Forecasting Models of Electricity Prices

Special Issue Editor
Javier Contreras

MDPI

Special Issue Editor
Javier Contreras
Universidad de Castilla
Spain

Editorial Office
MDPI AG
St. Alban-Anlage 66
Basel, Switzerland

This edition is a reprint of the Special Issue published online in the open access journal *Energies* (ISSN 1996-1073) from 2016–2017 (available at: http://www.mdpi.com/journal/energies/special_issues/forecast_model_electr_price).

For citation purposes, cite each article independently as indicated on the article page online and as indicated below:

Author 1; Author 2; Author 3 etc. Article title. *Journal Name*. **Year**. Article number/page range.

ISBN 978-3-03842-414-7 (PDF)
ISBN 978-3-03842-415-4 (Pbk)

Table of Contents

Chapter 1: Introduction

Chapter 2: Statistical time Series Analysis

Chapter 3: Fuzzy Logic, Artificial Neural Networks and Hybrid Methods

Chapter 4: Market Equilibrium and Fundamental Models

Chapter 5: Ensemble and Portfolio Decision Models

About the Guest Editor

Javier Contreras received his B.S. degree in Electrical Engineering from the University of Zaragoza, Zaragoza, Spain, in 1989; his M.Sc. degree in Electrical Engineering from the University of Southern California, Los Angeles, in 1992; and his Ph.D. degree in Electrical Engineering from the University of California, Berkeley, in 1997. Since 1998, he has been with the University of Castilla–La Mancha (UCLM), Ciudad Real, Spain, where he is currently Full Professor. Dr. Contreras has also been a visiting scholar at the University of Hong Kong and the University of Illinois at Urbana-Champaign. He has been a consultant for several electricity companies in Spain and has participated as principal investigator in national, European and international research projects. Dr. Contreras has focused his research towards a broad cross-disciplinary program in the area of price forecasting, electricity markets, renewable energy and operation and planning of electrical power systems. He is Fellow of IEEE.

Preface to "Forecasting Models of Electricity Prices"

The prediction of prices in the short-, medium- and long-term is very important for electric companies, retailers and consumers. In the short-term, a producer of electricity without the capability of altering market prices needs accurate price forecasts to achieve its optimal self-schedule of production and to derive a sensible bidding strategy in the electricity markets in which it operates. If the producers are to alter the market prices, they need information about their own effect on prices and the competitors' price bids. In the medium-term, a producer requires price forecasts several months in advance in order to sign energy contracts. Retailers and large consumers need price estimations for the same reasons as the producers, in order to maximize their utilities and to optimize their bids in the market. Hence, price forecasts represent fundamental information for all the agents acting in the electricity markets.

Since electricity prices possess singular features that are not present in other markets—weekly and daily seasonalities, price spikes, regime switching behavior, etc.—sophisticated prediction methods are required. This book presents the current state-of-the-art electricity price forecasting methods including statistical time series analysis, heuristic models, equilibrium models and portfolio methods, among others.

Javier Contreras

Guest Editor

Chapter 1:
Introduction

energies

MDPI

Editorial

Forecasting Models of Electricity Prices

Javier Contreras

Escuela Técnica Superior de Ingenieros Industriales, Universidad de Castilla—La Mancha,
Campus Universitario S/N, 13071 Ciudad Real, Spain; javier.contreras@uclm.es

Academic Editor: Enrico Sciubba
Received: 14 January 2017; Accepted: 20 January 2017; Published: 29 January 2017

This book contains the successful invited submissions [1–11] to a Special Issue of *Energies* on the subject area of "Forecasting Models of Electricity Prices".

The electric power industry has been in a transition from a centralized towards a deregulated production scheme since the early 1980s. Previous centralized schemes were based on electricity tariffs that were paid by the customers as a function of the aggregate cost of production. In the new unbundled scheme, price forecasting has become an important tool for electric companies and customers to decide on their production offers and demand bids and for regulators to characterize the degree of competition of the market.

Electricity prices have unique features that are not observed in other markets, such as weekly and daily seasonalities, on-peak vs. off-peak hours, price spikes, etc. The fact that electricity is not easily storable and the requirement of meeting the demand at all times makes the development of forecasting techniques a challenging issue.

This Special Issue includes the most important forecasting techniques applied to the forecasting of electricity prices, such as:

- Statistical time series models;
- Artificial Neural Networks;
- Wavelet transform models;
- Regime-switching Markov models;
- Fundamental market models;
- Equilibrium models;
- Ensemble and portfolio decision models.

The response to our call had the following statistics:

- Submissions (15);
- Publications (11);
- Rejections (4);
- Article types: Review Article (0); Research Article (11);

The authors' geographical distribution (published papers) is:

- China (3)
- Spain (3)
- Portugal (2)
- Denmark (1)
- Poland (1)
- Taiwan (1)

Energies **2017**, *10*, 160

Published submissions are related to a broad range of applications for load and price forecasting including classical Auto Regressive, heuristics, equilibrium methods, switching models, and combinations of them, among others.

We found the edition and the selection of papers for this book to be very inspiring and rewarding. We also thank the editorial staff and reviewers for their efforts and help during the process.

Conflicts of Interest: The authors declare no conflict of interest.

References

1. Bello, A.; Reneses, J.; Muñoz, A. Medium-Term Probabilistic Forecasting of Extremely Low Prices in Electricity Markets: Application to the Spanish Case. *Energies* **2016**, *9*, 193. [CrossRef]
2. Cheng, C.; Chen, F.; Li, G.; Tu, Q. Market Equilibrium and Impact of Market Mechanism Parameters on the Electricity Price in Yunnan's Electricity Market. *Energies* **2016**, *9*, 463. [CrossRef]
3. Jiang, P.; Liu, F.; Song, Y. A Hybrid Multi-Step Model for Forecasting Day-Ahead Electricity Price Based on Optimization, Fuzzy Logic and Model Selection. *Energies* **2016**, *9*, 618. [CrossRef]
4. Uniejewski, B.; Nowotarski, J.; Weron, R. Automated Variable Selection and Shrinkage for Day-Ahead Electricity Price Forecasting. *Energies* **2016**, *9*, 621. [CrossRef]
5. Osório, G.J.; Gonçalves, J.N.D.L.; Lujano-Rojas, J.M.; Catalão, J.P.S. Enhanced Forecasting Approach for Electricity Market Prices and Wind Power Data Series in the Short-Term. *Energies* **2016**, *9*, 693. [CrossRef]
6. Monteiro, C.; Ramirez-Rosado, I.J.; Fernandez-Jimenez, L.A.; Conde, P. Short-Term Price Forecasting Models Based on Artificial Neural Networks for Intraday Sessions in the Iberian Electricity Market. *Energies* **2016**, *9*, 721. [CrossRef]
7. Cheng, C.; Luo, B.; Miao, S.; Wu, X. Mid-Term Electricity Market Clearing Price Forecasting with Sparse Data: A Case in Newly-Reformed Yunnan Electricity Market. *Energies* **2016**, *9*, 804. [CrossRef]
8. Bello, A.; Bunn, D.; Reneses, J.; Muñoz, A. Parametric Density Recalibration of a Fundamental Market Model to Forecast Electricity Prices. *Energies* **2016**, *9*, 959. [CrossRef]
9. Lee, C.-M.; Ko, C.-N. Short-Term Load Forecasting Using Adaptive Annealing Learning Algorithm Based Reinforcement Neural Network. *Energies* **2016**, *9*, 987. [CrossRef]
10. Sánchez de la Nieta, A.A.; González, V.; Contreras, J. Portfolio Decision of Short-Term Electricity Forecasted Prices through Stochastic Programming. *Energies* **2016**, *9*, 1069. [CrossRef]
11. Neupane, B.; Woon, W.L.; Aung, Z. Ensemble Prediction Model with Expert Selection for Electricity Price Forecasting. *Energies* **2017**, *10*, 77. [CrossRef]

Chapter 2:
Statistical Time Series Analysis

energies

MDPI

Article

Medium-Term Probabilistic Forecasting of Extremely Low Prices in Electricity Markets: Application to the Spanish Case

Antonio Bello *, Javier Reneses and Antonio Muñoz

Institute for Research in Technology, Technical School of Engineering (ICAI), Universidad Pontificia Comillas, Madrid 28015, Spain; javier.reneses@iit.comillas.edu (J.R.); antonio.munoz@iit.comillas.edu (A.M.)
* Correspondence: antonio.bello@iit.comillas.edu; Tel.: +34-91-540-2800

Academic Editor: Javier Contreras
Received: 14 January 2016; Accepted: 3 March 2016; Published: 15 March 2016

Abstract: One of the most relevant challenges that have arisen in electricity markets during the last few years is the emergence of extremely low prices. Trying to predict these events is crucial for market agents in a competitive environment. This paper proposes a novel methodology to simultaneously accomplish punctual and probabilistic hourly predictions about the appearance of extremely low electricity prices in a medium-term scope. The proposed approach for making real *ex ante* forecasts consists of a nested compounding of different forecasting techniques, which incorporate Monte Carlo simulation, combined with spatial interpolation techniques. The procedure is based on the statistical identification of the process key drivers. Logistic regression for rare events, decision trees, multilayer perceptrons and a hybrid approach, which combines a market equilibrium model with logistic regression, are used. Moreover, this paper assesses whether periodic models in which parameters switch according to the day of the week can be even more accurate. The proposed techniques are compared to a Markov regime switching model and several naive methods. The proposed methodology empirically demonstrates its effectiveness by achieving promising results on a real case study based on the Spanish electricity market. This approach can provide valuable information for market agents when they face decision making and risk-management processes. Our findings support the additional benefit of using a hybrid approach for deriving more accurate predictions.

Keywords: electricity markets; medium-term electricity price forecasting; probabilistic forecasting; extremely low prices; spikes; hybrid approach

---

1. Introduction

In the current global context of the growing complexity of electricity markets, trying to predict electricity prices is essential for all market agents. However, this is not an easy task, since the price of electricity is far more volatile than other commodities. The presence of extremely high prices has been a recurrent phenomenon in markets worldwide. Nevertheless, the recent increasing deployment of non-dispatchable generation is also leading to the appearance of extremely low prices (zero or even negative prices depending on the considered regulatory framework).

This paper focuses on improving the understanding of the factors that contribute to the occurrence of these extreme price events and also their accurate forecasting with a medium-term scope. More specifically, the aim of this paper is to propose a novel methodology that allows one to predict not only the expected number of hours with very low prices in the medium term, but also the associated probability density function. The proposed methodology relies on a thorough in-sample analysis to adjust the models and an out-sample simulation approach to test its performance when making real *ex ante* forecasts.

The covered time horizon is from one month up to one year. In general, retailers and large consumers need reliable medium-term predictions to optimize their operation, as well as to properly negotiate in the short-term market and accomplish beneficial bilateral contracts. In addition, producers need medium-term predictions to optimize their generation programs and negotiate favorable bilateral and financial contracts. On the other hand, it is essential to anticipate the occurrence of these abnormally low priced hours, because this situation significantly increases the exposure of industry participants to price risk. Even in extreme cases, these unanticipated large changes in the spot price can lead to bankruptcies of energy companies if they are not prepared to tackle such risks. For this reason, an effective risk management support for the operation of electrical systems must also be able to foresee extremely low values.

The proposed methodology, which is currently in operation in one of the major Spanish electricity companies, is tested in a real case: the Spanish electricity market. The Spanish electricity market constitutes one of the most interesting cases in which the remarkable growth of renewable energy production frequently pushes the most expensive thermal power stations outside the generation program of the wholesale market. The consequent reduction in thermal production, coupled with a decline in the demand curve (especially in off-peak hours) due to the financial crisis and a low interconnection capacity to evacuate the surplus of non-dispatchable energy, causes at certain times a sharp reduction in the clearing price. Apart from the oversupply of generation technologies with zero opportunity cost (renewable energy sources (RES), run-of-the-river hydro and nuclear), an excess of gas (due to take or pay clauses) can make combined cycles have zero opportunity cost. The conjunction of these events causes the emergence of a scenario in which the matching of supply and demand is occurring at $0 €/MWh$ (note that in Spain, unlike other countries, such as Australia and Germany, negative prices are not allowed).

The main contributions of this paper can be summarized as follows:

1. A general methodology has been developed to make real *ex ante* forecasts (point and probabilistic) of extremely low prices for a mid-term horizon on an hourly basis. The methodology combines different forecasting methods and spatial interpolation techniques within a Monte Carlo simulation of multiple predicted scenarios for the considered risk factors.
2. The accuracy of a novel hybrid approach that integrates fundamental and behavioral information, logistic regressions, decision trees and multilayer perceptrons has been compared to the results obtained by means of a traditional Markov regime switching model and different naive methods. This comprehensive comparison has been carried out in both in-sample and out-of-sample datasets. It has also been examined if the use of periodic models helps to improve prediction capabilities.
3. The performance of the proposed methodology has been tested in a real-sized electricity system. Note that the empirical application presented in this paper is in a single price market that does not incorporate distribution network constraints in the market clearance. In the Spanish electricity market, the high complexity of the electricity price dynamics is mainly due to the huge penetration of renewable energy sources in the generation mix and the limited interconnection capacity with France. These aspects have been taken into account in all of the forecasting models presented in this paper. However, in order to extend the methodology to other markets, where locational marginal prices may exist, and for which this methodology could be applicable, the impact of variables related to local distributed generation should be taken into account (as [1] investigates in the electrical system in Italy). In this sense, another paper that presents the influence of distributed generation (DG) on congestion and locational marginal price (LMP) is [2].

The paper is structured as follows. After a state of the art review, Section 3 describes the methodology developed in the paper. Section 4 introduces the proposed forecasting techniques, as well as the in-sample results obtained. In Section 5, the case study and the real *ex ante* forecasting results are presented. Finally, the conclusions and the main contributions of the paper are summarized in Section 6.

2. Previous Work

Diverse models have been proposed in the literature to forecast electricity prices with different aims and time horizons. The wide number of forecasting techniques is likely to be grouped by various criteria that have been proposed in several studies [3,4]. According to [5], electricity price forecasting models include statistical and non-statistical models. The latter group, which is classified in more detail in [6], comprehends simulation models and equilibrium analysis models [7]. These approaches are preferred in a medium- to long-term horizon, as they can provide price predictions even when there are structural or regulatory changes in the market. However, as they are highly demanding computationally, they tend to group hours of similar characteristics. The latter makes that the forecasts not be as accurate as data-driven methods [8]. On the other hand, statistical methods, which rely on historical data, are useful for short-term price forecasting, but they degrade when are used for medium- or long-term horizons [9]. They include time series models and artificial intelligence techniques.

A great number of time series models has been successfully implemented. In this way, the ARIMA (autoregressive integrated moving average) models are the most representative, with different particularizations. Thus, there are references that accommodate the seasonality using the same set of parameters for all hours of the day [10,11]; and others that perform ARIMA model fitting (or its variants, AR or ARMA) for each time slot of the day [12,13]. Other generalizations of the ARIMA models are the so-called linear transfer function or transfer function models with ARIMA noise [14,15], which have the peculiarity of including past and present influence of other series. Other kinds of time series are the multiple-input multiple-output models, which predict the n-dimensional price vector in a single step [16]. Artificial intelligence techniques, which can be classified into artificial neural networks (ANN) [17], fuzzy logic and their combination, the neuro-fuzzy method [18], are more powerful for complex, nonlinear time series analysis than the rest of the statistical models. The methods presented before show a considerable ability to forecast the expected electricity prices under normal market conditions. So far, however, none of these techniques can effectively deal with spikes or extreme prices in electricity markets [19]. Among the first references that address these specific features of electricity prices is [20], where spikes are modeled by introducing large positive jumps together with a high speed of mean reversion. Other authors model spikes by allowing signed jumps [21]. According to [22], spike forecasting techniques can be classified into traditional and non-traditional approaches. Traditional approaches fall, broadly speaking, into three categories: (i) traditional autoregressive time series models; (ii) nonlinear time series models with particular emphasis on Markov-switching models; and (iii) continuous-time diffusion or jump diffusion models. Non-traditional approaches include artificial neural networks or other data-mining techniques.

Traditional autoregressive time series models treat spikes through Poisson and Bernoulli jump processes [23], the inclusion of thresholds [24] or the use of different multivariate error distributions [16]. Meanwhile, regime-switching models are the nonlinear extension of traditional time series. These models are capable of identifying the nonlinearities of the dynamics and distinguish the normal chaotic motion from the turbulent and spike regime. One of the most representative model of this class is the threshold autoregressive (TAR) one, which determines the regime by the value of an observable variable corresponding to a threshold value. In the case of including exogenous (fundamental) variables, TAR processes lead to the TARX model. An alternative is the self-exciting threshold autoregressive (SETAR) model, which arises when the threshold variable is taken as the lagged value of the price series itself [25]. Markov switching models are the most prominent among those in which the switching mechanism between the states cannot be determined by an observable variable. For the treatment of spikes, they suggest different states in which at least one is consistent with its appearance [26]. With regard to continuous-time diffusion processes, spikes are essentially captured by the combination of a Poisson jump component and an intensity parameter. This parameter can be constant [27] or can be driven by deterministic seasonal variables [28]. Recently, in [22], a nonlinear variant of the autoregressive conditional hazard model has been used to estimate the probability of a spike with a short-term horizon, and in [29], a spike component is predicted in the short term using a linear approximation based on consumption and wind.

Some other approaches are based on the namely nontraditional techniques, which include: decision trees and rule-based approaches; probability methods, such as Bayesian classifiers [30]; neural network (NN) methods, such as spiking NN [31]; example-based methods, such as k-nearest neighbors [19,32]; and SVM (support vector machine) [33].

To the best knowledge of the authors, no references have been published dealing with the problem of medium-term price spikes or extreme price forecasting. The proposed work is unique in the sense that it proposes to use several forecasting techniques for making both point and probabilistic medium-term prediction of extremely low prices with an hourly accuracy.

3. Methodology

Essentially, the steps of the methodology suggested in this paper are the following:

1. The choice of a threshold to define what is considered as an extreme low price event. This point is discussed in depth in Section 3.2. It is important to point out that the methodology is not materially affected by the choice of the threshold.
2. The selection of explanatory variables that contribute to explain the phenomenon of the emergence of very low prices from a perspective that takes into account the market behavior and their statistical significance. This is further discussed in Section 3.3.
3. The adjustment of a forecasting technique for predicting the occurrence of extremely low variables in terms of a probability value from actual market data (in-sample dataset). In Section 4, we detail all of the forecasting techniques that have been used and calibrated for this purpose. Due to the fact that in our study, the dependent variable (occurrence of extremely low prices) is dichotomous in nature, the potential models to apply for the analysis are restricted to binary choice models. The proposed models classify observations based on a cutoff value. If the probability predicted by the model is greater than this cutoff value, the observation will be classified as a normal price. Otherwise, it will be deemed as an extremely low price. The choice of this cutoff point is discretional, and it will influence the sensitivity and specificity, which vary inversely with the probability value chosen. These statistics, as well as the rest of the Cooper statistics [34], can be calculated from a contingency table (Table 1) as shown in Table 2. In this paper, the cutoff point was chosen so as to provide a balance between sensitivity and positive predictivity (*i.e.*, a failure to predict an actual extremely low price is penalized as heavily as a false alarm). As a result of this step, the parameters and the optimal cutoff value for each forecasting technique are obtained and will be used in the following stage.
4. The development of probabilistic real *ex ante* forecasts through cross scenario analysis, which is the basis of Section 5. In order to use Monte Carlo simulation to tackle uncertainty in the medium term, a large number of realizations of the model are needed, usually entailing a huge computational time and effort. In order to cope with this inconvenience, we have adapted an efficient method proposed in [35] for making market equilibrium models tractable (a practical implementation can be also found in [7]) to other forecasting techniques. This method, which is illustrated schematically in Figure 1, allows one to compute a huge number of simulations by decreasing the computational time and without a major loss of accuracy. As can be seen in the figure, the first step of the methodology consists of computing a reduced number of executions (m simulations) of each of the proposed forecasting models. As a result, we obtain m result matrices about the appearance or not of extremely low prices (the classification is made with the cutoff value previously estimated) in each specific hour of the simulation time horizon. These initial m simulations of the model are spatially placed in a hypercube of N dimensions according to the combinations of scenarios. More specifically, each dimension of the hypercube corresponds to an uncertain variable. For the sake of clarity, N is equal to three in Figure 1. Note that each risk factor is distinguished by its cumulative distribution function (CDF), and a particular scenario is

defined by the pertinent percentiles of the considered risk factors. Latin hypercube sampling with correlation control techniques has been used with the aim of having a well-sampled hypercube in which each scenario is used at least once in the m executions of the statistical models. In the second stage, a vast amount (M >> m) of correlated random scenario combinations of the risk factors is generated to establish those unobserved areas of the hypercube. Here, the correlation structure between the variables is determined by using historical data. In the third step, these unsimulated areas (M feasible matrices about the appearance of extremely low prices) of the hypercube are interpolated from the initial executions by means of an interpolator based on local regression that considers the spatial structure of these initial executions. Finally, as the scenario definition is random and considers the correlation structure between the uncertain variables, all of the scenarios can be considered to be equally probable, and thus, it is possible to make both point and probabilistic forecasts of the variables of interest.

Table 1. Contingency table.

Observed Price	Predicted Price		Marginal Totals
	Extremely Low	**Rest**	
Extremely low	a	b	a + b
Rest	c	d	c + d
Marginal Totals	a + c	b + d	a + b + c + d

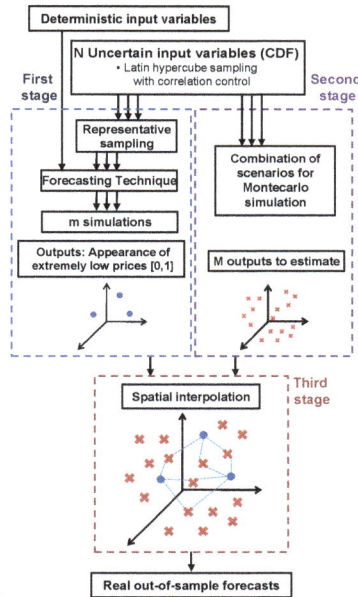

Figure 1. Global overview of Monte Carlo simulation with spatial interpolation techniques.

Table 2. Definitions of the Cooper statistics.

Statistic	Definition
Sensitivity: $a/(a + b)$	Proportion of the extremely low prices that the model predicts to be extreme
Specificity: $d/(c + d)$	Proportion of the normal prices that the model predicts correctly
Accuracy: $(a + d)/(a + b + c + d)$	Proportion of prices that the model classifies correctly
Positive predictivity (Pos. Pred.): $a/(a + c)$	Proportion of the prices predicted to be extremely low prices by the model that give positive results in observed prices
Negative predictivity (Neg. Pred.): $d/(b + d)$	Proportion of the prices predicted to be normal prices by the model that give negative results in observed prices
False positive rate: $c/(c + d)$	Proportion of the normal prices that are falsely predicted to be extreme by the model
False negative rate: $b/(a + b)$	Proportion of the extreme prices that are falsely predicted to be normal by the model

3.1. The Times Series Dataset

In this paper, a dataset from the Spanish day-ahead market, which comprises the period ranging between 1 January 2009 and 31 March 2012, is used. This period has been chosen because it is the moment that marked the inflexion point in relation to the appearance of extremely low prices in the Spanish market. However, the methodology would be equally extrapolated to other subsequent time periods. The complete data consisted of hourly spot prices and the actual production for each technology. The data corresponding to the Spanish market are available from the Iberian Energy Market Operator (OMIE [54]).

In order to thoroughly investigate the forecasting capability of each model, the data were divided into in-sample and out-of-sample datasets. The former set, which includes the training and testing sets, encompasses from 1 January 2009–30 November 2011. Thus, in Section 4, the generalization capabilities of the models with the actual data of the exogenous variables are carried out. In Section 5, an out-of-sample analysis for the period ranging between 1 December 2011 and 31 March 2012 with estimated scenarios of the explanatory variables is conducted. As no major structural or regulatory changes occurred during this period, it can be possible to capture the price dynamics by using a common statistical model.

A more detailed statistical analysis of the in-sample dataset is presented in Table 3. As noted, the distribution of electricity prices is not normal, presenting excess kurtosis and negative skewness. This means that excessively high or low prices have a higher probability of occurrence than in the case of a normal distribution. Moreover, prices below the average are more likely to occur than prices above the mean value.

Table 3. Statistical summary of market spot price.

Average	Standard Deviation	Skewness	Kurtosis	Min.	Quartile 1	Quartile 2	Quartile 3	Max.
41.041	13.280	−0.521	1.479	0.000	34.450	41.170	50.230	145.000

3.2. Extremely Low Prices Threshold

Due to the fact that market agents are not only interested in trying to model the incidence of zero prices, the objective of this section is the choice of a threshold for distinguishing extremely low prices from the rest. There are several approaches in the literature to classify whether an observation is extreme or not [36,37]. In fact, the choice of a reasonable threshold is still a subjective choice. In Spain, for instance, the threshold defining an extreme low price event is generally regarded between 10 and 15 €/MWh. In this article, a point of the characteristic modes of the distribution function has been used. In order to estimate the probability density function, the Epanechnikov kernel, which minimizes the asymptotic mean integrated squared error, has been used.

Figure 2 illustrates a graphical representation of the distribution function. This representation enables one to derive a threshold of 14.2 €/MWh. This point, which lies below the 5th percentile of the the unconditional distribution of the price series, fulfills two features. On the one hand, it is an inflection point of the density function on the left tail. On the other hand, it is a point that is away from the average a distance of more than two standard deviations. The last premise is consistent with that adopted in [19]. It should be noted that the methodology is not restricted by the choice of the threshold. The resulting indicator variable is coded 0 for extremely low prices and 1 for the remainder prices.

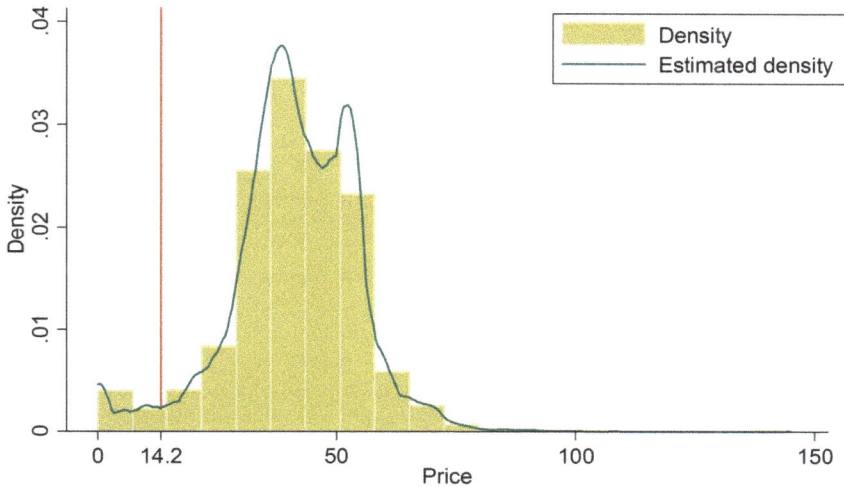

Figure 2. Density estimation

3.3. Explanatory Variables

Several studies have been conducted in order to detail the explanatory variables affecting spot market prices [28,38–40] and price spikes [41]. However, none of these references focus on the appearance of extremely low priced hours in the medium term. Therefore, in this paper, a critical point is not only the choice of the variables that help to better explain the aforementioned phenomenon, but also those factors that allow one to operationalize the forecasting model in a feasible way. This entails that it is possible to use only those variables that can be characterized with a reasonable accuracy at this medium-term time horizon. This is, for instance, the case of technologies that usually behave as price takers.

In the particular case of Spain, where most of the electricity is priced at the day-ahead hourly market, the price is set by the marginal generator bid. Among the power plants with the lowest short-opportunity costs are renewable energy sources (RES), nuclear and run-of-the-river hydro. On the contrary, the plants with the highest opportunity costs are gas, coal, head-dependent hydro and fuel oil. During the horizon of the study, RES were legislated under what was referred to as a special regime, promoted through a feed-in tariff system. Thus, apart from the special regime, the technologies that usually behave as price takers are run-of-the-river hydro and nuclear. Due to the zero variable cost of the former and the inflexibility of the latter, both of them usually bid at low prices to collect then the marginal price. It should be noted that it is well known that hydro conditions are one of the most important sources of uncertainty in the Spanish market. In this study, run-of-the-river hydro production has been considered as a reasonable approach for hydro conditions. This is because, run-of-the-river hydro production, unlike head-dependent hydro, behaves as a price taker, and therefore, it is expected to be relevant for the appearance of extremely low prices.

The selection of explanatory variables, which has been done taking into account the principle of parsimony, was based on backward elimination methods, which evaluate different statistics in order to control the exit of variables. Thus, the adjusted coefficient of determination and several model selection criteria (such as AIC and BIC) were measured.

Before constructing the proposed models, the presence of unit roots in the candidates for exogenous variables has been tested. To this end, augmented Dickey–Fuller has been used. For all considered explanatory variables, the absence of stationarity was strongly rejected beyond a 5% significance level. The absence of multicollinearity between the regressors has also been confirmed by applying the variance inflation factor.

4. Forecasting Techniques

The main target of this section is to compare several novel models with one of the most prominent models used in the prediction of spikes: Markov regime-switching models. In order to estimate their generalization capabilities, the in-sample data were divided into two different sets: 80% for training and 20% for testing. The validation set comprises the training and testing sets (from 1 January 2009– 30 November 2011). Note that this section analyzes the models' generalization capabilities with actual data of the exogenous variables.

It is well known that electricity prices exhibit seasonal fluctuations, which can be collected by including relevant explanatory variables. Still, despite a well-specified inclusion of demand in the models that could explain the weekly effects, a separate modeling for (1) working days, (2) Saturdays and (3) Sundays and holidays has also been carried out with the aim of testing if the forecasting performance is improved. The main benefit of this approach, which uses periodic models, is that it allows the model parameters to switch according to the different behaviors identified in the course of the week. Moreover, the construction of separate models has the advantage of selecting an optimum and specific threshold value for classifying an observation as a very low price or not depending on the predicted likelihood given by the statistical technique used in each case. This could be particularly relevant for the accuracy of the forecasts.

4.1. Logistic Regression

Logistic regression is a well-known supervised learning algorithm that can allow us to estimate the probability of the occurrence of extremely low prices, which is a dichotomous outcome. This technique is very useful to analyze the potential impact of the independent variables on the dependent variable. For each constructed model, the estimated coefficients of the explanatory variables are also shown. The Wald test was utilized to check their validity. The models goodness-of-fit is checked by means of the Cox and Snell R square and the Nagelkerke R square.

4.1.1. Model 1

As has been shown previously, the addressed classification problem is unbalanced with a proportion of hours with very low prices only accounting for 4.32%. For this reason, a model based on the traditional logistic regression procedures could sharply underestimate the probability of this rare event, and therefore, it could lead to erroneous results. One of the most popular techniques to correct these effects when the occurrence of the events is less than 5% was the bias correction method proposed in [42]. This procedure estimates the same logit model as the traditional one, but with an estimator that provides lower mean square error in the presence of rare events data for coefficients, probabilities and other quantities of interest.

Model 1 uses the explanatory variables referred to in Table 4 without taking into account the effects of work activity in a specific way. Table 5 and Table 6 show that the overall accuracy of the model is acceptable. Note that a naive model, for simple chance, would have a specificity and sensitivity of 50%. For example, the statistic R square is 65.6%, which is quite acceptable in predictive terms. According to Table 4, the estimated coefficient of the dependent variables, as shown by the Wald test, are significant from a statistical point of view.

Table 4. Variables in the equation: Model 1. ND, net demand.

Variable	Coefficient	S.E.	Wald	Signification
ND	0.0005860	0.0000176	1113.7107	0.000
H	−0.0016148	0.0000422	1463.4780	0.000
N	−0.0003917	0.0000587	44.5235	0.000
W	−0.0007231	0.0000226	1023.5278	0.000
CONST	1.7120790	0.4079335	17.6144	0.000

Table 5. Contingency table with a cutoff value of 0.65: Model 1.

Observed Price	Predicted Price		Correct Percentage
	Price [0,14.2]	Price > 14.2	
Price [0,14.2]	752	350	68.24
Price > 14.2	361	24,070	98.52
Global percentage			97.22

Table 6. Summary of Model 1.

−2 Log of the Likelihood	Cox and Snell R Square	Nagelkerke R Square
3501.401	0.196	0.656

4.1.2. Model 2

This model is an extension of the previous one. The novelty is that the effect of the working patterns in prices has been incorporated by using a periodic logistic regression model. Thus, a regression model is estimated for weekdays, another one for Saturdays and a different one for the holidays. Table 7 shows that the model goodness-of-fit is better, and the variation explained by the model is slightly higher than the previous one.

In accordance with Table 8, the explanatory variables are significant from a statistical point of view (except the constant term in the case of the working days). Table 9 shows the performance of the three models separately and for the global model. Furthermore, the optimal cutoff points that have been calculated are presented in parenthesis. As seen, this model presents a power for prediction slightly higher than Model 1.

Table 7. Summary of Model 2.

−2 Log of the Likelihood			Cox and Snell R Square			Nagelkerke R Square		
WORK	SAT	HOL	WORK	SAT	HOL	WORK	SAT	HOL
1954.41	464.10	981.57	0.18	0.22	0.23	0.69	0.70	0.59

Table 8. Variables in the equation: Model 2.

Variable	Coefficient ($\times 10^{-4}$)			Signification		
	WORK	**SAT**	**HOL**	**WORK**	**SAT**	**HOL**
ND	6.45	5.15	6.14	0.00	0.00	0.00
H	−16.41	−18.55	−16.55	0.00	0.00	0.00
N	−2.26	−6.16	−5.62	0.01	0.00	0.00
W	−7.50	−4.69	−7.99	0.00	0.00	0.00
CONST	−5531	38,342	32,099	0.33	0.00	0.00

Table 9. Cooper statistics of Model 2.

(%)	**WORK (0.66)**	**SAT (0.66)**	**HOL (0.61)**	**Global**
Sensitivity	69.85	78.31	61.54	68.97
Specificity	98.87	98.97	97.06	98.60
Positive Predictivity	70.06	78.31	61.11	68.97
Negative Predictivity	98.86	98.97	97.12	98.60
Accuracy	97.81	98.03	94.59	97.32

4.1.3. Model 3

This particular case is a variant of Model 2, not considering the correction proposed in [42]. Instead, it was decided to reduce the data with the aim that the number of low prices had a greater significance in the sample. In this way, according to Table 10, those intervals of data in which it is guaranteed the absence of extremely low prices in the dataset were eliminated. Hence, the sample was reduced by 42.3%. The model, although still slightly better than Model 1, cannot overcome the suitability of Model 2. Table 11 presents Cooper statistics broken down individually and globally. Note that there are categories with two values. The values on the left side refer to those corresponding to the simplified model, while the values on the right side refer to the correction made taking into account that the removed values of the training set have been successfully predicted. Meanwhile, Table 12 confirms the goodness-of-fit of each one of the regression models. Furthermore, Table 13 presents the coefficients associated with each explanatory variable, as well as the statistical significance of each one of them. As in Model 2, the constant term for weekdays is not significant.

Table 10. Intervals without very low prices.

Variable	WORK	SAT	HOL
ND	>32,500	>28,000	>26,000
ND′	>28,000	>23,000	>21,000
ND″	>13,000	>16,000	>9000

Table 11. Cooper statistics of Model 3.

(%)	**WORK (0.66)**	**SAT (0.66)**	**HOL (0.62)**	**Global**
Sensitivity	70.00	74.70	60.84	68.33
Specificity	97.53–98.84	98.73–98.98	95.45–96.73	97.44–98.57
Positive Predictivity	69.57	75.61	61.27	68.33
Negative Predictivity	97.58–98.86	98.67–98.93	95.37–96.67	97.44–98.57
Accuracy	95.48–97.78	97.53–98.00	91.79–93.91	95.26–97.27

Table 12. Summary of Model 3.

−2 Log of the Likelihood			Cox and Snell R Square			Nagelkerke R Square		
WORK	SAT	HOL	WORK	SAT	HOL	WORK	SAT	HOL
1948.50	460.18	972.42	0.26	0.23	0.27	0.64	0.70	0.55

Table 13. Variables in the equation: Model 3.

Variable	Coefficient ($\times 10^{-4}$)			Signification		
	WORK	SAT	HOL	WORK	SAT	HOL
ND	6.34	4.90	5.77	0.00	0.00	0.00
H	−16.32	−18.55	−16.55	0.00	0.00	0.00
N	−2.18	−5.98	−5.48	0.01	0.00	0.00
W	−7.44	−4.53	−7.88	0.00	0.00	0.00
CONST	−4721	41,169	36,893	0.41	0.00	0.00

4.1.4. Comparison between the Models

As seen, the three models have a quite acceptable predictive accuracy. Note that Model 1 showed slightly worst global performance than the other models. However, it presents the advantage that it is a much simpler model. The results obtained in this comparison demonstrate that the inclusion of weekly seasonality slightly improves the predictive capabilities of the forecasting methods. A remarkable fact is that there are discrepancies regarding the significance of the explanatory variables depending on the day of the week under consideration. Another finding to note is that the correction proposed in [42] is an effective tool when dealing with unbalanced problems. Finally, a greater difficulty in achieving effective predictions on Sundays and holidays has been found. This may be explained by a different market behavior during these days.

4.2. Decision Trees

Decision trees classify observations based on a set of decision rules, applied in a sequential manner. The probability of occurrence of extremely low prices is allocated to each end of a branch in the tree. The way of estimating probabilities does not require the assumption of specific probability distributions for the variables, which is an advantage of this methodology. In order to not over-fit the data, stopping rules control the growing process, and the over-fitted parts were pruned. In this paper, an ID3 (Iterative Dichotomiser 3) algorithm has been used, with splitting criteria based on the entropy [43].

4.2.1. Model 1

A global model ignoring the effects of weekly patterns has been built. According to Figure 3, the most representative variable is the hydro production. Note that under a scenario below the 88th centile, there is a negligible probability of occurrence of very low prices. Another interesting aspect is that nuclear production is not representative in this case. Table 14 shows the classification success rate of the tree. As seen, the model is able to accurately predict an acceptable number of the events that occurred during this period of time.

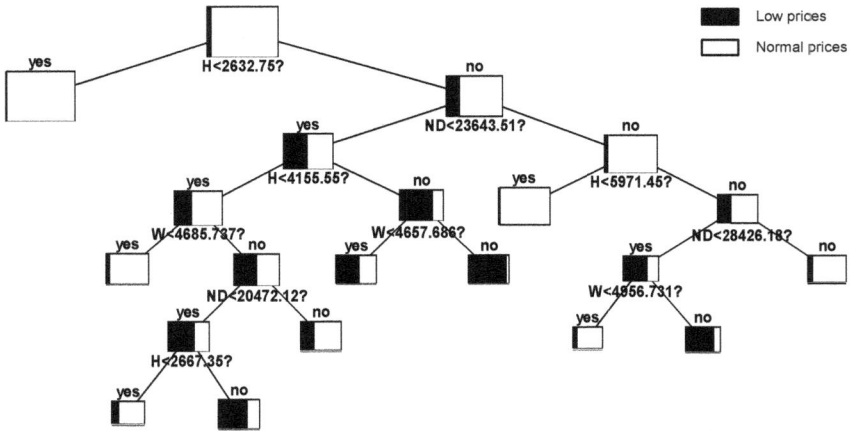

Figure 3. Decision tree: Model 1.

Table 14. Contingency table with a cutoff value of 0.85: Model 1.

Observed Price	Predicted Price		Correct Percentage
	Price [0,14.2]	Price > 14.2	
Price [0,14.2]	657	445	59.62
Price > 14.2	401	24,030	98.36
Global percentage			96.69

4.2.2. Model 2

In this case, the weekly seasonality has been taken into account by building three different trees. If we analyze the constructed trees in comparative terms, we can observe clear similarities and differences between them. First, it is clear that hydro production is the most relevant variable. Moreover, the scenarios that lead to hours with very low prices are similar. However, a more detailed individual analysis of each tree allows one to establish significant differences in the representativeness of the explanatory variables and characteristic values of the selected splits.

Figure 4 shows the tree for Saturdays. As can be seen in the figure, those hourly scenarios of hydro production that fall below 3767.2 MW have a negligible probability of occurrence of very low prices. One of the most remarkable facts is that, from a statistical point of view, hydro production seems to be even more important than in the other types of days, such as Sundays and holidays. Furthermore, nuclear production is irrelevant. The typical behavior of Sundays and holidays is reflected in Figure 5. An interesting aspect is that wind power is not significant. In contrast, nuclear production itself is useful in explaining the output's behavior. Finally, Figure 6 corresponds to working days. According to the figure, the main factors influencing the appearance of low priced hours are the hydro production and the system net demand.

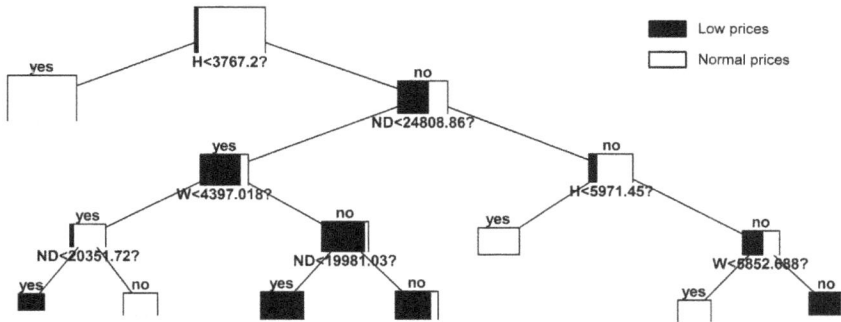

Figure 4. Decision tree: Saturdays.

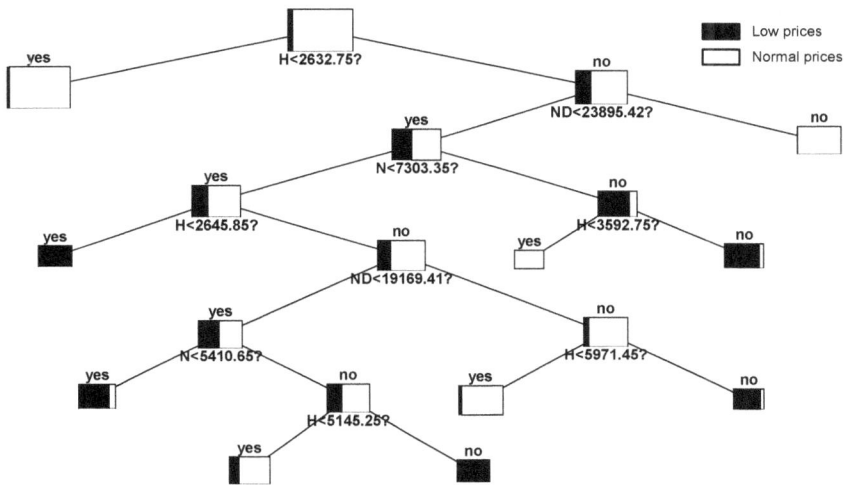

Figure 5. Decision tree: Sundays and holidays.

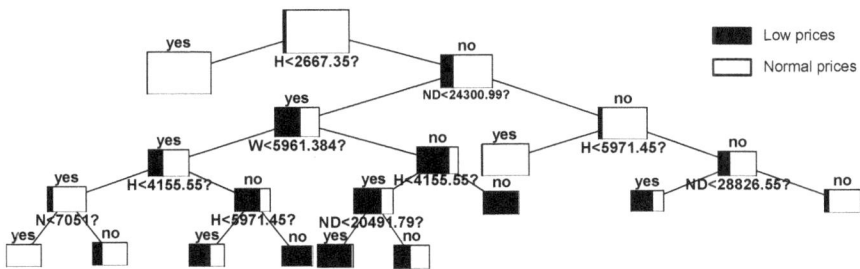

Figure 6. Decision tree: working days.

Table 15 shows the global performance of the model derived from each of the trees, which has been constructed separately. As the optimal cutoff value was set independently for each tree, it is not possible to achieve a balance between positive and negative predictivity.

Table 15. Contingency table: Model 2.

Observed Price	Predicted Price		Correct Percentage
	Price [0,14.2]	Price > 14.2	
Price [0,14.2]	669	433	60.71
Price>14.2	308	24,123	98.74
Global percentage			97.10

4.2.3. Comparison between the Models

Table 16 shows a comparison between the proposed models. The percentages of success in the training, test and validation sets are shown, as well as the optimal cutoff value for each tree. In general, both models are acceptable, since the overall accuracy is superior to that obtained by chance (95.68%). The second model performs slightly better. A remarkable aspect is that the tree constructed for Sundays and holidays is only able to identify 50% of the hours with very low prices.

Table 16. Comparison among the different decision trees.

Strategy	Decision Tree	Percentage of Success (%)			Cutoff
		Training	Test	Validation	
Without weekly effect	FULL	96.671	99.001	97.137	0.85
Weekly effect	WORK	97.449	99.353	97.830	0.75
	SAT	97.875	100	98.246	0.8
	HOL	95.427	96.590	95.660	0.76

Decision Tree	Cooper Statistics (%)				
	Sensitivity	Specificity	Pos. Pred.	Neg. Pred.	Accuracy
FULL	59.619	98.359	62.098	98.182	96.687
WORK	62.923	98.815	66.830	98.597	97.503
SAT	70.482	99.397	84.783	98.604	98.081
HOL	50.000	97.800	62.996	96.309	94.465

4.3. Multilayer Perceptron

Multilayer perceptrons constitute a useful tool for regression and classification [44]. In this case, its use is justified because we expect that there could exist a nonlinear relationship between the proposed inputs and output.

There are several problems associated with local minima and decisions over the size of the network to use. Thus, the use of this technique usually implies experimenting with different architectures, as the determination of the number of hidden layers and the number of hidden neurons in each hidden layer. It has been shown in practice that one hidden layer configuration is enough for most applications. For this reason, a topology with two layers of adaptive weights (a hidden layer and an output layer) has been selected.

The activation function of neurons is the hyperbolic tangent, both for the hidden and the output layer. With the aim of providing a probabilistic interpretation, the outputs have been scaled to the interval [0,1].

Regarding the proper number of hidden neurons to be used, a sweep computing the validation error was carried out to find its optimal value. As said, one of the main problems in the multilayer perceptron is getting stuck in local minima. This problem has been solved by initializing the weights with random values and by repeating the process several times. Before training, the inputs were normalized by a simple linear rescaling. The fitting criterion was the quadratic error minimization, as it penalizes large errors more than small ones.

Using this technique, two models were constructed: a single MLP for the whole set of data and three different MLP to take into account the weekly seasonality. In Table 17, the structure of the proposed models is summarized. In order to evaluate the forecast performance, the Cooper statistics are displayed for the optimal cutoff value that has been computed. It is evident that the prediction ability of this model is significantly improved if the weekly seasonality is included. In any case, each of the models based on MLP is clearly superior in terms of prediction performance to the other proposed techniques. Curiously, this model demonstrates a better ability to capture the dynamics of the very low prices on Sundays and holidays.

Table 17. Comparison of proposed models: MLP.

Strategy	MLP	Network Size	Cutoff	Cooper Statistics (%)			
				Sensitivity	Specificity	Pos. Pred.	Neg. Pred.
Without weekly effect	FULL	4-25-1	0.65	78.131	98.993	77.778	99.013
Weekly effect	WORK	4-24-1	0.64	82.308	99.311	81.930	99.329
	SAT	4-28-1	0.78	92.169	99.627	92.169	99.627
	HOL	4-28-1	0.59	91.259	99.318	90.941	99.345

4.4. Hybrid Approach

The hybrid approach presented in this section is a novel forecasting methodology that combines a fundamental market equilibrium model with the logistic regression approach implemented in Section 4.1.1, with the ultimate objective of benefiting from the advantages that each of them offers separately. Fundamental models, which are preferred in a medium- and long-term horizon, can provide useful insights for the analysis of the strategic behavior in electricity markets and constitute a valuable tool to represent the electricity market with its main technical and economic characteristics, especially when there are structural or regulatory changes in the market. However, as stated in [7], market equilibrium models fail when the aim is to capture the high-order moments of the probability distribution functions compared to the data-driven methods. This is because in the tails of the distribution, fundamentals are less important than behavioral factors. This naturally leads to complementing the fundamental approach with some of the statistical methods discussed before, which is one of the ultimate goals of this paper.

4.4.1. Fundamental Market Equilibrium Model

In the first step of the methodology, the operation and the behavior of the Spanish electric power system are fairly represented using a fundamental market equilibrium model based on conjectural variations as stated in [45,46]. The model, which is equivalent to the one used in [7], is formulated as a traditional cost-based optimization problem where each generation company i tries to maximize its own profit. In this model, the strategic behavior of each generation company i is represented by means of a parameter known as the conjectured-price response θ_i. This exogenous positive parameter has been valued by using historical data following [47,48]. This parameter, which measures the market power of the various companies taking part in the market, is the minus derivative of the electricity market price λ with respect to the production q_i of the generation company (Equation (1)).

$$\theta_i = -\frac{\partial \lambda}{\partial q_i} \geq 0, \qquad \forall i \tag{1}$$

As was shown in [45], the market equilibrium (note that under game theory, the market equilibrium is the point in which each market agent maximizes its own profit, but bearing in mind that the rest of the agents also maximize their profits) can be calculated by solving an equivalent quadratic optimization problem (Equations (2)–(4)):

$$\min_{q_i} \sum_i \overline{C}_i (q_i) \tag{2}$$

s.t.

$$\sum_i q_i = D \qquad : (\lambda) \tag{3}$$

$$\mathcal{H}(q_i) \geq 0 \tag{4}$$

The term $\overline{C}_i (q_i)$ is the so-called effective cost function of agent i, which is defined as:

$$\overline{C}_i (q_i) = C_i (q_i) + \theta_i \cdot \frac{q_i^2}{2}, \qquad \forall i \tag{5}$$

As seen, the effective cost function takes into account a linear or quadratic cost function $C_i (q_i)$ and a term that models the strategic behavior of the company i. Therefore, the minimization problem Equations (2)–(4) is a quadratic optimization problem that can be effectively solved using readily available commercial software. The decision variables of this problem are the dispatch of the generators, subject to the demand-balance equation (Equation (3)) and the technical constraints (Equation (4)) of the operation of all hydro and thermal groups (emission limits, variable costs, minimum and maximum power, efficiency, *etc.*).

Since in medium-term market equilibrium models, an hourly representation is not often used in practice because the size and resolution times increase considerably for a real-sized electricity systems, the hours within each month have been grouped into $l = 1,2,\ldots,16$ net demand levels (denoted as ND in this paper), or system states, by means of a k-means clustering process, as explained in [49]. System states consist of a number of hours in which market conditions are considered to be the same.

The use of system states as proposed in [49] is an alternative approach to the traditional representation based on load levels, which prevents the loss of chronological information between individual hours. This is very important for decision variables, such as the starting-up and the shutting-down of thermal groups. It should be noted that load levels, unlike system states, are only defined based on system demand. The consideration of the net demand for the computation of system states allows us to better represent the operation in power systems with a high penetration of renewable energy sources. Furthermore, as stated in [7], this novel approach based on system states enables one to reach a better forecasting accuracy and allows one to successfully capture the so-called stylized facts [50] of electricity prices. However, even in this complex model, there are many difficulties to properly account for the occurrence of extreme events, and that is why a complementary approach with a statistical model is needed.

4.4.2. Communication between the Models

For the hybridization of the models, the system marginal price for each state λ is firstly estimated by computing the dual variable of the power-balance constrain (Equation (3)) of the market equilibrium model. Hereinafter, the price is allocated to the hours that belong to the corresponding state, and it is used as an explanatory variable in the logistic regression model for rare events, which was detailed above in Section 4.1.1. As stated in Table 18, the market price λ shows the statistical significance thereof. Another factor to highlight is that the sign of the considered variables coincides with what would be expected *a priori*. The fact that the fundamental model seems not to add very much to the logistic model implies, as was expected, poor fundamental specification at low prices. However, as can be seen in Table 19, the obtained results suggest that this approach performs slightly better than the individual models presented in Section 4.1. This finding suggests that λ, which simultaneously captures the production cost and the strategic behavior of market agents, can provide useful insights to predict extremely low prices.

Table 18. Variables in the equation: hybrid model.

Variable	Coefficient	S.E.	Wald	Signification
ND	0.0007084	0.0000618	131.3093	0.000
H	−0.0007399	0.0004128	3.2121	0.073
N	−0.0003914	0.0002106	3.4540	0.063
W	−0.0008067	0.0000741	118.4120	0.000
λ	0.0510856	0.0111722	20.9082	0.000
CONST	2.5112069	1.5278314	2.7015	0.100

Table 19. Contingency table with a cutoff value of 0.66: hybrid model.

Observed Price	Predicted Price		Correct Percentage
	Price [0,14.2]	Price > 14.2	
Price [0,14.2]	796	306	72.23
Price >14.2	306	24,125	98.75
Global percentage			97.60

It should be stressed that the hybridization may also be performed with the rest of the techniques that have been previously presented. The major reasons why logistic regression has been used are based on the commitment between accuracy, transparency and simplicity of implementation that this technique has demonstrated. This is of great importance, since, as will be explained ahead in Section 5, the implementation of the methodology to make real predictions in the medium term requires simulating multiple scenarios of the variables subject to uncertainty, which is computationally highly intensive in real-sized electricity systems. Note that although the problem size using system states is much lower than the hourly representation, the model for the Spanish electricity market consists of more than 300,000 equations and 800,000 variables. The optimization problem is formulated in GAMS (General Algebraic Modeling System) and is solved using CPLEX 12.4. The resolution time is almost two minutes for just one realization of the risk factors using a PC with Intel Core Duo i7-4790 CPU @3.6 GHz CPU and 32.0 GB RAM.

4.5. Markov Regime-Switching Model

Markov regime-switching models (MRS) assume the existence of an unobserved variable representing the state or regime, which governs a given dataset at each point in time. The usefulness of MRS models for power market applications has already been recognized. However, their effectiveness for forecasting has been vaguely proven, and only lately has this issue been approached in the literature [39,41,51].

Markov switching models do not require a previous dating of which periods are considered extreme. Therefore, fixed imposed thresholds are not needed. The model is able to capture changes in the mean and the variance between state processes. The motions of the state variable between the regimes are governed by an underlying Markov process.

In this paper, one of the most popular specifications of MRS models in the energy economics literature is followed. Specifically, as proposed in [51,52], the specification includes two independent regimes (R_1 and R_2) and a mean-reverting heteroskedastic process for the base regime dynamics. In Equation (6), where the base regime is described, ϵ_t is supposed to be N(0,1)-distributed. On the other hand, a Gaussian distributed spike regime is assumed (Equation (7)).

$$X_{t,R_1} = \alpha_{R_1} + \beta_{R_1} X_{t-1} + \sigma_{R_1} \mid X_{t-1} \mid^{\gamma_{R_1}} \epsilon_t \qquad (6)$$

$$X_{t,R_2} \sim N(\alpha_{R_2}, \sigma_{R_2}) \qquad (7)$$

Following the recommendations provided in [51], the prices themselves instead of the log-prices are modeled. Moreover, the deseasonalization of prices is conducted in a similar way as was stated in [51]. Thus, an additive model is considered. Equation (8) represents that the hourly spot price Y_t can be decomposed into a stochastic part X_t and a predictable component (trend and seasonal component). On the one hand, for estimating the trend T_t, a wavelet filtering-annual-smoothing technique is used. On the other hand, weekly periodicity is considered for the seasonal component S_t. This component is removed by applying a variation of the moving average technique, using the median instead of the mean value. The reason is that the median is more robust than the mean in the presence of outliers. Figure 7 graphically shows the decomposition process, which has been performed.

$$Y_t = T_t + S_t + X_t \qquad (8)$$

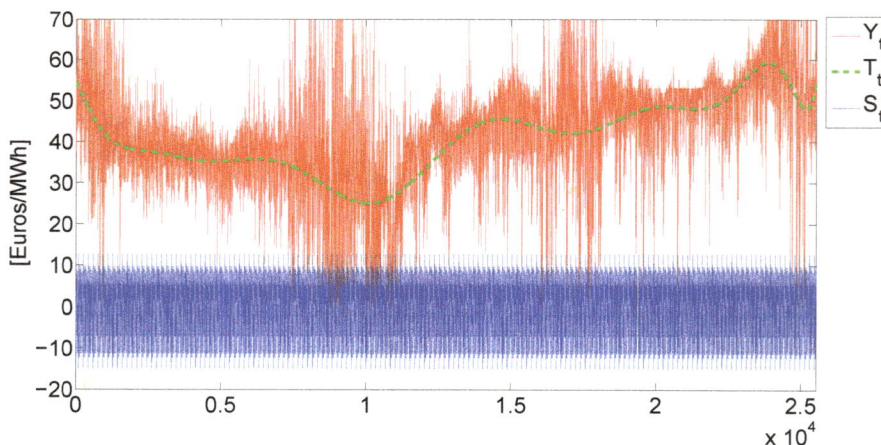

Figure 7. Decomposition of the price series using an additive model.

Next, MRS models are fitted to deseasonalized prices X_t. The calibration of parameters is accomplished by an iterative procedure based on the application of the expectation-maximization algorithm proposed in [53]. The estimated model parameters are given in Table 20. As can be seen, the parameters obtained for the base regime suggest a high speed of mean-reversion (which is represented by the parameter β_{R_i}) and that extremely low prices increase volatility more than extremely high prices (this is captured by the parameter γ_{R_i}). Regarding the probabilities q_{ii} of staying in the same regime, high values in both regimes are observed. This suggests that several consecutive observations in each regime will be appreciated, which represent an advantage in comparison to jump diffusion models.

Table 20. Parameter estimates and descriptive statistics of the Markov regime-switching (MRS) model.

R_i	Parameters				Statistics		
	α_{R_i}	β_{R_i}	$\sigma_{R_i}^2$	γ_{R_i}	$E(X_{t,R_i})$	q_{ii}	$P(R=i)$
Base	4.59	0.91	11,654.67	−0.96	50.62	0.98	0.84
Spike	47.38		259.01		47.372	0.89	0.16

Figure 8 shows the deseasonalized prices X_t and the spikes that have been identified. The lower picture displays the probabilities of being in the spike regime. The deseasonalized series has been shifted so that its minimum coincides with that of the original series Y_t.

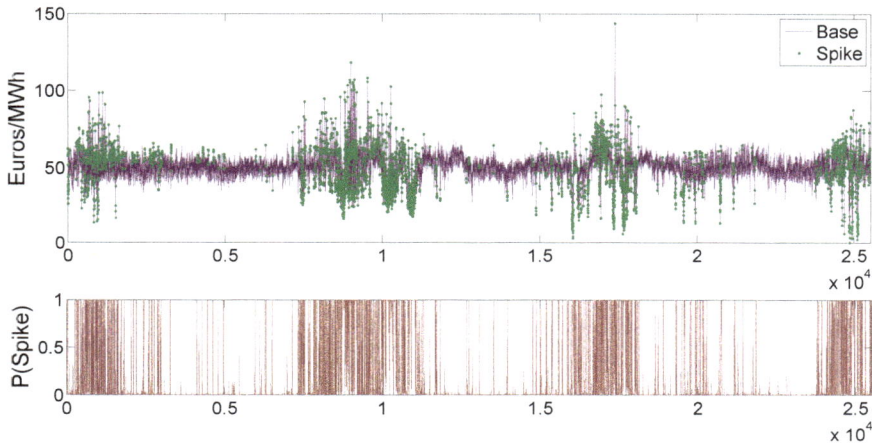

Figure 8. Calibration results for the MRS model.

In order to test the ability of the model to predict extremely low prices in the in-sample dataset, 5000 price trajectories have been simulated. The performance of the model is analyzed with two measures typically used in spike classification: sensitivity (85.75%) and positive predictivity (45.88%). Although the model is able to predict many of the extremely low prices correctly, this technique tends to classify many non-spikes as spikes. Therefore, it seems that this model has no advantages over the other proposed models.

5. Real *Ex Ante* Forecasts

This section is aimed at making real *ex ante* forecasts in a probabilistic way by using predicted scenarios of the risk factors and the parameters that have been estimated for the forecasting techniques presented in Section 4. On the one hand, Section 5.1 presents a description of how the last stage of the methodology explained in Section 3 is actually implemented with the proposed forecasting techniques. On the other hand, in Section 5.2, we briefly discuss the simulation with the MRS model, which is used as a benchmark. Finally, the presented case study is presented in Section 5.3.

It is important to highlight that the medium-term horizon is referred here to a forecasting scope that varies from one to two months. More specifically, if the primary objective is the prediction of extreme hourly prices for month m, the simulations are carried out in a single step in the first hour of month m-1.

5.1. Simulation with the Proposed Models

In order to make simulations with the proposed models, a multi-scenario analysis has been conducted. Therefore, the first step of this methodology is to generate scenarios for those random variables based on historical data. As was stated in Section 3, all risk factors are represented by their corresponding cumulative distribution function (CDF) in such a way that each scenario corresponds to a percentile. Different strategies have been used for the hybrid approach and the remaining techniques. This is because in the market equilibrium model, it is also important to incorporate the uncertainty related to fuel prices and the unavailability of thermal plants.

5.1.1. Logistic Regression, Decision Trees and Multilayer Perceptrons

In the presented case study, possible hourly realizations for water inflows (five scenarios), power demand (five scenarios) and wind production (55 scenarios) have been generated. In order to obtain a well-sampled spatial structure, representative percentiles ranging from the 1st to the 99th of its CDF have been chosen. Meanwhile, it is assumed that the international exchanges, as well as the production of the rest of the technologies belonging to the special regime are completely determined by their expected values. It should be noted that it is out of the scope of this work to understand how the prediction errors in the generation of scenarios contribute to the error of *ex ante* forecasts. Taking into account all of the possible combinations of the generated scenarios, a total of 1375 simulations have been performed with each forecasting technique. For each scenario and for each hour, the likelihood of the appearance of low priced hours has been computed. If this value is lower than the optimal cutoff value defined for each particular model in Section 4, the observation is classified as an extremely low price. Because of its practical interest, an additional variable that indicates the number of extremely low prices per month has also been constructed for every simulation on the basis of the sum of all hourly indicator variables. In the next step, a huge amount (more specifically, 100,000) of random scenario combinations of the percentiles of the well-known uncertain variables is generated in order to establish the unobserved areas of the hypercube, which have to be estimated by means of the spatial interpolator. Finally, probabilistic forecasts are calculated taking into account that all of the 100,000 scenarios are equiprobable.

5.1.2. Hybrid Approach

In the specific case of the hybrid approach, we have taken into account as medium-term risk factors, in addition to the three variables considered in Section 5.1.1, the natural gas prices (11 scenarios), the CO_2 emission allowance prices (11 scenarios), the coal prices (11 scenarios) and the unplanned unavailability of thermal power plants (three scenarios). As a result, there are 5,490,375 possible combinations of uncertain variables. In the initial stage, a representative sample of 1375 uncorrelated scenario combinations has been defined by means of Latin hypercube sampling as stated in Section 3. The next step is to perform 1375 simulations of the hybrid approach by including the well-sampled scenario combinations of the uncertain variables and the deterministic inputs. Finally, probabilistic forecasts are computed in a similar way to what is done in the previous section. The main difference is that in this case, the 100,000 random scenario combinations that have been generated for the Monte Carlo simulation present a correlated structure. This is particularly important, since it is well established that commodity prices are correlated, and it would be unrealistic to consider certain combinations when making the spatial interpolation. For the sake of clarity, a general outline of the methodology followed is provided in Figure 9.

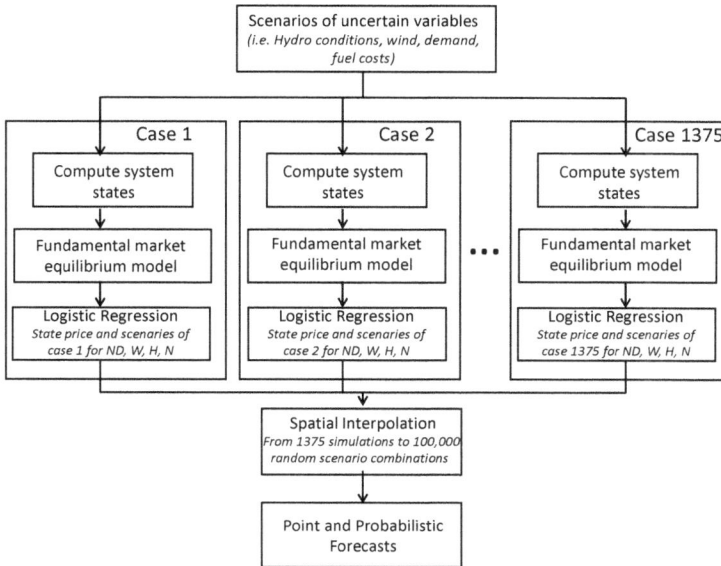

Figure 9. Global overview of the out-of-sample simulation with the hybrid approach.

5.2. Simulation with the Markov Regime-Switching Model

Several price trajectories are simulated in order to guarantee the stability of the results. Specifically, 5000 different paths have been used for each month. Then, using the simulated forecasts for the spot price, the corresponding probabilistic forecasts have been determined. Forecasting based on decomposition methods has been performed by extending each of the predictable components. The trend is the component that presents major problems, since wavelets are functions that are quite localized in time and space. In order to extend the signal, polynomial extrapolation or a spline fit might be utilized. In this case, as this component is closely related to expectations about fuel price levels, climate and consumption conditions, an adjusted linear model based on futures prices information being traded is used. This approach is suitable to properly internalize the expectations of all market agents. Regarding the seasonal component, it has been extended through the duplication of the last seasonal period. This can be considered as appropriate, since the seasonal component does not vary with time.

5.3. Case Study and Results Analysis

This section firstly assesses the capabilities of the proposed techniques to provide real *ex ante* point forecasts. The number of hours with very low prices per month has been selected, due to its interest in practical applications, as accuracy measure to evaluate the performance of the different approaches. For this assessment, a comparative study with two naive methods has been conducted. On the one hand, Naive 1 makes forecasts for month m by taking into account the proportion of extreme low prices that have taken place from 1 January 2009 to the last hour of month m-2. On the other hand, Naive 2 considers the proportion of events in similar months of the in-sample dataset.

In Table 21, the values of these measures are provided. Comparing the predictions of each proposed model with those that actually occurred, it can be concluded that the hybrid approach seems to be superior to the rest of models. This result suggests that the inclusion of the prediction of the market equilibrium price as an input of logistic regression in each scenario can provide useful information about the economic and technical characteristics of the market. This is even more important when possible structural changes can occur in the market. As can be seen, models based on logistic regression are able to achieve high levels of accuracy and slightly outperform multilayer perceptrons. Furthermore, note that MLP models perform significantly worse in the out-of-sample test. Regarding decision trees, it seems that they do not provide satisfactory results from the constructed scenarios. In turn, Markov regime-switching models show acceptable results. However, they have the well-known disadvantage of being very sensitive to the predictable component estimation. As seen, all models are successful when facing the naive test. It is also interesting that, unlike in the in-sample dataset, the prediction ability is not improved when periodic models are used.

Table 21. Number of hours with very low prices expected per month.

Month	Logistic Regression			Decision Trees		MLP		Hybrid	MRS	Naive 1	Naive 2	Actual
	M1	M2	M3	M1	M2	M1	M2					
December	3.92	3.87	3.81	0.00	0.00	0.77	9.84	**4.14**	3.31	31.81	69.00	6
January	**8.81**	7.44	7.51	0.00	0.05	9.67	16.99	7.89	3.50	32.11	77.33	9
February	10.04	9.94	10.14	0.00	0.10	9.13	14.27	**11.59**	2.48	29.24	43.85	11
March	20.96	18.65	18.56	0.00	0.13	21.06	23.07	**15.38**	4.53	30.65	98.14	11

Since probabilistic forecasts are crucial for an adequate risk management, the proposed methodology has also been used to compute the probability of an extremely low price for each hour in the forecast period. Figure 10 shows the probability of the appearance of extremely low prices, which has been estimated on an hourly basis during a representative month.

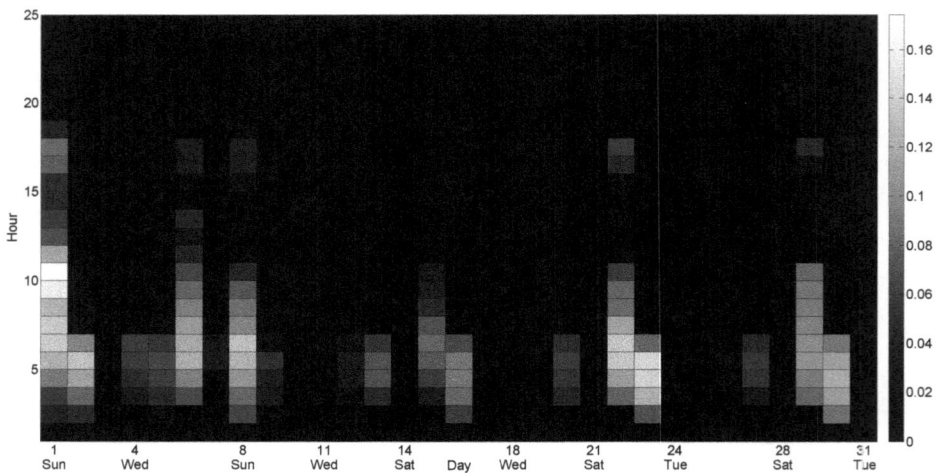

Figure 10. Probability of the appearance of extremely low prices predicted by the Logistic Regression M1 for January 2012.

Similarly, the proposed methodology has been applied to estimate the probability density function (PDF) associated with the number of hours with very low prices throughout the projection period. An illustrative example of the forecasted PDF for February 2012 by using the hybrid approach appears in Figure 11. In this figure, the dashed line represents the actual number of hours that occurred in the market. As shown, the distribution is unimodal and right-skewed. In this particular case, it is evident that the actual value always falls under the range of the most likely values. The real value (11 h) is near the mode value (8 h) and the expected value (11.59 h).

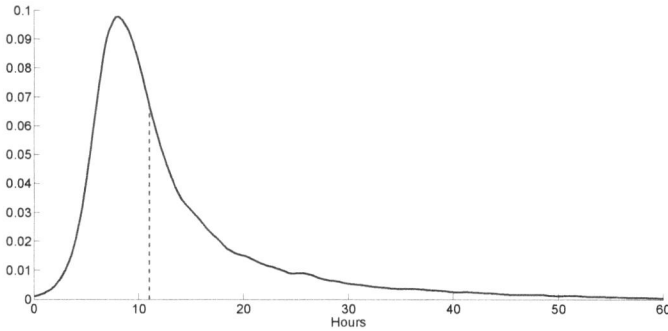

Figure 11. Probability density function for the number of hours with extremely low prices that has been predicted by the hybrid model for February 2012.

In order to compare the forecast quality of the proposed models, the Brier Score (BS) has been used. The BS is probably the most commonly-used verification measure for assessing the accuracy of binary probabilistic predictions. It is the mean squared error of the probability forecasts over the verification sample and it is expressed as:

$$BS = \frac{1}{S} \sum_{i=1}^{S} (p_i - o_i)^2 \tag{9}$$

where S is the sample size, p_i is the predicted probability of the event occurring according to the i-th hourly forecast and o_i is equal to one or zero, depending on whether the event subsequently occurred or not during that hour. The BS ranges from zero for a perfect forecast to one for the worst possible forecast.

With the objective of making it easier to interpret the results, two naive models have also been used as benchmarks. Naive 3 is based on the previous similar month, while Naive 4 relies on taking the historical values of month m-2 as forecasts of future prices for month m. A comparison of the results for the probabilistic estimates for all specifications is reported in Table 22. As can be seen, the obtained results suggest that the proposed hybrid methodology produces superior probabilistic forecasts than the rest of the alternative techniques. Naive techniques are clearly outperformed by all of the proposed procedures, which demonstrates the practical interest of the developed methodology. In this case, a slight increase in accuracy was obtained when considering different dynamics for each day of the week through the periodic models.

Table 22. Comparison of the proposed models in terms of the Brier Score (BS).

Logistic Regression			Decision Trees		MLP		Hybrid	MRS	Naive 3	Naive 4
M1	M2	M3	M1	M2	M1	M2				
0.01239	0.01213	0.01205	0.01264	0.01261	0.01246	0.01241	**0.01195**	0.01258	0.04611	0.03005

6. Conclusions

In this paper, a novel methodological approach to analyze and make real *ex ante* forecasts of the occurrence of extremely low prices in electricity markets with a medium-term horizon has been presented. The proposed methodology, which is a mixture of different forecasting techniques with a Monte Carlo simulation that integrates a spatial interpolation tool, is able to simultaneously perform punctual and probabilistic predictions with an hourly basis. The methodology has been specifically applied to the Spanish wholesale market, but may be extended equally to other electricity markets worldwide.

Logistic regression for rare events, decision trees, multilayer perceptrons and a novel hybrid approach, which is able to incorporate both fundamental and behavioral information, have been compared to a Markov regime switching model and several naive methods. Further research has been undertaken in order to evaluate whether periodic models, in which parameters switch according to the day of the week, can provide better prediction capabilities.

Overall, all of the proposed models present reasonable errors taking into account the complex nature of the phenomenon and substantially outperform naive techniques in both the in- and out-of-sample datasets. Encouraging results have been obtained from real *ex ante* forecasts of the distribution function of the exogenous variables used to predict the phenomenon. The results reveal that the integration of a market equilibrium model and logistic regression in a hybrid approach provides a significant improvement in the prediction accuracy in comparison to the individual models. We also found that the inclusion of a prior estimation of the market equilibrium price can provide valuable information when used as an input in a statistical technique, such as logistic regression, especially when there are structural or regulatory changes in the market. Logistic regression with the correction for rare events data has proven to be a simple, but effective tool enabling one to outperform multilayer perceptrons and decision trees in terms of forecasting accuracy. When the real explanatory variables are used, it is clear that MLP performs better than the other models, but it behaves significantly worse when making real *ex ante* forecasts. However, still, MLP is superior to decision trees. With respect to decision trees, they have shown that they can provide valuable information and offer great interpretability for probabilistic approaches.

Another interesting conclusion is that a meaningful improvement of the prediction capability is reached when considering different dynamics for working days, Saturdays, Sundays and holidays. Open research lines may include the extension of this methodology to extremely high priced hours and to other markets, where locational marginal prices may exist and for which the impact of variables related to local distributed generation should be taken into account.

Acknowledgments: The authors are very grateful to the Spanish Ministry of Economy and Competitiveness (Ministerio de Economía y Competitividad) for their funding (MTM2013-48462-C2-2-R).

Author Contributions: Antonio Bello conducted the research, while Javier Reneses and Antonio Muñoz supervised this paper.

Conflicts of Interest: The authors declare no conflict of interest.

Nomenclature

D	Electric demand
H	Run-of the-river hydro production
HOL	Sundays and holidays
IE	Difference between exports and imports
N	Nuclear energy production
ND	Net demand (D-SR)
ND'	Net demand (ND + IE-N)
ND''	Net demand (ND-W-N-H)
SAT	Saturdays
SR	Special regime energy production
W	Wind energy production
WORK	Working days

References

1. Desideri, U.; Proietti, S.; Arcioni, L. Analysis and Statistic Evaluation of Distributed Generation in Italy. In Proceedings of the ASME 7th Biennial Conference on Engineering Systems Design and Analysis, Manchester, UK, 19–22 July 2004 ; Volume 1, pp. 79–85.
2. Gautam, D.; Nadarajah, M. Influence of Distributed Generation on Congestion and LMP in Competitive Electricity Market. *Int. J. Electr. Comput. Energ. Electron. Commun. Eng.* **2010**, *4*, 605–612.
3. Weron, R. Electricity price forecasting: A review of the state-of-the-art with a look into the future. *Int. J. Forecast.* **2014**, *30*, 1030–1081.
4. González, A.; Muñoz, A.; García-González, J. Modeling and forecasting electricity prices with input/output hidden Markov models. *IEEE Trans. Power Syst.* **2005**, *20*, 13–24.
5. Li, G.; Liu, C.; Lawarree, J.; Gallanti, M.; Venturini, A. State-of-the-art of electricity price forecasting. In Proceedings of the International Symposium CIGRE/IEEE PES, New Orleans, LA, USA, 5–7 October 2005; pp. 110–119.
6. Ventosa, M.; Baíllo, A.; Ramos, A.; Rivier, M. Electricity market modeling trends. *Energy Policy* **2005**, *33*, 897–913.
7. Bello, A.; Reneses, J.; Muñoz, A.; Delgadillo, A. Probabilistic forecasting of hourly electricity prices in the medium-term using spatial interpolation techniques. *Int. J. Forecast.* **2015**, doi:10.1016/j.ijforecast.2015.06.002.
8. Aggarwal, S.K.; Saini, L.M.; Kumar, A. Electricity price forecasting in deregulated markets: A review and evaluation. *Int. J. Electr. Power Energy Syst.* **2009**, *31*, 13–22.
9. García-Martos, C.; Rodríguez, J.; Sánchez, M.J. Forecasting electricity prices and their volatilities using Unobserved Components. *Energy Econ.* **2011**, *33*, 1227–1239.
10. Contreras, J.; Espínola, R.; Nogales, F.; Conejo, A. ARIMA models to predict next-day electricity prices. *IEEE Trans. Power Syst.* **2003**, *18*, 1014–1020.
11. Conejo, A.; Plazas, M.; Espínola, R.; Molina, A. Day-Ahead Electricity Price Forecasting Using the Wavelet Transform and ARIMA Models. *IEEE Trans. Power Syst.* **2005**, *20*, 1035–1042.
12. Zareipour, H.; Canizares, C.; Bhattacharya, K.; Thomson, J. Application of Public-Domain Market Information to Forecast Ontario's Wholesale Electricity Prices. *IEEE Trans. Power Syst.* **2006**, *21*, 1707–1717.
13. García-Martos, C.; Rodríguez, J.; Sánchez, M. Mixed Models for Short-Run Forecasting of Electricity Prices: Application for the Spanish Market. *IEEE Trans. Power Syst.* **2007**, *22*, 544–552.
14. Nogales, F.; Contreras, J.; Conejo, A.; Espínola, R. Forecasting next-day electricity prices by time series models. *IEEE Trans. Power Syst.* **2002**, *17*, 342–348.
15. Nogales, F.J.; Conejo, A.J. Electricity price forecasting through transfer function models. *J. Op. Res. Soc.* **2005**, *57*, 350–356.
16. Panagiotelis, A.; Smith, M. Bayesian density forecasting of intraday electricity prices using multivariate skew t distributions. *Int. J. Forecast.* **2008**, *24*, 710–727.
17. Cruz, A.; Muñoz, A.; Zamora, J.L.; Espínola, R. The effect of wind generation and weekday on Spanish electricity spot price forecasting. *Electr. Power Syst. Res.* **2011**, *81*, 1924–1935.
18. Amjady, N. Day-Ahead Price Forecasting of Electricity Markets by a New Fuzzy Neural Network. *IEEE Trans. Power Syst.* **2006**, *21*, 887–896.
19. Lu, X.; Dong, Z.Y.; Li, X. Electricity market price spike forecast with data mining techniques. *Electr. Power Syst. Res.* **2005**, *73*, 19–29.
20. Deng, S. *Stochastic Models of Energy Commodity Prices and Their Applications: Mean-reversion with Jumps and Spikes*; University of California Energy Institute: Berkeley, CA, USA, 1998; pp. 1–35.
21. Escribano, A.; Peña, J.I.; Villaplana, P. Modeling electricity prices: International evidence. *Oxford Bulletin of Economics and Statistics* **2011**, *73*, 622–650.
22. Christensen, T.; Hurn, A.; Lindsay, K. Forecasting spikes in electricity prices. *Int. J. Forecast.* **2012**, *28*, 400–411.
23. Cuaresma, J.C.; Hlouskova, J.; Kossmeier, S.; Obersteiner, M. Forecasting electricity spot-prices using linear univariate time-series models. *Appl. Energy* **2004**, *77*, 87–106.
24. Misiorek, A.; Trueck, S.; Weron, R. Point and interval forecasting of spot electricity prices: Linear *vs.* non-linear time series models. *Stud. Nonlinear Dyn. Econom.* **2006**, *10*, 1–34.
25. Rambharat, B.R.; Brockwell, A.E.; Seppi, D.J. A threshold autoregressive model for wholesale electricity prices. *J. R. Stat. Soc. Ser. C* **2005**, *54*, 287–299.

26. Becker, R.; Hurn, S.; Pavlov, V. Modelling Spikes in Electricity Prices. *Econ. Rec.* **2007**, *83*, 371–382.
27. Cartea, A.; Figueroa, M.G. Pricing in electricity markets: a mean reverting jump diffusion model with seasonality. *Appl. Math. Financ.* **2005**, *12*, 313–335.
28. Knittel, C.R.; Roberts, M.R. An empirical examination of restructured electricity prices. *Energy Econ.* **2005**, *27*, 791–817.
29. Cerjan, M.; Matijas, M.; Delimar, M. Dynamic Hybrid Model for Short-Term Electricity Price Forecasting. *Energies* **2014**, *7*, 3304–3318.
30. Zhao, J.H.; Dong, Z.Y.; Li, X.; Wong, K.P. A general method for electricity market price spike analysis. In Proceedings of the IEEE Power Engineering Society General Meeting, San Francisco, CA, USA, 12–16 June 2005; pp. 286–293.
31. Sharma, V.; Srinivasan, D. A spiking neural network based on temporal encoding for electricity price time series forecasting in deregulated markets. In Proceedings of the 2010 International Joint Conference on Neural Networks (IJCNN), Barcelona, Spain, 18–23 July 2010; pp. 1–8.
32. Voronin, S.; Partanen, J. Price Forecasting in the Day-Ahead Energy Market by an Iterative Method with Separate Normal Price and Price Spike Frameworks. *Energies* **2013**, *6*, 5897–5920.
33. Zhao, J.H.; Dong, Z.Y.; Li, X. Electricity market price spike forecasting and decision making. *IET Gener. Transm. Distrib.* **2007**, *1*, 647–654.
34. Cooper, J.A.; Saracci, R.; Cole, P. Describing the validity of carcinogen screening tests. *Br. J. Cancer* **1979**, *39*, 87–89.
35. Dueñas, P.; Reneses, J.; Barquin, J. Dealing with multi-factor uncertainty in electricity markets by combining Monte Carlo simulation with spatial interpolation techniques. *IET Gener. Transm. Distrib.* **2011**, *5*, 323–331.
36. Lapuerta, C.; Wilson, P.; Moselle, B.; Carere, E. *Recommendations for the Dutch Electricity Market*; The Brattle Group, Ltd.: London, UK, 2001; pp. 1–40.
37. Bierbrauer, M.; Trück, S.; Weron, R. Modeling electricity prices with regime switching models; In Proceedings of the 4th International Conference on Computational Science (ICCS), Kraków, Poland, 6–9 June 2004; pp. 859–867.
38. Monteiro, C.; Fernandez-Jimenez, L.A.; Ramirez-Rosado, I.J. Explanatory Information Analysis for Day-Ahead Price Forecasting in the Iberian Electricity Market. *Energies* **2015**, *8*, 10464–10486.
39. Chen, D.; Bunn, D. The Forecasting Performance of a Finite Mixture Regime-Switching Model for Daily Electricity Prices. *J. Forecast.* **2014**, *33*, 364–375.
40. Karakatsani, N.; Bunn, D. Forecasting Electricity Prices: The Impact of Fundamentals and Time-Varying Coefficients. *Int. J. Forecast.* **2008**, *24*, 764–785.
41. Mount, T.D.; Ning, Y.; Cai, X. Predicting price spikes in electricity markets using a regime-switching model with time-varying parameters. *Energy Econ.* **2006**, *28*, 62–80.
42. King, G.; Zeng, L. Logistic Regression in Rare Events Data. *Polit. Anal.* **2001**, *9*, 137–163.
43. Quinlan, J.R. Discovering Rules by Induction from Large Collections of Examples. In *Expert Systems in the Micro-Electronic Age*; Michie, D., Ed.; Edinburgh University Press: Edinburgh, UK, 1979; pp. 168–201.
44. Bishop, C.M. *Neural Networks for Pattern Recognition*; Oxford University Press: New York, USA, 1995.
45. Barquín, J.; Centeno, E.; Reneses, J. Stochastic market equilibrium model for generation planning. In Proceedings of the International Conference on Probabilistic Methods Applied to Power Systems, Ames, IA, USA, 12–16 September 2004; pp. 367–372.
46. Centeno, E.; Reneses, J.; Barquín, J. Strategic Analysis of Electricity Markets Under Uncertainty: A Conjectured-Price-Response Approach. *IEEE Trans. Power Syst.* **2007**, *22*, 423–432.
47. Bunn, D.W. *Modelling Prices in Competitive Electricity Markets*, 1st ed.; Wiley: Chichester, UK, 2004.
48. Díaz, C.; Villar, J.; Campos, F.; Reneses, J. Electricity market equilibrium based on conjectural variations. *Electr. Power Syst. Res.* **2010**, *80*, 1572–1579.
49. Wogrin, S.; Dueñas, P.; Delgadillo, A.; Reneses, J. A New Approach to Model Load Levels in Electric Power Systems With High Renewable Penetration. *IEEE Trans. Power Syst.* **2014**, *29*, 2210–2218.
50. Weron, R. *Modeling and Forecasting Electricity Loads and Prices: A Statistical Approach*; Wiley: Chichester, UK, 2006.
51. Weron, R. Heavy-tails and regime-switching in electricity prices. *Math. Methods Op. Res. ZOR* **2009**, *69*, 457–473.

52. Janczura, J.; Weron, R. Efficient estimation of Markov regime-switching models: An application to electricity spot prices. *AStA Adv. Stat. Anal.* **2012**, *96*, 385–407.

53. Hamilton, J.D. Analysis of time series subject to changes in regime. *J. Econom.* **1990**, *45*, 39–70.

54. Iberian Energy Market Operator (OMIE). http://www.omie.es/en/inicio (accessed on 16 September 2015).

![energies logo] *energies*

MDPI

Article

Automated Variable Selection and Shrinkage for Day-Ahead Electricity Price Forecasting

Bartosz Uniejewski, Jakub Nowotarski and Rafał Weron *

Department of Operations Research, Wrocław University of Technology, 50-370 Wrocław, Poland;
uniejewskibartosz@gmail.com (B.U.); jakub.nowotarski@pwr.edu.pl (J.N.)
* Correspondence: rafal.weron@pwr.edu.pl; Tel.: +48-71-320-4525

Academic Editor: Javier Contreras
Received: 5 July 2016; Accepted: 29 July 2016; Published: 5 August 2016

Abstract: In day-ahead electricity price forecasting (EPF) variable selection is a crucial issue. Conducting an empirical study involving state-of-the-art parsimonious expert models as benchmarks, datasets from three major power markets and five classes of automated selection and shrinkage procedures (single-step elimination, stepwise regression, ridge regression, lasso and elastic nets), we show that using the latter two classes can bring significant accuracy gains compared to commonly-used EPF models. In particular, one of the elastic nets, a class that has not been considered in EPF before, stands out as the best performing model overall.

Keywords: electricity price forecasting; day-ahead market; autoregression; variable selection; stepwise regression; ridge regression; lasso; elastic net

JEL Classification: C14, C22, C51, C53, Q47

1. Introduction

Alongside short-term load forecasting, short-term electricity price forecasting (EPF) has become a core process of an energy company's operational activities [1]. The reason is quite simple. A 1% improvement in the mean absolute percentage error (MAPE) in forecasting accuracy would result in about 0.1%–0.35% cost reductions from short-term EPF [2]. In dollar terms, this would translate into savings of ca. $1.5 million per year for a typical medium-size utility with a 5-GW peak load [3].

As has been noted in a number of studies, be it statistical or computational intelligence, a key point in EPF is the appropriate choice of explanatory variables [1,4–11]. The typical approach has been to select predictors in an ad hoc fashion, sometimes using expert knowledge, seldom based on some formal validation procedures. Very rarely has an automated selection or shrinkage procedure been carried out in EPF, especially for a large set of initial explanatory variables.

Early examples of formal variable selection in EPF include Karakatsani and Bunn [12] and Misiorek [13], who used stepwise regression to eliminate statistically insignificant variables in parsimonious autoregression (AR) and regime-switching models for individual load periods. Amjady and Keynia [4] proposed a feature selection algorithm that utilized the mutual information technique. (for later applications, see, e.g., [11,14,15]). In an econometric setup, Gianfreda and Grossi [5] computed p-values of the coefficients of a regression model with autoregressive fractionally integrated moving average disturbances (Reg-ARFIMA) and in one step eliminated all statistically-insignificant variables. In a study concerning the profitability of battery storage, Barnes and Balda [16] utilized ridge regression to compute forecasts of the New York Independent System Operator (NYISO) electricity prices for a model with more than 50 regressors.

More recently, González et al. [17] used random forests to identify important explanatory variables among the 22 considered. Ludwig et al. [7] used both random forests and the least absolute shrinkage and selection operator (i.e., lasso or LASSO) as a feature selection algorithm to choose the relevant out of the 77 available weather stations. In a recent neural network study, Keles et al. [11] combined the *k-nearest-neighbor* algorithm with backward elimination to select the most appropriate input variables out of more than 50 fundamental parameters or lagged versions of these parameters. Finally, Ziel et al. [9,18] used the lasso to sparsify very large sets of model parameters (well over 100). They used time-varying coefficients to capture the intra-day dependency structure, either using B-splines and one large regression model for all hours of the day [9] or, more efficiently, using a set of 24 regression models for the 24 h of the day [18].

However, a thorough study involving state-of-the-art parsimonious expert models as benchmarks, data from diverse power markets and, most importantly, a set of different selection or shrinkage procedures is still missing in the literature. In particular, to our best knowledge, elastic nets have not been applied in the EPF context at all. It is exactly the aim of this paper to address these issues. We perform an empirical study that involves:

- nine variants of three parsimonious autoregressive model structures with exogenous variables (ARX): one originally proposed by Misiorek et al. [19] and later used in a number of EPF studies [13,18,20–27], one which evolved from it during the successful participation of TEAM POLAND in the Global Energy Forecasting Competition 2014 (GEFCom2014; see [28–30]) and an extension of the former, which creates a stronger link with yesterday's prices and additionally considers a second exogenous variable (zonal load or wind power),
- three two-year long, hourly resolution test periods from three distinct power markets (GEFCom2014, Nord Pool and the U.K.),
- nine variants of five classes of selection and shrinkage procedures: single-step elimination of insignificant predictors (without or with constraints), stepwise regression (with forward selection or backward elimination), ridge regression, lasso and three elastic nets (with $\alpha = 0.25$, 0.5 or 0.75),
- model validation in terms of the robust weekly-weighted mean absolute error (WMAE; see [1]) and the Diebold–Mariano (DM; see [31]) test

and draw statistically-significant conclusions of high practical value.

The remainder of the paper is structured as follows. In Section 2, we introduce the datasets. Next, in Section 3, we first discuss the iterative calibration and forecasting scheme, then describe the techniques considered for price forecasting: a simple naive benchmark, nine variants of three parsimonious ARX-type model structures and five classes of selection and shrinkage procedures. In Section 4, we summarize the empirical findings. Namely, we evaluate the quality of point forecasts in terms of WMAE errors, run the DM tests to formally assess the significance of differences in the forecasting performance and analyze variable selection for the best performing elastic net model. Finally, in Section 5 wrap up the results and conclude.

2. Datasets

The datasets used in this empirical study include three spot market time series. The first one comes from the Global Energy Forecasting Competition 2014 (GEFCom2014), the largest energy forecasting competition to date [28]. The dataset includes three time series at hourly resolution: locational marginal prices, day-ahead predictions of system loads and day-ahead predictions of zonal loads and covers the period 1 January 2011–14 December 2013; see Figure 1. The origin of the data has never been revealed by the organizers. The full dataset is now available as supplementary material accompanying [28] (Appendix A); however, during the competition, the information set was being extended on a weekly basis to prevent 'peeking' into the future. The dataset was preprocessed by the organizers and does not include any missing or doubled values.

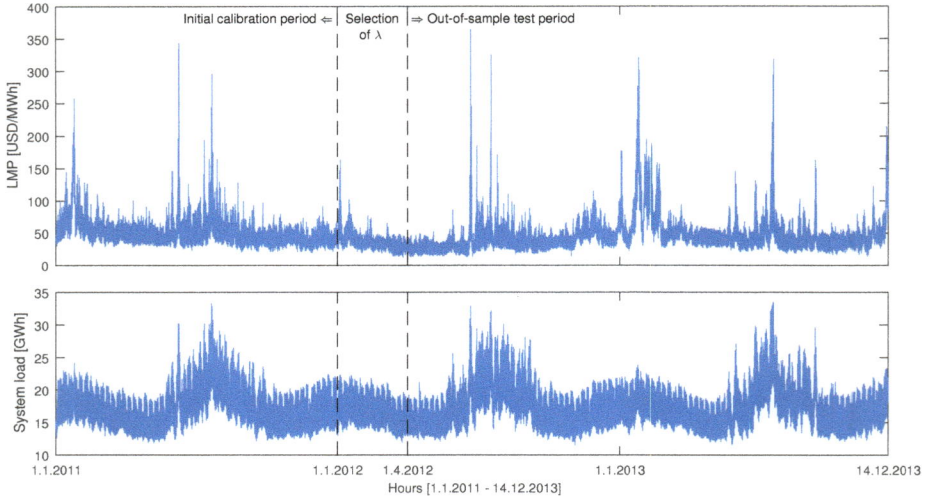

Figure 1. GEFCom2014 hourly locational marginal prices (LMP; top) and hourly day-ahead predictions of the system load (bottom) for the period 1 January 2011–14 December 2013. The day-ahead predictions of the zonal load are generally indistinguishable from those of the system load at this resolution; see Figure 8 in [28]. The vertical dashed lines mark the beginning of the 91-day period for selecting λ's (ridge regression, lasso, elastic nets) and the beginning of the 623-day long out-of-sample test period.

The second dataset comes from one of the major European power markets: Nord Pool (NP). It comprises hourly system prices, hourly consumption prognosis for four Nordic countries (Denmark, Finland, Norway and Sweden) and hourly wind prognosis for Denmark and covers the period 1 January 2013–29 March 2016; see Figure 2. The time series were constructed using data published by the Nordic power exchange Nord Pool (www.nordpoolspot.com) and preprocessed to account for missing values and changes to/from the daylight saving time, analogously as in [20] (Section 4.3.7). The missing data values (corresponding to the changes to the daylight saving/summer time; moreover, eight out of 28,392 hourly consumption figures were missing for Norway) were substituted by the arithmetic average of the neighboring values. The 'doubled' values (corresponding to the changes from the daylight saving/summer time) were substituted by the arithmetic average of the two values for the 'doubled' hour.

The third dataset comes from N2EX, the U.K. day-ahead power market operated by Nord Pool. It comprises hourly system prices for the period 1 January 2013–29 March 2016; see Figure 3. The time series was constructed using data published by Nord Pool (www.nordpoolspot.com) and, like the second dataset, preprocessed to account for changes to/from the daylight saving time. Note that the U.K. dataset includes only prices, as no day-ahead forecasts of fundamental variables were available to us. Hence, models calibrated to the U.K. data are 'pure price' models. To better see the effect of excluding fundamentals from forecasting models, we use the GEFCom2014 dataset twice, once with fundamentals (system and zonal load forecasts; to compare with the results for Nord Pool) and once without them.

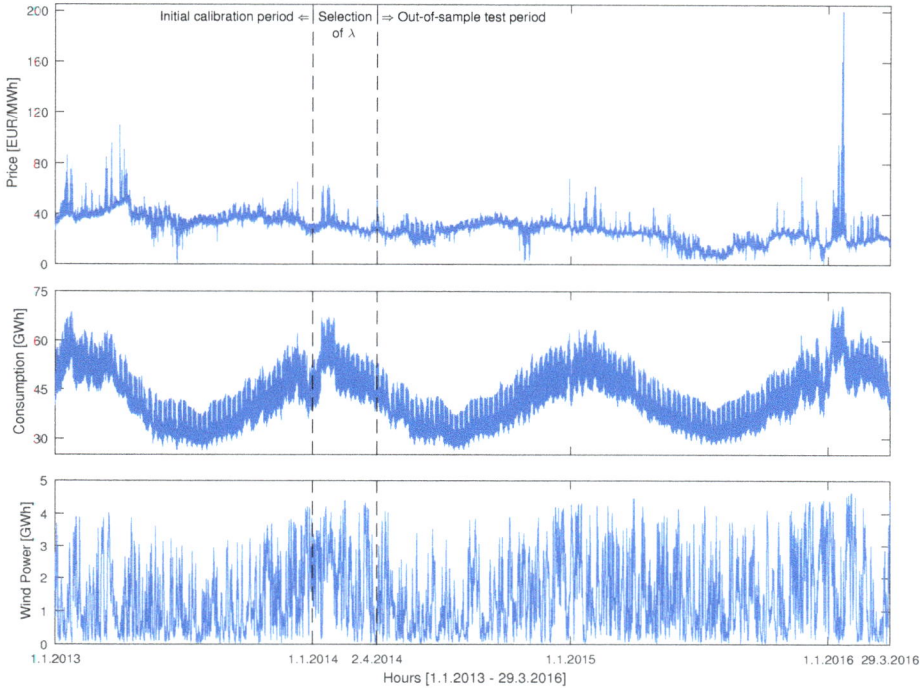

Figure 2. Nord Pool hourly system prices (top), hourly consumption prognosis (middle) and hourly wind power prognosis for Denmark (bottom) for the period 1 January 2013–29 March 2016. The vertical dashed lines mark the beginning of the 91-day period for selecting λ's (ridge regression, lasso, elastic net) and the beginning of the 728-day long out-of-sample test period.

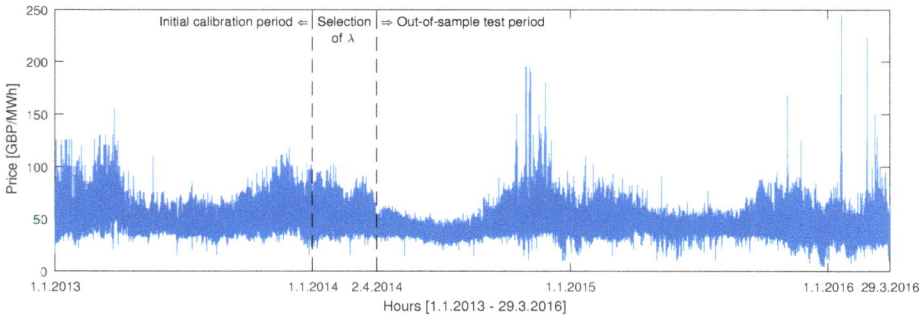

Figure 3. U.K. power market hourly system prices (Nord Pool's N2EX market) for the period 1 January 2013–29 March 2016. The vertical dashed lines mark the beginning of the 91-day period for selecting λ's (ridge regression, lasso, elastic net) and the beginning of the 728-day long out-of-sample test period.

3. Methodology

It should be noted that although we use here the terms short-term, spot and day-ahead interchangeably, the former two do not necessarily refer to the day-ahead market. Short-term EPF generally involves predicting 24 hourly (or 48 half-hourly) prices in the day-ahead market, cleared typically at noon on the day before delivery, i.e., 12–36 h before delivery, the adjustment markets, cleared a few hours before delivery, and the balancing or real-time markets, cleared minutes before delivery [32]. The spot market, especially in the literature on European electricity markets, is often used as a synonym of the day-ahead market. However, in the U.S., the spot market is another name for the real-time market, while the day-ahead market is called the forward market [20,33]. Furthermore, some markets in Europe nowadays admit continuous trading for individual load periods up to a few hours before delivery. With the shifting of volume from the day-ahead to intra-day markets, also in Europe, the term spot is more and more often being used to refer to the real-time markets [1].

Throughout this article, we denote by $P_{d,h}$ the electricity price in the day-ahead market for day d and hour h. Like many studies in the EPF literature [1], we use the logarithmic transform to make the price series more symmetric (see Figure 4) and compare with the top panels in Figures 1–3. We can do this since all considered datasets are positive-valued. However, this is not a very restrictive property. If datasets with zero or negative values were considered, we could work with non-transformed prices. Furthermore, we center the log-prices by subtracting their in-sample mean prior to parameter estimation. We do this independently for each hour $h = 1, ..., 24$:

$$p_{d,h} = \log(P_{d,h}) - \frac{1}{T} \sum_{t=1}^{T} \log(P_{t,h}),$$ (1)

where T is the number of days in the calibration window; hence, the missing intercept ($\beta_{h,0} \equiv 0$) in our autoregressive models; for model parameterizations, see Sections 3.2–3.4.

For all three markets, the day-ahead forecasts of the hourly electricity price are determined within a rolling window scheme, using a 365-day calibration window. First, all considered models are calibrated to data from the initial calibration period (i.e., 1 January 2011–31 December 2011 for GEFCom2014 and 1 January 2013–31 December 2013 for Nord Pool and the U.K.), and forecasts for all 24 h of the next day (1 January) are determined. Then, the window is rolled forward by one day; the models are re-estimated, and forecasts for all 24 h of 2 January are computed. This procedure is repeated until the predictions for the 24 h of the last day in the sample (14 December 2013 for GEFCom2014 and 29 March 2016 for Nord Pool and the U.K.) are made.

For models requiring calibration of the regularization parameter (i.e., λ), we use a setup commonly considered in the machine learning literature. Namely, we divide our datasets into estimation (365 days), validation (91 days or 13 full weeks) and test periods (623 days for GEFCom2014, 728 days for Nord Pool and the U.K.; respectively 89 and 104 full weeks). For each of the five models—ridge regression, lasso and elastic nets with $\alpha = 0.25, 0.50$ and 0.75—34 different 'sub-models' with 34 values of λ spanning the regularization parameter space (see Sections 3.4.3 and 3.4.4 for details) are estimated in the 91-day validation period directly following the last day of the initial calibration period; see Figures 1–3. For all hours of the day, only one value of λ is chosen for each of the five models: the one that yields the smallest $\overline{\text{WMAE}}$ error during this 91-day period; for error definitions, see Section 4.1. This value of λ is later used for computing day-ahead price forecasts in the whole out-of-sample test period. To ensure that all models are evaluated using the same data, predictions of all models are compared only in the out-of-sample test periods: 1 April 2012–14 December 2013 (623 days) for GEFCom2014 and 2 April 2014–29 March 2016 (728 days) for Nord Pool and the U.K. Obviously, such a simple procedure for the selection of the regularization parameter may not be optimal. Generally, better performance is to be expected from shrinkage models when λ is recalibrated at every time step. Such an approach has been recently taken by Ziel [18], who used the Bayesian information criterion to select one out of 50 values of λ for every day and every hour in the 969-day-long out-of-sample test period. The downside of such an approach is, however, the increased computational time.

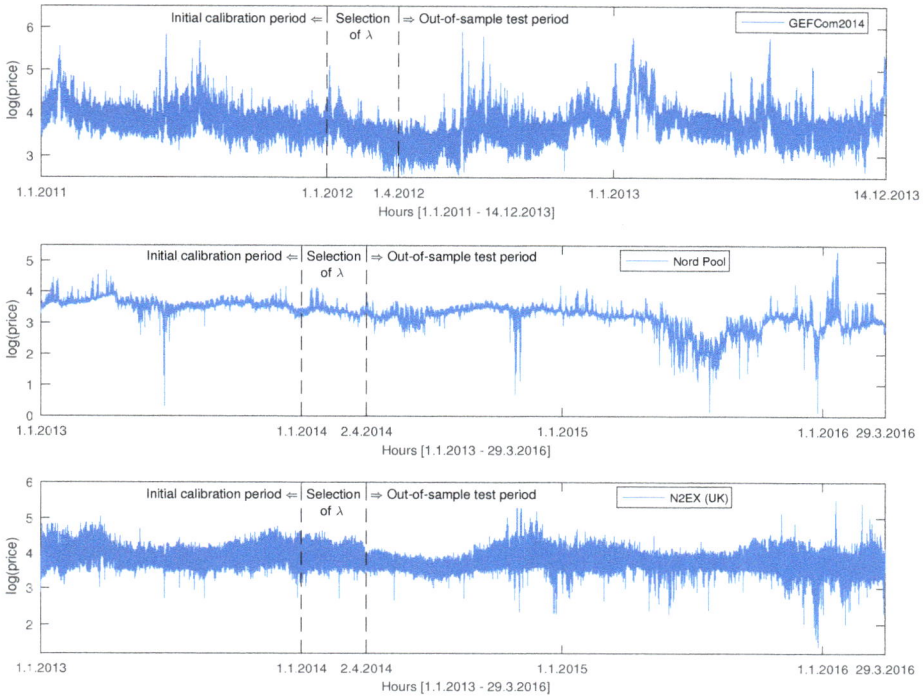

Figure 4. Global Energy Forecasting Competition 2014 (GEFCom2014) (top), Nord Pool (middle) and N2EX (U.K.; bottom) hourly log prices. As expected, the logarithmic transform makes the price series more symmetric. The vertical dashed lines mark the beginning of the 91-day period for selecting λ's (ridge regression, lasso, elastic nets) and the beginning of the out-of-sample test periods. Each day, the 365-day-long calibration window is rolled forward by 24 h; the models are re-estimated; and price forecasts for the 24 h of the next day are computed.

Our choice of the model classes is guided by the existing literature on short-term EPF. Like in [12,18,25–27,30], the modeling is implemented separately across the hours, leading to 24 sets of parameters for each day the forecasting exercise is performed. As Ziel [18] notes, when we compare the forecasting performance of relatively simple models implemented separately across the hours and jointly for all hours (like in [9,34–36]), the latter generally performs better for the first half of the day, whereas the former are better in the second half of the day. At the same time, models implemented separately across the hours offer more flexibility by allowing for time-varying cross-hour dependency in a straightforward manner. Hence, our choice of the modeling framework.

In the remainder of this section, we first define the benchmarks: a simple similar-day technique and a collection of parsimonious autoregressive models. Since the latter are usually built on some prior knowledge of experts, like in [18], we refer to them as expert models. Then, we move on to describe the selection and shrinkage procedures used in this study.

3.1. The Naïve Benchmark

The first benchmark, most likely introduced to the EPF literature in [34] and dubbed the naïve method, belongs to the class of similar-day techniques (for a taxonomy of EPF approaches, see, e.g., [1]). It proceeds as follows: the electricity price forecast for hour h on Monday is set equal to the price for the same hour on Monday of the previous week, and the same rule applies for Saturdays and Sundays; the electricity price forecast for hour h on Tuesday is set equal to the price for the same hour on Monday, and the same rule applies for Wednesdays, Thursdays and Fridays. As was argued in [34,35], forecasting procedures that are not calibrated carefully fail to outperform the naive method surprisingly often. We denote this benchmark by **Naive**.

3.2. Autoregressive Expert Benchmarks

The second benchmark is a parsimonious autoregressive structure originally proposed by Misiorek et al. [19] and later used in a number of EPF studies [18,20,21,23–27]. Within this model, the centered log-price on day d and hour h, i.e., $p_{d,h}$, is given by the following formula:

$$
\begin{aligned}
p_{d,h} = &\beta_{h,1} p_{d-1,h} + \beta_{h,2} p_{d-2,h} + \beta_{h,3} p_{d-7,h} + \beta_{h,4} p_{d-1}^{min} + \beta_{h,5} z_{d,h} \\
&+ \beta_{h,6} D_{Sat} + \beta_{h,7} D_{Sun} + \beta_{h,8} D_{Mon} + \varepsilon_{d,h},
\end{aligned} \tag{2}
$$

where the lagged log-prices $p_{d-1,h}$, $p_{d-2,h}$ and $p_{d-7,h}$ account for the autoregressive effects of the previous days (the same hour yesterday, two days ago and one week ago), while $p_{d-1}^{min} \equiv \min_{h=1,\dots,24}\{p_{d-1,h}\}$ is the minimum of the previous day's 24 hourly log-prices. The exogenous variable $z_{d,h}$ refers to the logarithm of hourly system load or Nordic consumption for day d and hour h (actually, to forecasts made a day before, see Section 2). The three dummy variables—D_{Sat}, D_{Sun} and D_{Mon}—account for the weekly seasonality. Finally, the $\varepsilon_{d,h}$'s are assumed to be independent and identically distributed (i.i.d.) normal variables. We denote this autoregressive benchmark by **ARX1** to reflect the fact that the load (or consumption) forecast is used as the exogenous variable in Equation (2). The corresponding model with $\beta_{h,5} \equiv 0$, i.e., with no exogenous variable, is denoted by **AR1**. The **ARX1** and **AR1** models, as well as all autoregressive structures considered in Sections 3.2 and 3.3, are estimated in this study with least squares (LS), using MATLAB's regress.m function.

In what follows, we also consider two variants of Equation (2) that treat holidays as special days:

$$
\begin{aligned}
p_{d,h} = &\beta_{h,1} p_{d-1,h} + \beta_{h,2} p_{d-2,h} + \beta_{h,3} p_{d-7,h} + \beta_{h,4} p_{d-1}^{min} + \beta_{h,5} z_{d,h} \\
&+ \beta_{h,6} D_{Sat} + \beta_{h,7} D_{Sun} + \beta_{h,8} D_{Mon} + \beta_{h,9} D_{Hol} + \varepsilon_{d,h},
\end{aligned} \tag{3}
$$

and that additionally utilize the fact that prices for early morning hours depend more on the previous day's price at midnight, i.e., $p_{d-1,24}$, than on the price for the same hour, as recently noted in [18,29]:

$$
\begin{aligned}
p_{d,h} = &\beta_{h,1} p_{d-1,h} + \beta_{h,2} p_{d-2,h} + \beta_{h,3} p_{d-7,h} + \beta_{h,4} p_{d-1}^{min} + \beta_{h,5} z_{d,h} \\
&+ \beta_{h,6} D_{Sat} + \beta_{h,7} D_{Sun} + \beta_{h,8} D_{Mon} + \beta_{h,9} D_{Hol} + \beta_{h,10} p_{d-1,24} + \varepsilon_{d,h}.
\end{aligned} \tag{4}
$$

We denote Models (3) and (4) by **ARX1h** and **ARX1hm**, respectively. Similarly, corresponding models with $\beta_{h,5} \equiv 0$ are denoted by **AR1h** and **AR1hm**. Note, that when forecasting the electricity price for the last load period of the day, i.e., $p_{d,24}$, models with suffix **hm** reduce to models with suffix **h** (this is true for all models considered in Section 3.2).

In Equations (3) and (4), D_{Hol} is a dummy variable for holidays. The holidays were identified using the Time and Date AS (www.timeanddate.com/holidays) web page: U.S. federal holidays (for GEFCom2014), national holidays in Norway (for Nord Pool) and public holidays, bank holidays and major observances in the U.K. (option 'Holidays and some observances').

The third benchmark is an extension of the **ARX1** model, which takes into account the experience gained during the GEFCom2014 competition that it may be beneficial to use different model structures for different days of the week, not only different parameter sets [29]. Hence, the multi-day ARX model (denoted later in the text by **mARX1**) is given by the following formula:

$$
\begin{aligned}
p_{d,h} = &\left(\sum_{i \in I} \beta_{h,1,i} D_i \right) p_{d-1,h} + \beta_{h,2} p_{d-2,h} + \beta_{h,3} p_{d-7,h} + \beta_{h,4} p_{d-1}^{min} + \beta_{h,5} z_{d,h} \\
&+ \beta_{h,6} D_{Sat} + \beta_{h,7} D_{Sun} + \beta_{h,8} D_{Mon} + \beta_{h,11} D_{Mon} p_{d-3,h} + \varepsilon_{d,h},
\end{aligned} \tag{5}
$$

where $I \equiv \{0, Sat, Sun, Mon\}$, $D_0 \equiv 1$ and the term $D_{Mon}p_{d-3,h}$ accounts for the autoregressive effect of Friday's prices on the prices for the same hour on Monday. Note that to some extent, this structure resembles periodic autoregressive moving average (PARMA) models, which have seen limited use in EPF [37,38]. Like for the **ARX1** model, also for **mARX1**, we consider two variants:

- **mARX1h**, which treats holidays as special days, i.e., with the $\beta_{h,9}D_{Hol}$ term in Equation (5),
- and **mARX1hm**, which additionally implements the dependence on the previous day's price at midnight, i.e., with the $\beta_{h,9}D_{Hol}$ and $\beta_{h,10}p_{d-1,24}$ terms in Equation (5).

The corresponding price only models, i.e., with $\beta_{h,5} \equiv 0$, are denoted by **mAR1**, **mAR1h** and **mAR1hm**.

Misiorek et al. [19] noted that the minimum of the previous day's 24 hourly prices was the best link between today's prices and those from the entire previous day. Their analysis, however, was limited to one small dataset (California CalPXprices, 3–9 April 2000) and only one simple function at a time (maximum, minimum, mean or median of the previous day's prices). To check if using more than one function leads to a better forecasting performance, we introduce a benchmark, which is an extension of the **ARX1** model that takes into account not only the minimum (p_{d-1}^{min}), but also the maximum (p_{d-1}^{max}) and the mean (p_{d-1}^{avg}) of the previous day's 24 hourly prices. Additionally, we include a second exogenous variable ($y_{d,h}$), which is taken as either the logarithm of the day-ahead zonal load forecast (GEFCom2014) or of the Danish wind power prognosis. The resulting **ARX2** model is given by the following formula:

$$
\begin{aligned}
p_{d,h} = {} & \beta_{h,1}p_{d-1,h} + \beta_{h,2}p_{d-2,h} + \beta_{h,3}p_{d-7,h} + \beta_{h,A}p_{d-1}^{min} + \beta_{h,5}z_{d,h} \\
& + \beta_{h,6}D_{Sat} + \beta_{h,7}D_{Sun} + \beta_{h,8}D_{Mon} \\
& + \beta_{h,11}p_{d-1}^{max} + \beta_{h,12}p_{d-1}^{avg} + \beta_{h,13}y_{d,h} + \varepsilon_{d,h}.
\end{aligned}
\tag{6}
$$

Like for the **ARX1** and **mARX1** models, also for **ARX2**, we consider two variants:

- **ARX2h** with the $\beta_{h,9}D_{Hol}$ term in Equation (6),
- and **ARX2hm** with the $\beta_{h,9}D_{Hol}$ and $\beta_{h,10}p_{d-1,24}$ terms in Equation (6).

The corresponding price only models, i.e., with $\beta_{h,5}, \beta_{h,13} \equiv 0$, are denoted by **AR2**, **AR2h** and **AR2hm**.

3.3. Full Autoregressive Model

Finally, we define a much richer autoregressive model that includes as special cases all expert models discussed in Section 3.2 and call it the **full ARX** or **fARX** model. We consider all regressors that, in our opinion, posses a non-negligible predictive power. The **fARX** model is similar in spirit to the general autoregressive model defined by Equation (2) in [18]. However, there are some important differences between them. On one hand, **fARX** includes exogenous variables and a much richer seasonal structure. On the other, it does not look that far into the past and concentrates only on days $d-1, d-2, d-3$ and $d-7$. The **fARX** model is given by the following formula:

$$
\begin{aligned}
p_{d,h} = {} & \sum_{i=1}^{24} \left(\beta_{h,i}p_{d-1,i} + \beta_{h,i+24}p_{d-2,i} + \beta_{h,i+48}p_{d-3,i} \right) + \beta_{h,73}p_{d-7,h} \\
& + \sum_{j=1}^{3} \left(\beta_{h,j+73}p_{d-j}^{min} + \beta_{h,j+76}p_{d-j}^{max} + \beta_{h,j+79}p_{d-j}^{avg} \right) \\
& + \beta_{h,83}z_{d,h} + \beta_{h,84}z_{d-1,h} + \beta_{h,85}z_{d-7,h} + \beta_{h,86}y_{d,h} \\
& + \sum_{k=1}^{7} \beta_{h,86+k}D_k + \sum_{k=1}^{7} \beta_{h,93+k}D_kz_{d,h} + \sum_{k=1}^{7} \beta_{h,100+k}D_kp_{d-1,h} + \varepsilon_{d,h},
\end{aligned}
\tag{7}
$$

where $D_1 \equiv D_{Sat}, D_2 \equiv D_{Sun}, ..., D_7 \equiv D_{Fri}$ are dummies for the seven days of the week (we treat holidays as the eighth day of the week, hence $D_1 = ... = D_7 = 0$ for holidays). The price only variant, **fAR**, is obtained by setting to zero all coefficients of the terms involving exogenous variables, i.e., $\beta_{h,i} \equiv 0$, for $i = 83, ..., 86, 94, ..., 100$.

Although we fit the **fARX** model to power market data and evaluate its forecasting performance, the main reason for including it in this study is to use it as the baseline model for the selection and shrinkage procedures discussed in Section 3.4. For this purpose, let us write the **fARX** model in a more compact form:

$$p_{d,h} = \sum_{i=1}^{n} \beta_{h,i} X_{d,h,i} + \varepsilon_{d,h}, \tag{8}$$

where $X_{d,h,i}$'s are the $n = 107$ regressors in Equation (7) and $\beta_{h,i}$'s are their coefficients.

3.4. Selection and Shrinkage Procedures

All autoregressive models considered in Sections 3.2 and 3.3 are estimated in this study with least squares (LS). However, there are many alternatives to using LS in multi-parameter models, in particular [39]:

- variable or subset selection, which involves identifying a subset of predictors that we believe to be influential, then fitting a model using LS on the reduced set of variables,
- shrinkage (also known as regularization), which fits the full model with all predictors using an algorithm that shrinks the estimated coefficients towards zero, which can significantly reduce their variance.

Depending on what type of shrinkage is performed, some of the coefficients may be shrunk to zero itself. As such, some shrinkage methods, like the lasso, de facto perform variable selection. It should be noted, however, that variable selection (or model sparsity) is beneficial for interpretability and faster simulation of model trajectories; for reducing the forecasting errors, only the shrinkage property is required.

3.4.1. Single-Step Elimination of Insignificant Predictors

This subset selection procedure is a simple alternative to stepwise regression discussed in Section 3.4.2 and has been used, for instance, in [5]. The idea is to fit the full regression model, in our case **fARX**, then in a single step, set to zero all statistically insignificant coefficients. We use MATLAB's regress.m function with the commonly-used 5% significance level. Setting to zero all coefficients in Equation (7) whose 95% confidence intervals (CI) include zero yields the **ssARX** model for a particular day and hour (the **ssAR** model is obtained analogously from **fAR**; see Section 3.3). This procedure can be conducted by imposing some additional constraints, for instance, leaving in the model all coefficients of the basic **ARX1** (or **AR1**) benchmark. This yields the **ssARX1** and **ssAR1** models. Of course, the most commonly-used significance level of 5% may not be optimal. We have additionally checked the performance of 90% and 97.5% CI. It turns out that the overall ranking of the **ssAR**-type models does not change much. However, **ssARX** and **ssAR** perform slightly better for the 90% CI, while **ssARX1** and **ssAR1** either for the 95% or the 97.5% CI.

3.4.2. Stepwise Regression

Although very fast, the single-step elimination may remove too many explanatory variables at once and lead to a poorly-performing subset of predictors. On the other hand, selecting the best subset from among all 2^n subsets of the n predictors is not computationally feasible for large n. Even if doable, it may lead to overfitting. For these reasons, stepwise methods, which explore a far more restricted set of models, are attractive alternatives to best subset selection [39]. In the context of EPF, they have been used, for instance, in [12,13,40].

There are two basic procedures: forward selection and backward elimination. Forward stepwise selection begins with a model containing no predictors and then iteratively adds variables to the model. At each step, the variable that gives the greatest additional improvement to the fit is added to the model, and the procedure continues until all important predictors are in the model. We use MATLAB's stepwisefit.m function, which computes the *p*-value of an *F*-statistic at each time step to test models with and without a potential term. If a variable is not currently in the model, the null hypothesis is that it would have a zero coefficient if added to the model. If there is sufficient evidence to reject the null hypothesis, that variable may be added to the model (we use stepwisefit's default 5% significance level for adding variables; naturally, this could be further fine tuned as for the single-step elimination procedures). In a given step, the function adds the variable with the smallest *p*-value. We denote the resulting models by **fsARX** and **fsAR**.

Backward stepwise elimination (or selection) begins with the full model containing all *n* variables, i.e., **fARX** or **fAR**, and then iteratively removes the least useful predictor, one at a time. MATLAB's stepwisefit.m function computes the null hypothesis that a given variable has a zero coefficient. If there is insufficient evidence to reject the null hypothesis, the variable may be removed from the model (we use stepwisefit's default 10% significance level for removing variables). In a given step, the function removes the variable with the largest *p*-value. We denote the resulting models by **bsARX** and **bsAR**.

3.4.3. Ridge Regression

Ridge regression is a regularization method introduced in statistics by Hoerl and Kennard [41]. To our best knowledge, apart from a limited study of Barnes and Balda [16] in the context of evaluating the profitability of battery storage, the method has not been used for EPF. Ridge regression is very similar to least squares, except that the β_i's in (8) are not estimated by minimizing the residual sum of squares (RSS), but by RSS penalized by a quadratic shrinkage factor:

$$\hat{\beta}^{ridge} = \underset{\beta_{h,i}}{\arg\min} \left\{ RSS + \lambda \sum_{i=1}^{n} \beta_{h,i}^2 \right\} \equiv \underset{\beta_{h,i}}{\arg\min} \left\{ \sum_{d,h \in T} \left(p_{d,h} - \sum_{i=1}^{n} \beta_{h,i} X_{d,h,i} \right)^2 + \lambda \sum_{i=1}^{n} \beta_{h,i}^2 \right\}, \quad (9)$$

where *T* represents the calibration period and $\lambda \geq 0$ is a tuning or regularization parameter, to be determined separately. Note that for $\lambda = 0$, we get the standard LS estimator; for $\lambda \to \infty$, all $\beta_{h,i}$'s tend to zero; while for intermediate values of λ, we are balancing two ideas: minimizing the RSS and shrinking the coefficients towards zero (and each other).

Ridge regression produces a different set of coefficient estimates for each value of λ; hence, selecting a good value for λ is critical. Cross-validation provides a simple way to tackle this problem [39]. We choose a grid of λ values (here: 34 equally-spaced values spanning the range from 1–100; if $\lambda \in \{94, 97, 100\}$ was selected, we additionally checked another set of 34 equally-spaced values spanning the range from 101–200) and using MATLAB's ridge.m function (we scale the regressors) compute the prediction errors for each value of the tuning parameter in the 91-day validation period; see Section 2. We then select λ for which the $\overline{WMAE}$ error (for the definition, see Section 4.1) is the smallest and use it for computing day-ahead price forecasts in the whole out-of-sample test period. The resulting model is denoted in the text by **RidgeX** or **Ridge** when the baseline model is **fAR**.

3.4.4. Lasso and Elastic Nets

Ridge regression has one unwanted feature when it comes to interpretation and model identification. Unlike stepwise regression, which will generally select models that involve just a subset of the variables, ridge regression will include all *n* predictors in the final model [39]. The quadratic shrinkage factor in Equation (9) will shrink all $\beta_{h,i}$'s towards zero, but it will not set any of them exactly to zero. In 1996, Tibshirani [42] proposed the least absolute shrinkage and selection operator (i.e., lasso or LASSO) that overcomes this disadvantage. It is the only shrinkage procedure that has been applied in EPF to a larger extent, however only in the last two years [7,9,18,25,43].

The lasso is a shrinkage method just like ridge regression. However, it uses a linear penalty factor instead of a quadratic one:

$$\hat{\beta}^{lasso} = \underset{\beta_{h,i}}{\operatorname{argmin}} \left\{ \text{RSS} + \lambda \sum_{i=1}^{n} |\beta_{h,i}| \right\}. \tag{10}$$

This subtle change makes the solutions nonlinear in $p_{d,h}$, and there is no closed form expression as in the case of ridge regression. Because of the nature of the shrinkage factor in Equation (10), making λ sufficiently large will cause some of the coefficients to be exactly zero [44]. Thus, the lasso de facto performs variable selection, just like the methods discussed in Sections 3.4.1 and 3.4.2. As in ridge regression, selecting a good value of λ for the lasso is critical. Here, we use MATLAB's lasso.m function and a grid of exponentially-decreasing λ's (the largest just sufficient to produce all $\beta_i = 0$; the function also automatically scales the regressors). We then select λ for which the $\overline{\text{WMAE}}$ error (for the definition, see Section 4.1) in the 91-day validation period is the smallest. The resulting model is denoted in the text by **LassoX**, or **Lasso** when the baseline model is **fAR**.

The lasso does not handle highly-correlated variables very well. The coefficient paths tend to be erratic and can sometimes show wild behavior [44]. This is not a critical issue for forecasting, but for interpretation and model identification, this has more serious consequences. In 2005, Zou and Hastie [45] proposed the elastic net, a new regularization and variable selection method that can be seen as an extension of ridge regression and the lasso. It often outperforms the lasso, while exhibiting a similar sparsity of representation. The elastic net uses a mixture of linear and quadratic penalty factors:

$$\hat{\beta}^{EN} = \underset{\beta_{h,i}}{\operatorname{argmin}} \left\{ \text{RSS} + \lambda \left(\frac{1-\alpha}{2} \sum_{i=1}^{n} \beta_{h,i}^2 + \alpha \sum_{i=1}^{n} |\beta_{h,i}| \right) \right\}, \tag{11}$$

where $\alpha \in [0,1]$. When $\alpha = 1$, the elastic net reduces to the lasso, and with $\alpha = 0$, it becomes ridge regression. The $\frac{1}{2}$ in the quadratic part of the elastic net penalty in Equation (11) leads to a more efficient and intuitive soft-thresholding operator in the optimization; the original formulation in [45] did not include the $\frac{1}{2}$ scaling. Note also that every elastic net problem can be rewritten as a lasso problem on augmented data. Hence, for fixed λ and α, the computational difficulty of the elastic net solution is similar to the lasso problem [44].

Compared to the lasso and ridge regression, the elastic net has an additional mixing parameter that has to be determined. It can be set on subjective grounds, as we do here, or optimized within a cross-validation scheme. We use MATLAB's lasso.m function (with a grid of exponentially-decreasing λ's; the function also automatically scales the regressors) and three values of the mixing parameter, $\alpha = 0.25, 0.50$ and 0.75. This yields six elastic net models:

- **EN25X**, **EN50X** and **EN75X** when the baseline model is **fARX**,
- and **EN25**, **EN50** and **EN75** when the baseline model is **fAR**,

that span the space between ridge regression (**RidgeX**, **Ridge**) and lasso models (**LassoX**, **Lasso**).

4. Empirical Results

We now present day-ahead forecasting results for the three considered datasets: GEFCom2014 hourly locational marginal prices, Nord Pool hourly system prices and U.K. hourly system prices. We use long, two-year out-of-sample test periods to make sure the obtained results are reliable (for the GEFCom2014 dataset, the test period is shorter: 623 days; see Figure 1). Recall from Section 2 that the models are re-estimated on a daily basis. Price forecasts $\hat{P}_{d+1,1}, ..., \hat{P}_{d+1,24}$ for all 24 h of the next day are determined at the same point in time, and the 365-day calibration window is rolled forward by one day.

4.1. Performance Evaluation in Terms of WMAE

Following [21,24,30,35], we compare the models in terms of the weekly-weighted mean absolute error (WMAE) loss function, which is a robust measure similar to MAPE, but with the absolute error normalized by the mean weekly price to avoid the adverse effect of negative and close to zero electricity spot prices. We evaluate the forecasting performance using weekly time intervals, each with $24 \times 7 = 168$ hourly observations. For each week $w = 1, ..., w_{max}$ in the out-of-sample test period, we calculate the error for each model as:

$$\text{WMAE}_w = \frac{1}{\bar{P}_{168}} \text{MAE}_w = \frac{1}{168 \cdot \bar{P}_{168}} \sum_{d=\text{Mon}}^{\text{Sun}} \sum_{h=1}^{24} |P_{d,h} - \hat{P}_{d,h}|, \qquad (12)$$

where $P_{d,h}$ is the actual price for hour h (not the centered log-price $p_{d,h}$), $\hat{P}_{d,h}$ is the model predicted price for that hour, $\bar{P}_{168} = \frac{1}{168} \sum_{d=\text{Mon}}^{\text{Sun}} \sum_{h=1}^{24} P_{d,h}$ is the mean price for a given week and $w_{max} = 89$ for GEFCom2014 and 104 for Nord Pool and the U.K. Next, we aggregate these errors into one mean value over all weeks in the out-of-sample test period:

$$\overline{\text{WMAE}} = \frac{1}{w_{max}} \sum_{w=1}^{w_{max}} \text{WMAE}_w. \qquad (13)$$

Note that we also analyzed the forecasts using the weekly root mean square error (see [1] (Section 3.3)), but the results were qualitatively the same and are omitted here due to space limitations.

In Table 1, we report $\overline{\text{WMAE}}$ errors for the three considered datasets and the 20 model types. We use the GEFCom2014 dataset twice: once we fit ARX-type models to the complete dataset with exogenous variables (system and zonal load; left part of the table) and once we fit AR-type models to the dataset without them (right part of the table). This allows us to compute the decrease in WMAE when exogenous variables are added to the model (the last column in Table 1). Several important conclusions can be drawn:

- All models beat the **Naive** benchmark and, except for the **fAR** model and the U.K. data, by a large margin. In particular, the improvement from using elastic nets can be as much as 5%! This indicates that they all are highly efficient forecasting tools.
- When we exclude single-step elimination without constraints (**ssAR/X**) and backward selection (**bsAR/X**) models, the selection and shrinkage methods generally outperform the expert benchmarks. In particular, the elastic net model with $\alpha = 0.75$ (i.e., closer in terms of α to the lasso than to ridge regression) beats every expert model, except **mAR1hm** for the U.K. data, where it is second best.
- The latter comment leads us to the next conclusion that adding the price for the last load period of the day, $p_{d-1,24}$, to the expert models improves their performance greatly. This fact has been recognized in the EPF literature only very recently [18,25,29] and apparently requires more attention. To see this, compare the models with suffix **m** to those without it. In particular, **mAR1hm** is the overall best performing model for the U.K. dataset and **ARX2hm** is the third best model for the Nord Pool dataset.
- Somewhat surprisingly, the full ARX model performs poorly. For the U.K. dataset, it is nearly as bad as the **Naive** benchmark. In all four cases (three datasets + GEFCom2014 without exogenous variables), it is worse than the overall best model and the best performing elastic net (**EN75/X**) by at least 1.4%. Given that a 1% improvement in MAPE translates into savings of ca. $1.5 million per year for a typical medium-size utility [2,3], this observation is of high practical value. Yet, from a statistical perspective, this finding is not that surprising. The **fARX** model has 107 parameters, which have to be calibrated to only 365 observations. Increasing the length of the calibration window should lead to a better performance of the full model.

Table 1. Mean values of the weekly-weighted mean absolute errors, i.e., $\overline{\text{WMAE}}$ defined by Equation (13), over all 89 weeks of the GEFCom2014 or all 104 weeks of the Nord Pool and U.K. out-of-sample test periods. $\overline{\text{WMAE}}$ errors are reported in percent, with standard deviation in parentheses. A heat map is used to indicate better ($\rightarrow$ green) and worse ($\rightarrow$ red) performing models. $\overline{\text{WMAE}}$ errors for the best performing model for each dataset are emphasized in bold. The last column presents the decrease in $\overline{\text{WMAE}}$ when exogenous variables are added to the model (**AR** $\rightarrow$ **ARX**; for the GEFCom2014 dataset). The bottom rows compare the performance across model classes.

	ARX-type			AR-type		AR - ARX
	GEFCom	Nord Pool		GEFCom	N2EX (UK)	GEFCom
Naive	14.708 (0.975)	11.141 (0.778)	Naive	14.708 (0.975)	9.767 (0.310)	0.000
Expert benchmarks						
ARX1	11.069 (0.639)	9.739 (0.614)	AR1	11.183 (0.701)	8.384 (0.253)	0.114
ARX1h	11.072 (0.639)	9.693 (0.616)	AR1h	11.181 (0.704)	8.389 (0.253)	0.109
ARX1hm	10.976 (0.617)	8.673 (0.516)	AR1hm	11.062 (0.657)	8.229 (0.247)	0.086
mARX1	11.102 (0.621)	9.482 (0.601)	mAR1	11.320 (0.696)	8.258 (0.253)	0.218
mARX1h	11.105 (0.622)	9.461 (0.602)	mAR1h	11.322 (0.699)	8.270 (0.254)	0.218
mARX1hm	10.974 (0.598)	8.461 (0.518)	mAR1hm	11.168 (0.644)	**8.098** (0.246)	0.195
ARX2	10.742 (0.575)	8.878 (0.546)	AR2	11.331 (0.700)	8.290 (0.253)	0.589
ARX2h	10.739 (0.575)	8.826 (0.546)	AR2h	11.333 (0.704)	8.288 (0.253)	0.594
ARX2hm	10.625 (0.565)	8.206 (0.485)	AR2hm	11.070 (0.656)	8.237 (0.249)	0.444
Full ARX model						
fARX	10.911 (0.507)	10.131 (0.708)	fAR	12.279 (0.602)	9.724 (0.334)	1.368
Selection and shrinkage methods						
ssARX	10.669 (0.577)	8.861 (0.537)	ssAR	12.061 (0.644)	9.344 (0.270)	1.393
ssARX1	9.894 (0.548)	8.409 (0.507)	ssAR1	11.343 (0.641)	8.395 (0.261)	1.449
fsARX	9.876 (0.502)	8.130 (0.502)	fsAR	11.193 (0.592)	8.563 (0.272)	1.317
bsARX	10.449 (0.502)	9.421 (0.599)	bsAR	11.968 (0.582)	9.252 (0.301)	1.519
RidgeX	9.777 (0.544)	8.972 (0.479)	Ridge	10.775 (0.653)	8.237 (0.260)	0.998
LassoX	9.476 (0.516)	8.419 (0.503)	Lasso	10.722 (0.609)	8.125 (0.253)	1.246
EN75X	9.475 (0.517)	**8.056** (0.489)	EN75	10.708 (0.610)	8.124 (0.253)	1.233
EN50X	**9.473** (0.518)	8.287 (0.496)	EN50	10.688 (0.611)	8.121 (0.253)	1.215
EN25X	9.474 (0.522)	8.529 (0.503)	EN25	**10.650** (0.613)	8.113 (0.253)	1.176
Comparisons						
Expert - Best	1.152	0.150	Expert - Best	0.412	0.000	
fARX - Best	1.438	2.075	fAR - Best	1.629	1.626	
Naive - Best	5.235	3.086	Naive - Best	4.058	1.670	

- Among the selection and shrinkage methods, the lasso and elastic nets tend to outperform single-step elimination (**ssAR/X/1**), stepwise regression (**fsAR/X, bsAR/X**) and even ridge regression (**Ridge/X**). Only for the Nord Pool dataset, the **fsARX** forward selection model is better than the lasso and two elastic nets.

4.2. Diebold–Mariano Tests

In order to formally investigate the advantages from using selection and shrinkage methods, we apply the Diebold–Mariano (DM; see [31]) test for significant differences in the forecasting performance. Since predictions for all 24 h of the next day are made at the same time using the same information set, forecast errors for a particular day will typically exhibit high serial correlation. Therefore, like [24,30,46], we conduct the DM tests for each of the 24 load periods separately, using absolute error losses of the model forecast:

$$L(\varepsilon_t) = |\varepsilon_t| = |P_{d,h} - \hat{P}_{d,h}|. \tag{14}$$

For each pair of models and for each hour independently, we calculate the loss differential series:

$$d_t = L(\varepsilon_t^{model_X}) - L(\varepsilon_t^{model_Y}). \tag{15}$$

We perform two one-sided DM tests at the 5% significance level: (i) a test with the null hypothesis $H_0: E(d_t) \leq 0$, i.e., the outperformance of the forecasts of $model_Y$ by those of $model_X$; and (ii) the complementary test with the reverse null $H_0^R: E(d_t) \geq 0$, i.e., the outperformance of the forecasts of $model_X$ by those of $model_Y$. Note that, like in [24,30,46], we assume here forecasts for consecutive days, hence loss differentials are not serially correlated. For the better performing models, this is a generally valid assumption.

In Figures 5 and 6, we summarize the DM results for all test cases (three datasets + GEFCom2014 without exogenous variables). Namely, we sum the number of significant differences in forecasting performance across the 24 h and use a heat map to indicate the number of hours for which the forecasts of a model on the X-axis are significantly better than those of a model on the Y-axis. Two extreme cases—(i) the forecasts of a model on the X-axis are significantly better for all 24 h of the day and (ii) the forecasts of a model on the X-axis are not significantly better for any hour—are indicated by white and black squares, respectively. Naturally, the diagonal (white crosses on black squares) should be ignored, as it concerns the same model on both axes. Columns with many non-black squares (the more green or white the better) indicate that the forecasts of a model on the X-axis are significantly better than the forecasts of many of its competitors. Conversely, rows with many non-black squares mean that the forecasts of a model on the Y-axis are significantly worse than the forecasts of many of its competitors. For instance, for the GEFCom2014 dataset and ARX-type models displayed in the left panel of Figure 5, the white row for the **Naive** benchmark indicates that the forecasts of this simple model are significantly worse than the forecasts of all of its competitors for all 24 h, while the black column for the **Naive** benchmark means that not a single competitor produces significantly worse forecasts than **Naive**, even for a single hour of the day.

The obtained DM-test results support our observations from Section 4.1 on WMAE errors. Again, we can conclude that applying the lasso or one of the elastic nets improves forecasting accuracy. Especially for the GEFCom2014 dataset (both for ARX- and AR-type models), these variable selection schemes lead to models that yield significantly better forecasts than those of the expert models (see the white columns for **Lasso/X**, **EN75/X**, **EN50/X** and **EN25/X** in the left panels of Figures 5 and 6), while their predictions are never outperformed by any of the competitors (see the black rows for these four models). For the Nord Pool and U.K. datasets, the results are not that clear cut, but still there are many more green or white squares in the columns than in the rows corresponding to these four selection schemes.

Again, **EN75/X** stands out as the best performing model overall. For the GEFCom2014 test cases, it always leads to significantly better forecasts than any of the expert benchmarks. For the Nord Pool dataset, its forecasts are significantly better for 10–23 h of the day and significantly worse for at most 2 h (only for models with suffix **m**: **mARX1hm** and **ARX2hm**, 2 h, and **ARX1hm**, 1 h). Finally, for the U.K. dataset, the results are the least convincing. **EN75/X** yields significantly better forecasts for 4–12 h of the day and significantly worse for at most 2 h (only for **mAR**-type models: **mAR1** and **mAR1h**, 2 h, and **mAR1hm**, 1 h).

Now, let us look in detail at the performance for each hour of the day. In Figure 7, we provide a graphical representation of the DM test statistic for four models and all considered datasets. The models include: the best overall **EN75/X** model and three benchmarks (**Naive, mARX1hm/mAR1hm** and **fAR/X**). For the GEFCom2014 dataset, **EN75/X** clearly beats all benchmarks across all hours. The situation for the remaining two datasets would be nearly the same if it was not for the early morning hours (Hours 6 and 7 for Nord Pool and Hour 8 for the U.K.), when the expert benchmarks yield significantly better predictions. This is somewhat surprising, since the morning peak comes a bit later in both markets. Perhaps looking at variables selected by the elastic net algorithm will provide more insight.

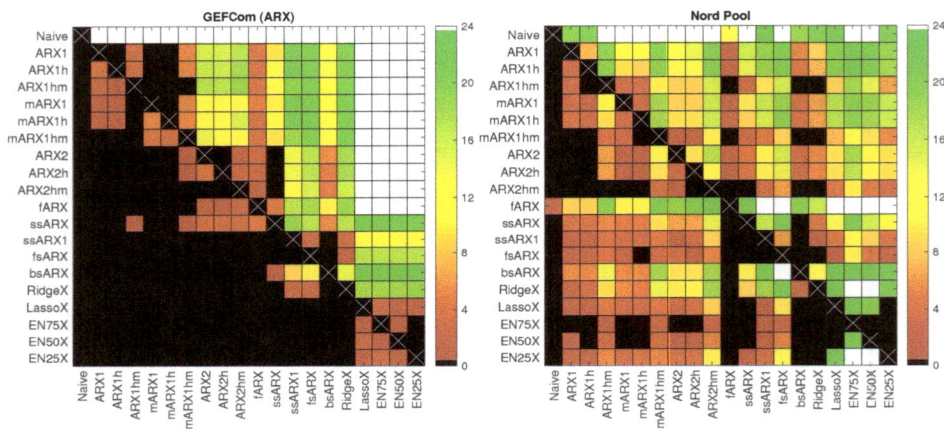

Figure 5. Results for conducted one-sided Diebold–Mariano tests at the 5% level for autoregressive model structures with exogenous variables (ARX)-type models and two datasets: GEFCom2014 (left panel) and Nord Pool (right panel). We sum the number of significant differences in forecasting performance across the 24 h and use a heat map to indicate the number of hours for which the forecasts of a model on the X-axis are significantly better than those of a model on the Y-axis. A white square indicates that forecasts of a model on the X-axis are better for all 24 h, while a black square that they are not better for a single hour.

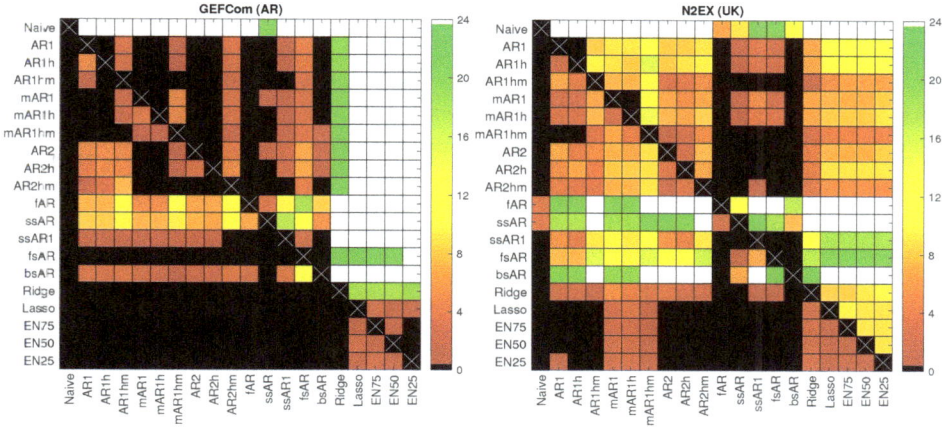

Figure 6. Results for conducted one-sided Diebold–Mariano tests at the 5% level for AR-type models and two datasets: GEFCom2014 (left panel) and N2EX (U.K.; right panel). We sum the number of significant differences in forecasting performance across the 24 h and use a heat map to indicate the number of hours for which the forecasts of a model on the X-axis are significantly better than those of a model on the Y-axis. A white square indicates that forecasts of a model on the X-axis are better for all 24 h, while a black square that they are not better for a single hour.

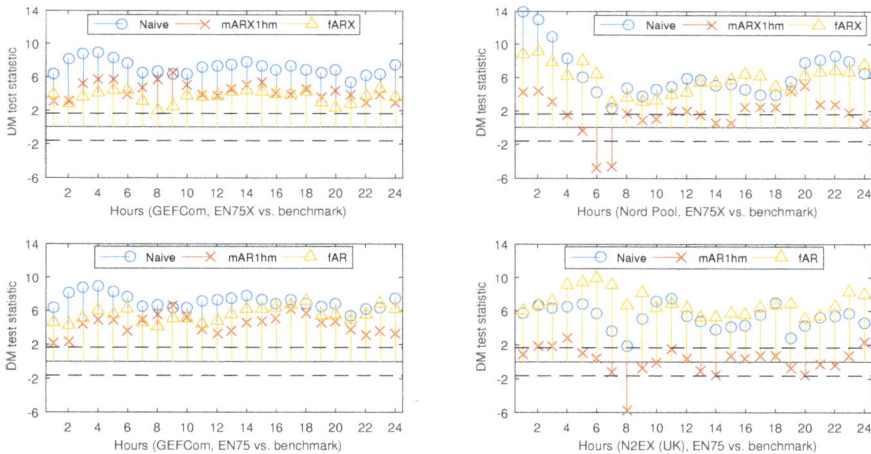

Figure 7. Results for the conducted one-sided Diebold–Mariano tests at the 5% significance level for selected ARX-type models and the GEFCom2014 and Nord Pool datasets (top panels) and selected AR-type models and the GEFCom2014 and N2EX (U.K.) datasets (bottom panels). The tests were conducted separately for each of the 24 h. The figures report the value of the test statistic for each test, as well as two thresholds (dashed lines in the plots). The lower one refers to null hypothesis $H_0:E(d_t) \leq 0$, i.e., the outperformance of the forecasts of **EN75/X** by those of a given benchmark (**Naive, mARX1hm/mAR1hm, fAR/X**). The upper threshold refers to the complementary test with the reverse null, i.e., $H_0:E(d_t) \geq 0$ or the outperformance of the forecasts of a given benchmark by those of **EN75/X**. Only points lying below (or above) the dashed threshold lines are significant at the 5% level. See also Figures 5 and 6.

49

4.3. Variable Selection

In Tables 2 and 3, we provide the number of days in the out-of-sample test period for which a given $\beta_{h,i}$ was selected for the best performing elastic net model, i.e., **EN75/X**. The maximum number of days is 623 ($= 7 \times 89$ weeks) for GEFCom2014 and 728 ($= 7 \times 104$ weeks) for Nord Pool and N2EX (U.K.). A heat map is used to indicate more ($\rightarrow$ green) and less ($\rightarrow$ red) commonly-selected $\beta_{h,i}$'s. The $\beta_{h,i}$'s are numbered as in Equation (7). Note that $\beta_{h,83}, ..., \beta_{h,86}, \beta_{h,94}, ..., \beta_{h,100} \equiv 0$ in the **EN75** model; see Table 3. Several interesting conclusions can be drawn:

- There is no single variable that is always used, regardless of the dataset, hour of the day or the day in the out-of-sample test period. The closest to 'perfection' is the day-ahead load forecast for the predicted hour, i.e., $z_{d,h}$ (see Row 83 in Table 2). Surprisingly, this dependence on the load forecast is stronger than the autoregressive effect (see the next bullet point). This may be a hint that the load-price relationship should be given more attention and that functionals of load-related (or other fundamental) variables should be included in EPF models, like in [10].
- As expected, the price 24 h ago, i.e., $p_{d-1,h}$, is an influential variable; see the diagonals in Rows 1–24 in both Tables. However, it is not only the same hour a day earlier, but also the neighboring hours. The diagonal is less visible around mid-day, and for Nord Pool, it almost disappears except for the late night hours. The latter may be to some extent due to the importance of wind in this market and the explanatory power of the day-ahead wind prognosis for the predicted hour.
- As recently observed in [18,29], the price for Hour 24, i.e., $p_{d-1,24}$, is an influential variable. Somewhat surprisingly, sometime between 7–9 a.m. and 9–11 p.m., Hour 22, i.e., $p_{d-1,22}$, becomes more important. What is more surprising, these late night hours are generally more often selected than the same hour a day ago, i.e., $p_{d-1,h}$. These observations require more thorough studies. Nevertheless, our limited results suggest that these late hour variables should be taken into account when constructing expert models.
- Clearly, the least important variables for all markets are the daily average prices over the last three days, i.e., p_{d-j}^{avg} for $j = 1, 2, 3$, which are almost never selected. There are some exceptions, though, for the GEFCom2014 dataset and the **EN75** model; see Table 3. Of the two other aggregated variables, p_{d-j}^{max} is slightly more influential than p_{d-j}^{min}, which contradicts the observations of Misiorek et al. [19] and may suggest its use in expert models instead of the minimum.
- If prices from days $(d - 2)$ or $(d - 3)$ are ever selected, it is only for hours around midnight the day before (i.e., $p_{d-2,23}$, $p_{d-2,24}$, $p_{d-3,1}$) or similar hours (i.e., the diagonals in Rows 25–48 and 49–72). On the other hand, the same hour one week ago, i.e., $p_{d-7,h}$, has a high explanatory power (see Row 73 for all datasets), which justifies its use in expert models [18–23,30].
- Finally, the weekly dummies (Rows 87–93), the dummy-linked load forecasts (Rows 94–100 in Table 2 only) and the dummy-linked last day's prices (Rows 101–107) are generally selected for the **EN75/X** model. This may be an indication that the weekly seasonality requires better modeling than offered by typically-used expert models.

Table 2. The number of days in the out-of-sample test period for which a given $\beta_{h,i}$ (see Equation (7)) was selected for the **EN75X** model. The maximum number of days is 623 (= 7×89 weeks) for GEFCom2014 and 728 (= 7×104 weeks) for Nord Pool. A heat map is used to indicate more ($\rightarrow$ green) and less ($\rightarrow$ red) commonly-selected $\beta_{h,i}$'s.

Table 3. The number of days in the out-of-sample test period for which a given $\beta_{h,i}$ (see Equation (7)) was selected for the **EN75** model. The maximum number of days is 623 (=7 × 89 weeks) for GEFCom2014 and 728 (=7 × 104 weeks) for N2EX (U.K.). A heat map is used to indicate more ($\rightarrow$ green) and less ($\rightarrow$ red) commonly-selected $\beta_{h,i}$'s. Note that $\beta_{h,83}, ..., \beta_{h,86}, \beta_{h,94}, ..., \beta_{h,100} \equiv 0$ in the **EN75** model.

5. Conclusions

A key point in electricity price forecasting (EPF) is the appropriate choice of explanatory variables. The typical approach has been to select predictors in an ad hoc fashion, sometimes using expert knowledge, but very rarely based on formal selection or shrinkage procedures. However, is this the right approach? Can the application of automated selection and shrinkage procedures to large sets of explanatory variables lead to better forecasts than those of the commonly-used expert models?

Conducting an empirical study involving state-of-the-art parsimonious autoregressive structures as benchmarks, datasets from three major power markets and five classes of automated selection and shrinkage procedures (single-step elimination, stepwise regression, ridge regression, lasso and elastic nets), we have addressed these important questions. To this end, we have compared the predictive performance of 20 types of models over three two-year-long out-of-sample test periods in terms of the robust weekly-weighted mean absolute error (WMAE) and tested the statistical significance of the results using the Diebold–Mariano [31] test.

We have shown that two classes of selection and shrinkage procedures—the lasso and elastic nets—lead to on average better performance than any of the considered expert benchmarks. On the other hand, single-step elimination, stepwise regression and ridge regression are not recommended for EPF as they do not yield significant accuracy gains compared to well-structured parsimonious autoregressive models. The lasso has been recently shown to perform well in EPF [9,18], but it is the more flexible elastic net that stands out as the best performing model overall. Given that both are automated procedures that do not require advanced expert knowledge or supervision, our results may have far reaching consequences for the practice of electricity price forecasting.

We have also looked at variables selected by the elastic net algorithm to gain insights for constructing efficient parsimonious models. In particular, we have confirmed the high explanatory power of the load forecasts for the target hour, of last day's prices for the same or neighboring hours and of the price for the same hour a week earlier. Somewhat surprisingly, we have found that not only the last available data point (price for Hour 24), but also prices for Hours 21–23 of the previous day should be considered when building expert models.

Acknowledgments: The study was partially supported by the National Science Center (NCN, Poland) through Grants 2015/17/B/HS4/00334 (to Bartosz Uniejewski and Rafał Weron) and 2013/11/N/HS4/03649 (to Jakub Nowotarski).

Author Contributions: All authors conceived of and designed the forecasting study. Bartosz Uniejewski performed the numerical experiments. Jakub Nowotarski and Rafał Weron supervised the experiments. All authors analyzed the results and contributed to writing the paper.

Conflicts of Interest: The authors declare no conflict of interest.

References

1. Weron, R. Electricity price forecasting: A review of the state-of-the-art with a look into the future. *Int. J. Forecast.* **2014**, *30*, 1030–1081.
2. Zareipour, H.; Canizares, C.A.; Bhattacharya, K. Economic impact of electricity market price forecasting errors: A demand-side analysis. *IEEE Trans. Power Syst.* **2010**, *25*, 254–262.
3. Hong, T. Crystal Ball Lessons in Predictive Analytics. *EnergyBiz Mag.* **2015**, 35–37.
4. Amjady, N.; Keynia, F. Day-ahead price forecasting of electricity markets by mutual information technique and cascaded neuro-evolutionary algorithm. *IEEE Trans. Power Syst.* **2009**, *24*, 306–318.
5. Gianfreda, A.; Grossi, L. Forecasting Italian electricity zonal prices with exogenous variables. *Energy Econ.* **2012**, *34*, 2228–2239.
6. Maciejowska, K. Fundamental and speculative shocks, what drives electricity prices? In Proceedings of the 11th International Conference on the European Energy Market (EEM14), Kraków, Poland, 28–30 May 2014; doi:10.1109/EEM.2014.6861289.
7. Ludwig, N.; Feuerriegel, S.; Neumann, D. Putting Big Data analytics to work: Feature selection for forecasting electricity prices using the LASSO and random forests. *J. Decis. Syst.* **2015**, *24*, 19–36.

8. Monteiro, C.; Fernandez-Jimenez, L.A.; Ramirez-Rosado, I.J. Explanatory information analysis for day-ahead price forecasting in the Iberian electricity market. *Energies* **2015**, *8*, 10464–10486.

9. Ziel, F.; Steinert, R.; Husmann, S. Efficient modeling and forecasting of electricity spot prices. *Energy Econ.* **2015**, *47*, 89–111.

10. Dudek, G. Multilayer perceptron for GEFCom2014 probabilistic electricity price forecasting. *Int. J. Forecast.* **2016**, *32*, 1057–1060.

11. Keles, D.; Scelle, J.; Paraschiv, F.; Fichtner, W. Extended forecast methods for day-ahead electricity spot prices applying artificial neural networks. *Appl. Energy* **2016**, *162*, 218–230.

12. Karakatsani, N.; Bunn, D. Forecasting electricity prices: The impact of fundamentals and time-varying coefficients. *Int. J. Forecast.* **2008**, *24*, 764–785.

13. Misiorek, A. Short-term forecasting of electricity prices: Do we need a different model for each hour? *Medium Econom. Toepass.* **2008**, *16*, 8–13.

14. Amjady, N.; Keynia, F. Electricity market price spike analysis by a hybrid data model and feature selection technique. *Electr. Power Syst. Res.* **2010**, *80*, 318–327.

15. Voronin, S.; Partanen, J. Price forecasting in the day-ahead energy market by an iterative method with separate normal price and price spike frameworks. *Energies* **2013**, *6*, 5897–5920.

16. Barnes, A.K.; Balda, J.C. Sizing and economic assessment of energy storage with real-time pricing and ancillary services. In Proceedings of the 2013 4th IEEE International Symposium on Power Electronics for Distributed Generation Systems (PEDG), Rogers, AR, USA, 8–11 July 2013; doi:10.1109/PEDG.2013.6785651.

17. González, C.; Mira-McWilliams, J.; Juárez, I. Important variable assessment and electricity price forecasting based on regression tree models: Classification and regression trees, bagging and random forests. *IET Gener. Transm. Distrib.* **2015**, *9*, 1120–1128.

18. Ziel, F. Forecasting Electricity Spot Prices Using LASSO: On Capturing the Autoregressive Intraday Structure. *IEEE Trans. Power Syst.* **2016**, doi:10.1109/TPWRS.2016.2521545.

19. Misiorek, A.; Trück, S.; Weron, R. Point and interval forecasting of spot electricity prices: Linear vs. non-linear time series models. *Stud. Nonlinear Dyn. Econom.* **2006**, *10*, doi:10.2202/1558-3708.1362.

20. Weron, R. *Modeling and Forecasting Electricity Loads and Prices: A Statistical Approach*; John Wiley & Sons: Chichester, UK, 2006.

21. Weron, R.; Misiorek, A. Forecasting spot electricity prices: A comparison of parametric and semiparametric time series models. *Int. J. Forecast.* **2008**, *24*, 744–763.

22. Serinaldi, F. Distributional modeling and short-term forecasting of electricity prices by Generalized Additive Models for Location, Scale and Shape. *Energy Econ.* **2011**, *33*, 1216–1226.

23. Kristiansen, T. Forecasting Nord Pool day-ahead prices with an autoregressive model. *Energy Policy* **2012**, *49*, 328–332.

24. Nowotarski, J.; Raviv, E.; Trück, S.; Weron, R. An empirical comparison of alternate schemes for combining electricity spot price forecasts. *Energy Econ.* **2014**, *46*, 395–412.

25. Gaillard, P.; Goude, Y.; Nedellec, R. Additive models and robust aggregation for GEFCom2014 probabilistic electric load and electricity price forecasting. *Int. J. Forecast.* **2016**, *32*, 1038–1050.

26. Maciejowska, K.; Nowotarski, J.; Weron, R. Probabilistic forecasting of electricity spot prices using factor quantile regression averaging. *Int. J. Forecast.* **2016**, *32*, 957–965.

27. Nowotarski, J.; Weron, R. To combine or not to combine? Recent trends in electricity price forecasting. *ARGO* **2016**, *9*, 7–14.

28. Hong, T.; Pinson, P.; Fan, S.; Zareipour, H.; Troccoli, A.; Hyndman, R.J. Probabilistic energy forecasting: Global Energy Forecasting Competition 2014 and beyond. *Int. J. Forecast.* **2016**, *32*, 896–913.

29. Maciejowska, K.; Nowotarski, J. A hybrid model for GEFCom2014 probabilistic electricity price forecasting. *Int. J. Forecast.* **2016**, *32*, 1051–1056.

30. Nowotarski, J.; Weron, R. On the importance of the long-term seasonal component in day-ahead electricity price forecasting. *Energy Econ.* **2016**, *57*, 228–235.

31. Diebold, F.X.; Mariano, R.S. Comparing predictive accuracy. *J. Bus. Econ. Stat.* **1995**, *13*, 253–263.

32. Garcia-Martos, C.; Conejo, A. Price forecasting techniques in power systems. In *Wiley Encyclopedia of Electrical and Electronics Engineering*; Wiley: Chichester, UK, 2013; pp. 1–23, doi:10.1002/047134608X.W8138.

33. Burger, M.; Graeber, B.; Schindlmayr, G. *Managing Energy Risk: An Integrated View on Power and Other Energy Markets*; Wiley: Chichester, UK, 2007.

34. Nogales, F.J.; Contreras, J.; Conejo, A.J.; Espinola, R. Forecasting next-day electricity prices by time series models. *IEEE Trans. Power Syst.* **2002**, *17*, 342–348.
35. Conejo, A.J.; Contreras, J.; Espínola, R.; Plazas, M.A. Forecasting electricity prices for a day-ahead pool-based electric energy market. *Int. J. Forecast.* **2005**, *21*, 435–462.
36. Paraschiv, F.; Fleten, S.E.; Schürle, M. A spot-forward model for electricity prices with regime shifts. *Energy Econ.* **2015**, *47*, 142–153.
37. Broszkiewicz-Suwaj, E.; Makagon, A.; Weron, R.; Wyłomańska, A. On detecting and modeling periodic correlation in financial data. *Physica A* **2004**, *336*, 196–205.
38. Bosco, B.; Parisio, L.; Pelagatti, M. Deregulated wholesale electricity prices in Italy: An empirical analysis. *Int. Adv. Econ. Res.* **2007**, *13*, 415–432.
39. James, G.; Witten, D.; Hastie, T.; Tibshirani, R. *An Introduction to Statistical Learning with Applications in R*; Springer: New York, NY, USA, 2013.
40. Bessec, M.; Fouquau, J.; Meritet, S. Forecasting electricity spot prices using time-series models with a double temporal segmentation. *Appl. Econ.* **2016**, *48*, 361–378.
41. Hoerl, A.E.; Kennard, R.W. Ridge regression: Biased estimation for nonorthogonal problems. *Technometrics* **1970**, *12*, 55–67.
42. Tibshirani, R. Regression shrinkage and selection via the lasso. *J. Royal Stat. Soc. B* **1996**, *58*, 267–288.
43. Ziel, F. Iteratively reweighted adaptive lasso for conditional heteroscedastic time series with applications to AR-ARCH type processes. *Comput. Stat. Data Anal.* **2016**, *100*, 773–793.
44. Hastie, T.; Tibshirani, R.; Wainwright, M. *Statistical Learning with Sparsity: The Lasso and Generalizations*; CRC Press: Philadelphia, PA, USA, 2015.
45. Zou, H.; Hastie, T. Regularization and Variable Selection via the Elastic Nets. *J. Royal Stat. Soc. B* **2015**, *67*, 301–320.
46. Bordignon, S.; Bunn, D.W.; Lisi, F.; Nan, F. Combining day-ahead forecasts for British electricity prices. *Energy Econ.* **2013**, *35*, 88–103.

Article

![energies logo] MDPI

Mid-Term Electricity Market Clearing Price Forecasting with Sparse Data: A Case in Newly-Reformed Yunnan Electricity Market

Chuntian Cheng *, Bin Luo, Shumin Miao and Xinyu Wu

Institute of Hydropower System & Hydroinformatics, Dalian University of Technology, Dalian 116024, China; luobin@mail.dlut.edu.cn (B.L.); miaoshumin@mail.dlut.edu.cn (S.M.); wuxinyu@dlut.edu.cn (X.W.)
* Correspondence: ctcheng@dlut.edu.cn; Tel.: +86-411-8470-8468

Academic Editor: Javier Contreras
Received: 16 June 2016; Accepted: 27 September 2016; Published: 8 October 2016

Abstract: For the power systems, for which few data are available for mid-term electricity market clearing price (MCP) forecasting at the early stage of market reform, a novel grey prediction model (defined as interval GM(0, N) model) is proposed in this paper. Over the traditional GM(0, N) model, three major improvements of the proposed model are: (i) the lower and upper bounds are firstly identified to give an interval estimation of the forecasting value; (ii) a novel whitenization method is then established to determine the definite forecasting value from the forecasting interval; and (iii) the model parameters are identified by an improved particle swarm optimization (PSO) instead of the least square method (LSM) for the limitation of LSM. Finally, a newly-reformed electricity market in Yunnan province of China is studied, and input variables are contrapuntally selected. The accuracy of the proposed model is validated by observed data. Compared with the multiple linear regression (MLR) model, the traditional GM(0, N) model and the artificial neural network (ANN) model, the proposed model gives a better performance and its superiority is further ensured by the use of the modified Diebold–Mariano (MDM) test, suggesting that it is suitable for mid-term electricity MCP forecasting in a data-sparse electricity market.

Keywords: market clearing price (MCP); mid-term; grey prediction model; sparse data; particle swarm optimization (PSO)

1. Introduction

Electricity price is one of the most important parameters of deregulated and competitive electricity markets; wherein, the electricity market clearing price (MCP) is the final outcome of the market bid price and generally determined by the aggregated demand and supply bid curves, which exists when an electricity market is clear of shortage and surplus. On 15 March 2015, the Communist Party of China (CPC) Central Committee and State Council's opinions on further deepening the reforms of the electric power system [1] marked the beginning of new electricity market reform in China, after Investment Decentralization in 1986, Unbundling of Government Admin and Business Operation in 1997 and Unbundling of Generation and Transmission in 2002 [2]. Unlike mature electricity markets in other countries, to guarantee the security and stability operation at the early stage of reform, the Chinese electricity market mainly comprises the mid- and long-term exchanges so far. Therefore, the mid-term electricity MCP forecasting has become one of the most concerning issues, as it is essential for resource reallocation, maintenance scheduling, financial risk reducing, bilateral contracting, budgeting and planning purposes [3].

Over the last 15–20 years, electricity price forecasting (EPF) has been carried out with various methods, bringing a variety of success. Till now, there is no consensus criterion on the classification of these models [4,5]. To our knowledge, they can be mainly classified into the following categories: (1) equilibrium and simulation models, such as the Nash–Cournot model [6–8], production-cost models (PCM) [9], strategic PCM (SPCM) [10], and so on; (2) statistical methods, such as multiple regression [11], exponential smoothing [12,13], time series [14–16], wavelet transform (WT) [17–19], AR-type models (such as auto-regressive moving average (ARMA) [16], auto-regressive integrated moving average (ARIMA) [20,21], seasonal ARIMA (SARIMA) [22]), and so on; (3) artificial intelligence models, such as artificial neural network (ANN) [23,24], support vector machine (SVM) [25,26] and data mining models [27,28]; and (4) hybrid models, which combine two or more single methods or models described above [29–31]. According to these research works, most works are focused on short-term EPF, commonly known as day-ahead EPF. In contrast, not much has been performed in the mid-term horizon [3,32–35]. Compared to short-term forecasting, mid-term forecasting is much more challenging. This is because the prediction horizon is much longer, and the price may not be contiguous to the immediate past periods, which means the trend information of the immediate past is not as valuable as for short-term forecasting. In addition, the influence factors of mid-term EPF are various and, worse, some are unavailable. When it comes to the newly-reformed electricity market in China, being at the early stage, another bigger challenge is how to build accurate models with relatively few available samples, which is beyond the capability of most forecasting models above. Appropriate prediction models need to be studied.

In systems theory, if the description of a system is completely known, it is known as a white system. In contrast, a black-box system means that it is completely unknown. When in an intermediate state, it is called a grey system [36]. The grey system theory (GST) was firstly proposed by Deng in 1982 [37], and several grey prediction models were created later. For the grey model, it is characterized by its prominent capability in modeling with small samples and poor information, which commonly exist in the natural world. Thus, it has been successfully employed in various fields and demonstrated satisfactory results, such as social [38,39], economic [40,41], agricultural [42], industrial [43], etc. The electricity MCP is a synthetic variable of the market mechanism with many uncertainties, obviously in accordance with the characteristics of GST, so the electricity MCP forecasting is adapted to be solved through grey prediction models. Furthermore, as the mid-term electricity MCP is not closely related to the immediate past periods and affected by multiple factors, a widely-used static multi-variable grey prediction model in GST named GM(0, N) is concerned in this paper. However, to directly forecast the mid-term electricity MCP, there are some difficulties or disadvantages in the traditional GM(0, N) model, including:

1. The input variables of the model would affect the forecasting result directly. Thus, the selection of appropriate influence factors of electricity MCP is extremely important, but also tough, as it varies from different electricity markets.
2. The mid-term electricity MCP is obviously volatile in nature and out of a monotonous trend. The forecasting accuracy is unsatisfactory when the value of the next forecasting point is away from the fitting model constructed only with known data.
3. The least square method (LSM) used in the traditional GM(0, N) model to identify parameters is effective on condition of the existence of the inverse of matrix $B^T B$, but this does not work when $B^T B$ is a singular matrix in some cases.

To overcome these problems and achieve a successful implementation for mid-term electricity MCP forecasting with few available data, a novel grey prediction model defined as the interval GM(0, N) model is proposed in this paper. The main contributions of this work can be summarized as follows:

1. Based on depth analysis of the newly-reformed electricity market, the influence factors of electricity MCP are studied, and input variables are carefully selected according to three aspects: supply factors, demand factors and supplemented factors.

2. In the proposed interval GM(0, N) model, two improved GM(0, N) models are included, respectively estimating the upper and lower bounds of the forecasting value. Firstly, to reduce randomness and increase smoothness, all input sequences (not containing values of the next forecasting point) including the MCP sequence and factor sequences are ranked in accordance with the ascending order of the MCP sequence. Then, one of the factor sequences is selected as the benchmark ranking sequence according to its ranking order and correlation with the MCP sequence. The position of the forecasting point in the ranked MCP sequence thus can be determined by sorting the benchmark ranking sequence, which contains the factor value of the forecasting point. Finally, two neighboring (upper and lower) points of the forecasting point in the ranked MCP sequence are regarded as two virtual values to construct two new MCP sequences, which are respectively used as characteristic sequences for two improved GM(0, N) models, obtaining the forecasting interval. In the two improved GM(0, N) models, the input sequences used for building the model include not only known data, but also the virtual MCP and predicted influence factors of the forecasting point.
3. Based on the forecasting interval, a novel whitenization method considering correlations between electricity MCP and influence factors is established to determine the definite forecasting value.
4. The parameters of the GM(0, N) model are identified by an improved particle swarm optimization (PSO) instead of LSM.
5. The performance of the proposed model has been validated by applying it to the newly-reformed Yunnan electricity market. Further comparisons between the proposed model and other models, including the multiple linear regression (MLR) model, the traditional GM(0, N) model and the artificial neural network (ANN) model, are carefully discussed and also evaluated by using the modified Diebold–Mariano (MDM) test. The results indicate that the proposed model is an effective means of mid-term electricity MCP forecasting with sparse data.

The remainder of this paper is organized as follows. In Section 2, the method and theory are introduced. Section 3 is devoted to the detailed presentation of the proposed model. In Section 4, electricity market conditions are analyzed, and input variables are selected. In Section 5, the case study and forecasting results are carefully discussed. Finally, Section 6 outlines the main conclusions.

2. Method and Theory

2.1. Principle of the Traditional GM(0, N) Model

Set $X_1^{(0)} = \{x_1^{(0)}(1), x_1^{(0)}(2), \cdots, x_1^{(0)}(n)\}$ as the raw characteristic sequence (here, the electricity MCP sequence); n is the data length. Set $X_i^{(0)} = \{x_i^{(0)}(1), x_i^{(0)}(2), \cdots, x_i^{(0)}(n)\}$, $i \in I, I = \{2, 3, \cdots, N\}$ as the relevant factor sequences (here, the influence factors of mid-term electricity MCP); N is the number of all sequences, and the total number of factor sequences is $N - 1$. Therefore, the AGO (accumulated generation operation) sequence $X_i^{(1)} = \{x_i^{(1)}(1), x_i^{(1)}(2), \cdots, x_i^{(1)}(n)\}$ is firstly generated by Equation (1).

$$x_i^{(1)}(k) = \sum_{m=1}^{k} x_i^{(0)}(m), \ (k = 1, 2, \cdots, n; \ i = 1, 2, \cdots, N) \tag{1}$$

Then, call formula Equation (2):

$$x_1^{(1)}(k) = \sum_{i=2}^{N} b_i x_i^{(1)}(k) + a, \ (k = 1, 2, \cdots, n) \tag{2}$$

as the traditional GM(0, N) model. Where, b_i and a are identified parameters, which can be solved by the least square method (LSM), if the inverse of matrix $B^T B$ exists. That is:

$$P_{0N} = [b_2, b_3, \cdots, b_N, a]^T = \left(B^T B\right)^{-1} B^T y_N \tag{3}$$

$$\text{where, } B = \begin{bmatrix} x_2^{(1)}(1) & x_3^{(1)}(1) & \cdots & x_N^{(1)}(1) & 1 \\ x_2^{(1)}(2) & x_3^{(1)}(2) & \cdots & x_N^{(1)}(2) & 1 \\ \vdots & \vdots & \vdots & \vdots & \vdots \\ x_2^{(1)}(n) & x_3^{(1)}(n) & \cdots & x_N^{(1)}(n) & 1 \end{bmatrix}, y_N = [x_1^{(1)}(1), x_1^{(1)}(2), \cdots, x_1^{(1)}(n), 1]^T.$$

Thus, we can get the next forecasting value for the characteristic sequence by Equation (4).

$$\begin{cases} \hat{x}_1^{(0)}(1) = x_1^{(0)}(1) \\ \hat{x}_1^{(0)}(k+1) = \hat{x}_1^{(1)}(k+1) - \hat{x}_1^{(1)}(k), \ k = 1, 2, \cdots, n \end{cases} \tag{4}$$

where $\hat{x}_1^{(1)}(k+1)$ is the forecasting value of $x_1^{(1)}(k+1)$; $\hat{x}_1^{(0)}(k+1)$ is the forecasting value of $x_1^{(0)}(k+1)$.

2.2. Limitation and Requirement for Implementation

2.2.1. Limitation of Least Square Method (LSM) in the Traditional GM(0, N) Model

In the traditional GM(0, N) model, parameters b_i and a are identified by LSM, shown in Equation (3). However, this is conditioned on the existence of the inverse of matrix $B^T B$.

Proof: The determinant of $B^T B$ is $Det(B^T B)$; the inverse matrix of $B^T B$ exists only when:

$$Det(B^T B) \neq 0 \tag{5}$$

that is when the B is a column full rank matrix. That is when, for any constant $u \in (-\infty, +\infty)$,

$$X_i^{(1)} + u X_j^{(1)} \neq \vec{0}, \ (\forall i, j \in I, i \neq j) \tag{6}$$

otherwise, the inverse of matrix $B^T B$ does not exist.

From the equations above, it can be observed that the limitation of LSM depends on the input dataset. However, Equation (6) does not hold for all cases, especially with the increasing length of available data. Therefore, the LSM cannot be adapted to identify parameters b_i and a for the GM(0, N) model in some cases. To avoid any mistakes caused by this limitation and to ensure the general applicability of the model, the parameter identification method should be improved.

2.2.2. Requirements for Input Variables

Although the great feature of the GM(0, N) model is allowing the randomness of data sequences and modeling with small samples and poor conditions, the volatile nature of the electricity price has a negative effect on forecasting results. Despite the fact that AGO has been applied to convert the sequences to a monotonic increasing trend, the modeling accuracy is still limited by directly using raw known data. For better accuracy, the input data need to be pre-processed. Furthermore, the related influence factors of electricity MCP are various, and input variables should be carefully selected, as they also have an impact on the forecasting result.

2.3. Performance Evaluation

2.3.1. Checking Method of Grey Prediction Models

To evaluate the fitting precision of the proposed grey prediction model, two main parameters in the after-test residue checking method, posterior-error (C) and micro-error-probability (P), are adopted, respectively defined as:

$$C = \frac{S_2}{S_1} \tag{7}$$

$$P = probability\{|\Delta x_1^{(0)}(k) - \Delta \overline{x}_1^{(0)}(k)| < 0.6745S_1\} \tag{8}$$

where:

$$\begin{cases} S_1 = \sqrt{\frac{1}{n-1}\sum\limits_{k=1}^{n}\left[x_1^{(0)}(k) - \overline{x}_1^{(0)}(k)\right]^2} \\ S_2 = \sqrt{\frac{1}{n-1}\sum\limits_{k=1}^{n}\left[\Delta x_1^{(0)}(k) - \Delta \overline{x}_1^{(0)}(k)\right]^2} \\ \overline{x}_1^{(0)}(k) = \frac{1}{n}\sum\limits_{k=1}^{n} x_1^{(0)}(k) \\ \Delta x_1^{(0)}(k) = x_1^{(0)}(k) - \hat{x}_1^{(0)}(k) \\ \Delta \overline{x}_1^{(0)}(k) = \frac{1}{n}\sum\limits_{k=1}^{n} \Delta x_1^{(0)}(k) \end{cases} \tag{9}$$

The fitting precision grade is shown in Table 1.

Table 1. Reference for the fitting precision grade.

Parameters	Fitting Precision Grade			
	Good	Qualified	Just	Unqualified
C	<0.35	0.35–0.50	0.50–0.65	≥0.65
P	>0.95	0.80–0.95	0.70–0.80	≤0.70

2.3.2. Performance Evaluation of Forecasting Models

For evaluating electricity MCP forecasting values, mean absolute error (MAE), mean squares error (MSE) and mean absolute percentage error (MAPE) are the most widely-used measurements, respectively shown in Equations (10)–(12). The smaller the MAE, MSE and MAPE, the better forecasting performance the model shows.

$$MAE = \frac{1}{n}\sum_{k=1}^{n}|x_1^{(0)}(k) - \hat{x}_1^{(0)}(k)| \tag{10}$$

$$MSE = \frac{1}{n}\sum_{k=1}^{n}\left[x_1^{(0)}(k) - \hat{x}_1^{(0)}(k)\right]^2 \tag{11}$$

$$MAPE = \frac{1}{n}\sum_{k=1}^{n}\left|\frac{x_1^{(0)}(k) - \hat{x}_1^{(0)}(k)}{x_1^{(0)}(k)}\right| \times 100\% \tag{12}$$

To further judge the forecasting performances of different forecasting models from a statistical point of view, a statistical evaluation method named modified Diebold–Mariano (MDM) test [44] is also proposed. The MDM test is an extension of the Diebold–Mariano test [45] and has been widely used in forecasting research. The MDM test statistic for the *h*-step ahead forecast is specified as:

$$MDM = \left[\frac{n + 1 - 2h + n^{-1}h(h-1)}{n}\right]^{\frac{1}{2}}\left[V(\overline{d})\right]^{-\frac{1}{2}}\left[\overline{d}\right] \tag{13}$$

where $V(\overline{d}) = \left[n^{-1}\left(\gamma_0 + 2\sum\limits_{k=1}^{h-1}\gamma_k\right)\right]$ is the variance of $\overline{d}$; $\gamma_k = n^{-1}\sum\limits_{t=k+1}^{n}\left(d_t - \overline{d}\right)\left(d_{t-k} - \overline{d}\right)$ is the *k*-th autocovariance of d_t; $\overline{d} = n^{-1}\sum\limits_{t=1}^{n} d_t$ is the mean of d_t; $d_t = L(e_{1,t}) - L(e_{2,t})$, $t = 1,2,3,\cdots.n$ is the loss differentials; $L(e_{i,t})$ is the loss function of forecast error $e_{i,t}$, $i = 1,2$; $e_{i,t} = x_t - \hat{x}_{i,t}$, x_t denotes the actual data series, and $\hat{x}_{i,t}$ denotes the forecasting series of forecasting model *i*. In our study, the popular loss functions used in power systems, i.e., the MAE and MSE loss functions, are adopted.

Under the null hypothesis of the equal forecasting performances of forecasting models, the MDM test statistic follows a *t*-distribution with $(n-1)$ degrees of freedom, so that tests can be carried out. As the detailed principle of the MDM test can be easily found in related papers, we will not repeat it here.

3. Novel Interval GM(0, N) Model

Sorting is a common way to get a descending (or ascending) sequence. However, the position of the forecasting point in the MCP sequence cannot be obtained, as its actual value is unknown. For this, one of the factor sequences is selected as the benchmark ranking sequence to determine the position of the forecasting point in the MCP sequence. Then, the actual value of the forecasting point is respectively replaced by its two neighboring (upper and lower) values in the MCP sequence. Two new MCP sequences thus are constructed, which are respectively used as the characteristic sequence for two improved GM(0, N) models, obtaining the upper and lower bounds of the forecasting value (i.e., forecasting interval). Furthermore, the model parameters are identified by an improved PSO instead of LSM. Finally, the definite forecasting value is believed to be related to this forecasting interval and further determined by a novel whitenization method. In this section, the implementation of this improved grey prediction model (defined as the interval GM(0, N) model) is given in detail.

3.1. Selection of the Benchmark Ranking Sequence

The factor sequence that shares the most similarities with the characteristic sequence is selected to help with determining the position of the forecasting point in the MCP sequence. The chosen factor sequence is called the benchmark ranking sequence, and the steps are given below.

Step 1. Get factor sequences that have the same (or reverse) ranking order as the characteristic sequence, defined as $X_i^{sub(0)}$. Two sequences $X = (x_1, \cdots, x_k, \cdots, x_n)$ and $Y = (y_1, \cdots, y_k, \cdots, y_n)$ are regarded as having the same (or reverse) ranking order, if:

(1) The lengths of X and Y are equal.
(2) A new sequence $\tilde{X} = \{x_1, \cdots, x_k, \cdots, x_n\}$ is obtained by ranking X in ascending order of value. Corresponding with the same index of the subscript, $\tilde{Y} = \{y_1, \cdots, y_k, \cdots, y_n\}$ is formed. For any k, there is $y_k - y_{k-1} \geq 0$ (or < 0).

Similarly, the factor sequences $X_i^{sub(0)}$ can be obtained, $i \in I^{sub}$, $I^{sub} \subset I$.

Step 2. The correlation coefficient $r_{1,i}$ between $X_1^{(0)}$ and $X_i^{(0)}$ is calculated by correlation analysis, shown in Equation (14).

$$r_{1,i} = r(X_1^{(0)}, X_i^{(0)}) = \frac{\sum\limits_{k=1}^{n} (x_1^{(0)}(k) - \overline{x}_1^{(0)})(x_i^{(0)}(k) - \overline{x}_i^{(0)})}{\sqrt{\sum\limits_{k=1}^{n} (x_1^{(0)}(k) - \overline{x}_1^{(0)}) \sum\limits_{k=1}^{n} (x_i^{(0)}(k) - \overline{x}_i^{(0)})}}, \ (i \in I) \tag{14}$$

where $\overline{x}_i^{(0)} = \frac{1}{n} \sum\limits_{k=1}^{n} x_i^{(0)}(k)$ is the mean value of $X_i^{(0)}$. Then, the benchmark ranking sequence (define its index in factor sequences as τ) is determined by the following rules: if $I^{sub} \neq \emptyset$, τ is the index when $r_{1,\tau} = \underset{i \in I^{sub}}{\text{MAX}}\{r_{1,i}\}$, else if $I^{sub} = \emptyset$, τ is the index when $r_{1,\tau} = \underset{i \in I}{\text{MAX}}\{r_{1,i}\}$. Therefore, the characteristic sequence and factor sequences are ranked in accordance with the ascending or descending order of the benchmark ranking sequence. In what follows, sequences mean ranked sequences.

3.2. Calculation for Forecasting Interval

By comparing the corresponding factor value of the forecasting point in the benchmark ranking sequence, the position of the forecasting point can be obtained, which is also regarded as the position in the ranked MCP sequence. Suppose the position is k; the ranked MCP sequence that contains the forecasting point is:

$$X_1^{(0)} = \{x_1^{(0)}(1), x_1^{(0)}(2), \cdots, x_1^{(0)}(k-1), \varnothing(k), x_1^{(0)}(k+1), \cdots, x_1^{(0)}(n+1)\} \tag{15}$$

where the $\varnothing(k)$ is the point to be forecasted and the data length is changed from n to $n+1$. Defining the lower- and upper-bound virtual values of $x_1^{(0)}(k)$ as $x_1^{low(0)}(k)$ and $x_1^{up(0)}(k)$, respectively. Therefore, the lower- and upper-bound MCP sequences can be constructed, respectively named as $X_1^{low(0)}$ and $X_1^{up(0)}$, and expressed as:

$$\begin{cases} X_1^{low(0)} = \{x_1^{(0)}(1), x_1^{(0)}(2), \cdots, x_1^{(0)}(k-1), x_1^{low(0)}(k), x_1^{(0)}(k+1), \cdots, x_1^{(0)}(n+1)\} \\ X_1^{up(0)} = \{x_1^{(0)}(1), x_1^{(0)}(2), \cdots, x_1^{(0)}(k-1), x_1^{up(0)}(k), x_1^{(0)}(k+1), \cdots, x_1^{(0)}(n+1)\} \end{cases} \tag{16}$$

two neighboring (lower and upper) points of $x_1^{(0)}(k)$ in the ranked MCP sequence, i.e., $x_1^{(0)}(k-1)$ and $x_1^{(0)}(k+1)$, are used as the virtual values. Thus, Equation (16) is:

$$\begin{cases} X_1^{low(0)} = \{x_1^{(0)}(1), x_1^{(0)}(2), \cdots, x_1^{(0)}(k-1), x_1^{(0)}(k-1), x_1^{(0)}(k+1), \cdots, x_1^{(0)}(n+1)\} \\ X_1^{up(0)} = \{x_1^{(0)}(1), x_1^{(0)}(2), \cdots, x_1^{(0)}(k-1), x_1^{(0)}(k+1), x_1^{(0)}(k+1), \cdots, x_1^{(0)}(n+1)\} \end{cases} \tag{17}$$

Then, the $X_1^{low(0)}$ and $X_1^{up(0)}$ are respectively used as characteristic sequences for two improved GM(0, N) models, correspondingly offering the lower- and upper-bound forecasting values, defined as $\hat{x}_1^{low(0)}(k)$ and $\hat{x}_1^{up(0)}(k)$. Especially, when the position is at the beginning (or ending) of the MCP sequence, there only exists an upper-bound (or lower-bound) MCP sequence. In this case, a one-sided interval is obtained. In the improved GM(0, N) model, all model parameters are identified by an improved PSO (detailed in Section 3.4), and input sequences include not only known data, but also the virtual MCP and predicted factors of the forecasting point.

3.3. Definite Forecasting Value Determination

When the forecasting interval $\left[\hat{x}_1^{low(0)}(k), \hat{x}_1^{up(0)}(k)\right]$ is gained, the purpose of the whitenization method in GST is to obtain the definite forecasting value. Generally, the expression is:

$$\hat{x}_1^{(0)}(k) = \alpha \hat{x}_1^{low(0)}(k) + (1-\alpha) \hat{x}_1^{up(0)}(k) \tag{18}$$

where α is a coefficient that represents the proportion of $\hat{x}_1^{low(0)}(k)$ to $\hat{x}_1^{(0)}(k)$. This means that the definite forecasting value is a linear combination of the two boundary values. The most used is the equal proportion method, namely $\alpha = 0.5$. However, this may result in additional error because the gap between $\hat{x}_1^{(0)}(k)$ and $\hat{x}_1^{low(0)}(k)$ is not equal to that between $\hat{x}_1^{(0)}(k)$ and $\hat{x}_1^{up(0)}(k)$ in most cases. To give a more detailed explanation to coefficient α and to calculate it accurately, a novel whitenization method is needed. Defining the gap between $x_i^{(0)}(k)$ and $x_i^{(0)}(k-1)$ as $\Delta x_i^{(0)}(k)$, the absolute error between $\hat{x}_1^{(0)}(k)$ and $\hat{x}_1^{low(0)}(k)$ as $\Delta \hat{x}_1^{low(0)}(k)$, that is:

$$\begin{cases} \Delta x_i^{(0)}(k) = x_i^{(0)}(k) - x_i^{(0)}(k-1), i = \{1, 2, \cdots, N\} \\ \Delta \hat{x}_1^{low(0)}(k) = \left|\hat{x}_1^{(0)}(k) - \hat{x}_1^{low(0)}(k)\right| \end{cases} \tag{19}$$

as the $\hat{x}_1^{low(0)}(k)$ is simulated from the sequence $X_1^{low(0)}$ in which the lower-bound value $x_1^{(0)}(k-1)$ is used as the replacement value of point k; the gap $\Delta x_1^{(0)}(k) = x_1^{(0)}(k) - x_1^{(0)}(k-1) = x_1^{(0)}(k) - x_1^{low(0)}(k)$ is the root of the existence of $\Delta \hat{x}_1^{low(0)}(k)$. The amount of $\Delta x_1^{(0)}(k)$ certainly reflects the impact of

$\hat{x}_1^{low(0)}(k)$ on $\hat{x}_1^0(k)$. Similarly, $\Delta x_1^{(0)}(k+1)$ reflects the impact of $\hat{x}_1^{up(0)}(k)$ on $\hat{x}_1^0(k)$. Thus, the coefficient α, which represents the proportion of $\hat{x}_1^{low(0)}(k)$ to $\hat{x}_1^{(0)}(k)$, is related with $\Delta x_1^{(0)}(k)$ and $\Delta x_1^{(0)}(k+1)$ and can be calculated by Equation (20). A schematic diagram is also shown in Figure 1. Especially with the one-sided interval as explained in Section 3.2, the definite forecasting value is only related with $\Delta x_1^{(0)}(k)$ or $\Delta x_1^{(0)}(k+1)$, and the value of α respectively equals zero and one when the position is the beginning and ending.

$$\alpha = \frac{\Delta x_1^{(0)}(k+1)}{\Delta x_1^{(0)}(k) + \Delta x_1^{(0)}(k+1)} \tag{20}$$

However, $x_1^{(0)}(k)$ is unknown; thus, $\Delta x_1^{(0)}(k)$ and $\Delta x_1^{(0)}(k+1)$ are also unknown. Therefore, coefficient α cannot be calculated by Equation (20) directly. In this paper, considering the correlations between electricity MCP and influence factors, a novel whitenization method is proposed. The value of coefficient α is calculated by correlation coefficient weighted gaps of factor sequences, expressed in Equation (21).

$$\begin{cases} \alpha_i = \dfrac{\Delta x_i^{(0)}(k+1)}{\Delta x_i^{(0)}(k) + \Delta x_i^{(0)}(k+1)} \\ \alpha = \sum\limits_{i=2}^{N} \left(\dfrac{|r_{1,i}|}{\sum\limits_{i=2}^{N} |r_{1,i}|} \times \alpha_i \right) \end{cases} , i = \{2,3,\dots,N\} \tag{21}$$

where $r_{1,i}$ is the correlation coefficient between $X_1^{(0)}$ and $X_i^{(0)}$ expressed in Equation (14).

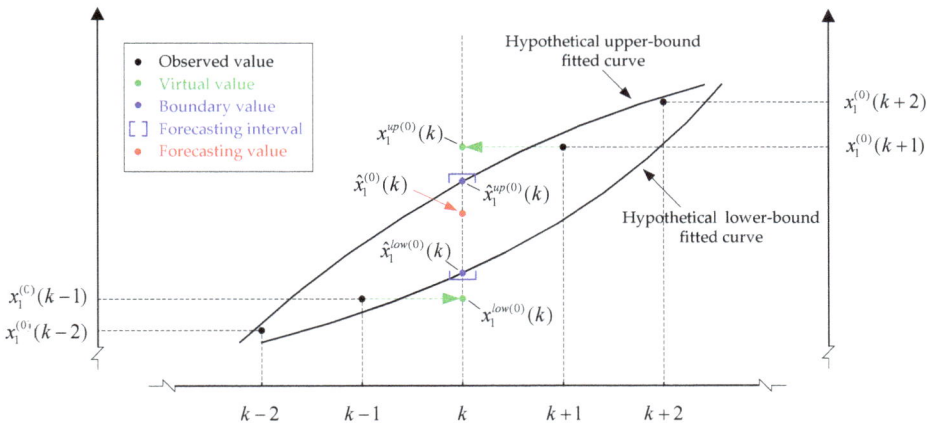

Figure 1. The schematic diagram of the whitenization method.

3.4. Parameters Identification by Improved Particle Swarm Optimization (PSO)

As mentioned in Section 2.1, model parameters to be identified are b_i, $i \in \{2,3,\cdots,N\}$ and a. Under given input data including the characteristic sequence and factor sequences, the goal is to seek a vector of these parameters, which provides the best forecasting results. The optimal model can be formulated as follows:

$$\min f(b_i,a) = MAPE = \frac{1}{n+1}\sum_{k=1}^{n+1} \left| \frac{x_1^{(0)}(k) - \hat{x}_1^{(0)}(k)}{x_1^{(0)}(k)} \right| \times 100\%$$
$$\text{subject to} \quad b_i \in (-\infty, +\infty), \ i \in \{2,3,\cdots,N\} \tag{22}$$
$$a \in (-\infty, +\infty)$$

where $n + 1$ is the data length of the characteristic sequence (containing the forecasting point).

Particle swarm optimization (PSO) is an evolutionary algorithm introduced by Kennedy and Eberhart in 1995 [46]. Due to its simplicity of implementation and ability to quickly converge to a reasonably good solution, PSO has been very widely used in many fields. However, for multimodal functions or high-dimensional problems, the canonical PSO tends to be trapped in premature convergence and provides a poor solution. Therefore, a PSO based on time-varying parameters is used in this paper to identify parameters for the proposed interval GM(0, N) model. It has been proven that the time-varying PSO can significantly improve algorithm performance in terms of the global search ability and the convergence speed [47–50]. The procedure of time-varying PSO is described in detail as follows.

Step 1. Initialize particles. The number of particles is set to 60 after repeated testing. The position vector of the t-th particle is defined as $X_t = \{x_{t1}, x_{t2}, \cdots, x_{tN}\}$ and the velocity vector as $V_t = \{v_{t1}, v_{t2}, \cdots, v_{tN}\}$, where $t = 1, 2, 3, \cdots, 60$. The positions of particles correspond to the solution of b_i, $i \in \{2, 3, \cdots, N\}$ and a. For faster convergence, the initial particles are constructed with the assistance of the multiple linear regression (MLR) model. The MLR model takes the form:

$$y = \beta X + \varepsilon \tag{23}$$

where y, X are respectively the AGO characteristic sequence $X_1^{(1)}$ and factor sequences $X_i^{(1)}$, $i = \{2, 3, \cdots, N\}$. Suppose the solution is $X^{ini} = \{\beta, \varepsilon\} = \{x_1^{ini}, x_2^{ini}, \cdots, x_N^{ini}\}$, then:

$$x_{td} = x_d^{ini} + r(x_{max} - x_{min}) + x_{min}, t = 1, 2, 3, \cdots, 60, d = 1, 2, 3, \cdots, D \tag{24}$$

where $x_{max} = \underset{1 \le d \le N}{\text{MAX}}\{|x_d^{ini}|\}$, $x_{min} = -\underset{1 \le d \le N}{\text{MAX}}\{|x_d^{ini}|\}$, r is a random number in interval $[0, 1]$, d is the index of parameters and D is the dimension of particle, $D = N$.

Step 2. Evaluate fitness for each particle and further identify the best position (defined as $P_t^k = \{p_{t1}^k, p_{t2}^k, \cdots, p_{tD}^k\}$) of each particle and the best position (defined as $P_g^k = \{p_{g1}^k, p_{g2}^k, \cdots, p_{gD}^k\}$) of the swarm. Up to the k-th evolution generation, let p_{tbest} denote the best fitness of the t-th particle. For each particle, if its current fitness value is better than p_{tbest}, then set this value as p_{tbest} and its position as P_t^k. Similarly, the particle with the best fitness of the swarm is identified, and its position is defined as P_g^k.

Step 3. Update the position and velocity for each particle, shown in Equation (25).

$$\begin{cases} v_{td}^{k+1} = \omega v_{td}^k + c_1 r_1 (p_{td}^k - x_{td}^k) + c_2 r_2 (p_{gd}^k - x_{td}^k) \\ x_{td}^{k+1} = x_{td}^k + v_{td}^{k+1} \end{cases} \tag{25}$$

where ω is the inertia weight; $d = 1, 2, 3, \cdots, D$; $t = 1, 2, 3, \cdots, 60$; k is the current evolution generation; c_1, c_2 are two acceleration coefficients in which c_1 is named the cognitive learning rate and c_2 is named the social learning rate; r_1, r_2 are two random numbers in the interval $[0, 1]$.

ω, c_1 and c_2 are three main parameters in PSO, which can significantly affect the performance of PSO. It has been demonstrated that a relatively larger value of ω is beneficial for global search, and a smaller one is good for local search [51]. Therefore, Shi [47] introduced a linear decreasing inertia weight, and later, Chatterjee [48] developed the trajectory of ω according to a nonlinear function, which is also used in this article, shown in Equation (26).

$$\omega = \omega_{start} - (\omega_{start} - \omega_{end})\left(\frac{g}{G_{max}}\right)^2 \tag{26}$$

where ω_{start} and ω_{end} are the initial value and end value, respectively. g and G_{max} represent the present and maximum evolution generation, respectively. Generally, $\omega_{start} = 0.9$, $\omega_{end} = 0.4$. G_{max} is set as constant 500 in this paper. In addition, Eberhart [46] described that a relatively high cognitive learning rate c_1 is beneficial for particles to wander through the entire search space, and a large social learning rate c_2 performs better in local search. Therefore, Ratnaweera [49] and Wang [50] proposed time-varying acceleration coefficients, which can be mathematically represented as follows:

$$c_1 = c_{1,max} + (c_{1,min} - c_{1,max})(\frac{g}{G_{max}}) \tag{27}$$

$$c_2 = c_{2,min} + (c_{2,max} - c_{2,min})(\frac{g}{G_{max}}) \tag{28}$$

where, $c_{1,min}$, $c_{1,max}$, $c_{2,min}$ and $c_{2,max}$ are constant factors. To find out the best ranges of c_1 and c_2, series numerical simulations are also carried out (detailed in Section 5.1).

Step 4. Repeat Steps 2–4 until the iteration reaches its maximum value. Therefore, the best fitness value can be confirmed, and its corresponding particle is obtained. The position of this particle is the optimal solution for parameters b_i and a.

3.5. Summary of Calculation Process

As described above, the calculation process of our proposed interval GM(0, N) model can be summarized as below in Figure 2.

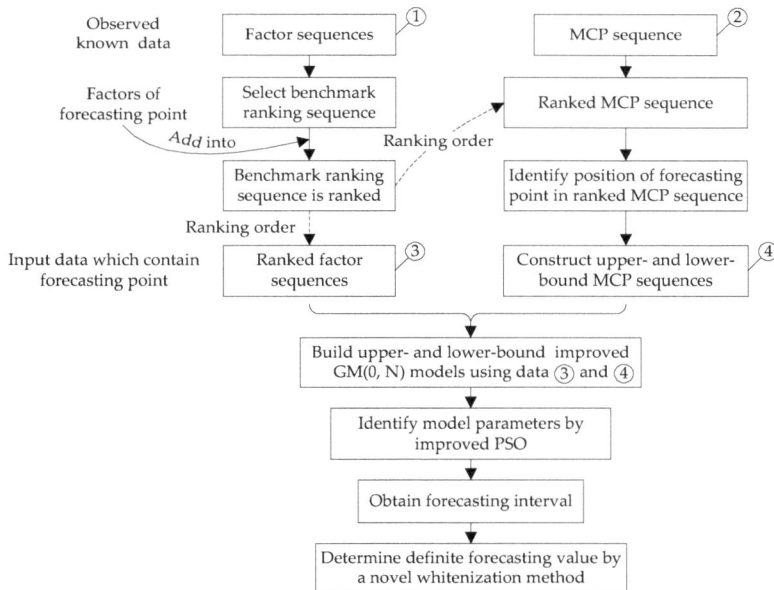

Figure 2. Flow chart of the calculation process of the proposed interval GM(0, N) model.

4. Study Area and Influence Factors

4.1. Newly-Reformed Yunnan Electricity Market

The Chinese government is starting its implementation of further electricity market reform in several piloting provinces, and Yunnan province is one of them [52]. The insufficient electricity demand and great surplus of clean energy are the main reasons for Yunnan to promote the reform of its electricity sector. At the early stage of electricity market reform, to guarantee the stability and security operation of the power grid, some units are treated as "must run units", which do not participate in the market competition. Meanwhile, to meet the requirements of environmental policies, electricity generated by clean and renewable energy sources should be fully acquired in a competitive market, such as wind, solar and small hydro. The generation contracts are made in monthly electricity market. The process and key times of Yunnan's monthly electricity market are shown in Figure 3. At the closure of bidding, the Yunnan Power Exchange Center (YPEC) aggregates the submitted seller and buyer bids. Then, the market supply and demand curves are generated, and the electricity MCP is calculated.

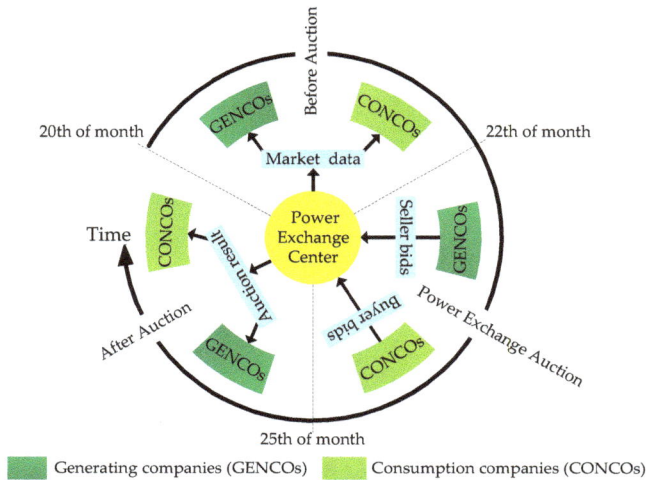

Figure 3. The process and key times of Yunnan's monthly electricity market.

4.2. Identification of Input Variables

The biggest difference between electricity and other commodities is that it cannot be stored in large quantities, which leads to a high volatility of electricity price. Many factors may determine or affect electricity price, such as historical price, historical/forecasted load, weather conditions, macroeconomic policies, competitors' irrational behaviors, and so on [5,10,53]. Till now, there are no consensus factors for price forecasting. Generally, the researchers tend to utilize past experience, also taking into consideration the unique economic environments of different markets, in selecting the input variables for their respective models. Therefore, we need to understand Yunnan's supply and demand conditions and select appropriate input variables. In this paper, factors that cannot be quantified and expressed in numerical values are not considered, such as macroeconomic policies and competitors' irrational behaviors.

Yunnan has extremely rich hydropower resources, and three of the 13 hydropower bases in China are located here. About 74.2% of the total installed capacity is hydropower, with 14.0% of small hydropower and 60.2% of medium/large hydropower. Besides hydropower, thermal, wind and solar power are also included, accounting for 16.2%, 8.0% and 1.5% of the total installed capacity respectively. Over 77% of the yearly generation in Yunnan is medium/large hydropower based; around 10% is generated by thermal power; and the remaining 13% is a mixture of production types including small hydro, wind and solar power. By the end of 2015, Yunnan's installed capacity and electricity generation of each energy resource are shown in Figures 4 and 5. Yunnan is also an important electricity supplier for China's west-to-east electricity transmission project, Laos and Vietnam. In 2015, about 44% of the total energy production has been exported. Therefore, its electricity demand can be classified into two types: one is provincial electricity demand, and the other is export electricity demand.

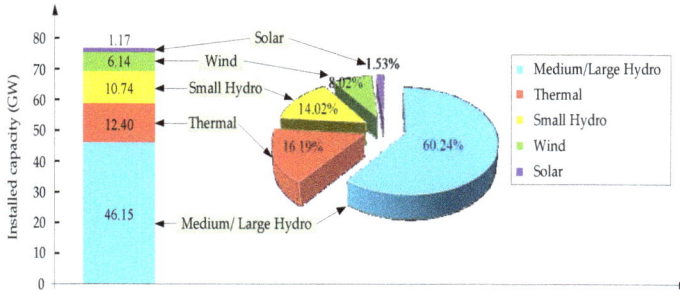

Figure 4. The installed capacity mix of Yunnan at the end of 2015.

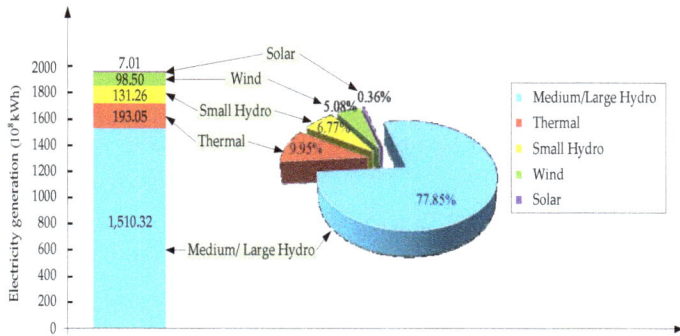

Figure 5. The electricity generation mix of Yunnan at the end of 2015.

Based on the experience, electricity supply and demand conditions of the Yunnan market, three types of influence factors are considered.

4.2.1. Supply Factors

The medium/large hydropower and thermal power are the main energy sources of Yunnan, which would directly affect the electricity MCP. As clean and renewable energy sources, the wind power, small hydropower and solar power all have the priority right to be purchased, which would increase the competition between other generators without the priority right and affect the electricity MCP, as well. Therefore, the available energy of each energy resource is concerned, and the energy production is considered as the input variable on the supply factors' side. It is noted that the factors that can further influence energy production are not included. The reasons for this are: (1) compared with energy production, the factors that can influence energy production are having relatively indirect influences on the market price; to a certain degree, their influences on the market price have already been reflected in that of energy production; and (2) the factors that influence energy production can be various, and some of them are unavailable. Take hydropower as an example: its energy production is strongly correlated with runoffs, reservoir's capacity and level, unit maintenance, flood prevention requirements, operation rules, and so on. It seems extremely difficult to consider all of these factors; the adopted energy production is a synthetic effect of these on electricity MCP. Besides, the portion that does not participate in the market competition (such as must run units) has been excluded from the available energies.

4.2.2. Demand Factors

The electricity MCP is obviously related to the market demand. Currently, in Yunnan, more than 40% of generated electricity is transmitted to outside areas through mid- or long-term contracts. The amount of export electricity has an influence on the competition of provincial electricity consumers and should affect the electricity MCP. For the provincial electricity demand, about 90% is provided in the monthly electricity market. The export electricity demand and provincial electricity demand are the main demand factors and selected as input variables.

4.2.3. Supplemented Factors

In Yunnan's developing electricity market, there are only two types of participants so far: the generating companies (GENCOs) and consumption companies (CONCOs). The number of GENCOs and CONCOs reflects the competition in the supply and demand sides. With more market participants, the market tends to be a perfectly competitive market, and the electricity MCP changes. Here, the number of GENCOs and CONCOs in the electricity market is used as supplemented input variables.

The input variables for electricity MCP forecasting considered in this article are listed in the following Table 2. Meanwhile, we should notice that the selection of appropriate input variables plays an important role in ensuring accuracy. For different electricity markets, the energy source compositions and economic environments are obviously different, resulting in specific input variables. The proposed model is promising to be applied to other electricity markets, but the input variables should be carefully re-identified before modeling.

Table 2. The input variables for electricity market clearing price (MCP) forecasting.

Index	Types	Input Variables	Symbols
1		Energy production of medium/large hydro power	F_1
2		Energy production of thermal power	F_2
3	Supply factors	Energy production of small hydro power	F_3
4		Energy production of wind power	F_4
5		Energy production of solar power	F_5
6	Demand factors	Export electricity demand	F_6
7		Provincial electricity demand	F_7
8	Supplemented	Number of GENCOs	F_8
9	factors	Number of CONCOs	F_9

4.3. Data Collection and Normalization

For the case study, actual monthly data including MCP sequence and factor sequences in the Yunnan electricity market, from April 2015–April 2016 are used. The data shown in Table 3 are normalized by the min-max normalization (*Nor.*), which is formulated as:

$$X' = \frac{X - X_{min}}{X_{max} - X_{min}} \tag{29}$$

where X_{max} and X_{min} are the maximum and minimum values of the X sequence, respectively. Correspondingly, the inverse normalization transformation (*Inv.*) is:

$$X = X'(X_{max} - X_{min}) + X_{min} \tag{30}$$

Table 3. Monthly MCP and influence factors of the Yunnan electricity market from April 2015–April 2016.

Month	MCP	F_1	F_2	F_3	F_4	F_5	F_6	F_7	F_8	F_9
April 2015	0.9174	0.2015	0.8689	0.0374	0.3354	0.0945	0.2972	0.7269	0.3462	0.0044
May 2015	0.4128	0.3630	0.8592	0.0290	0.2925	0.0199	0.5315	0.7343	0.2308	0.0000
June 2015	0.0494	0.5985	0.3532	0.3125	0.1701	0.0000	0.6965	0.8876	0.3846	0.1027
July 2015	0.0000	1.0000	0.3647	0.8816	0.0000	0.0995	0.9156	0.8774	0.6154	0.1754
August 2015	0.0105	0.9298	0.2992	1.0000	0.0241	0.0995	1.0000	0.9368	0.5000	0.1789
September 2015	0.0296	0.7841	0.0000	0.8854	0.0377	0.0945	0.9657	1.0000	0.2308	0.2170
October 2015	0.0303	0.7627	0.7197	0.5264	0.2543	0.1592	0.8949	0.9885	0.2692	0.2365
November 2015	0.1692	0.3667	0.7306	0.3354	0.3747	0.2985	0.3693	0.5218	0.2692	0.2294
December 2015	0.4726	0.3381	1.0000	0.2895	0.4741	0.3483	0.3101	0.5220	0.0000	0.1798
January 2016	0.9259	0.1890	0.6754	0.1360	0.6410	0.5323	0.2268	0.8476	1.0000	0.8335
February 2016	1.0000	0.0000	0.4339	0.0000	0.8095	0.6169	0.0000	0.0000	0.9231	0.6740
March 2016	0.7791	0.2034	0.5140	0.0053	1.0000	1.0000	0.2783	0.4350	0.9231	0.7741
April 2016	0.7330	0.2547	0.5546	0.0031	0.6599	0.7761	0.2992	0.7086	1.0000	1.0000

5. Case Study

5.1. An Example for Forecasting

To detail the forecasting procedure of the proposed interval GM(0, N) model, a forecasting example for April 2016 is given. In this case study, the proposed model is trained by data from April 2015–March 2016. The procedure is described as follows.

Step 1. According to Section 3.1, the influence factor F_1 is selected as the benchmark ranking sequence because it has nearly the reverse ranking order and the highest correlation coefficient with the MCP sequence. The correlation coefficient is 0.9059. Then, all data sequences that contain the forecasting point are ranked in accordance with the descending order of F_1, partly shown in Table 4.

Table 4. Ranked monthly MCP and the influence factors that contain the forecasting point.

Month	MCP	F_1	F_2	F_3	F_4	F_5	F_6	F_7	F_8	F_9
$\cdots$	-	-	-	$\cdots$	-	-	-	$\cdots$		
May 2015	0.4128	0.3630	0.8592	0.0290	0.2925	0.0199	0.5315	0.7343	0.2308	0.0000
December 2015	0.4726	0.3381	1.0000	0.2895	0.4741	0.3483	0.3101	0.5220	0.0000	0.1798
April 2016	$\varnothing(k)$	0.2547	0.5546	0.0031	0.6599	0.7761	0.2992	0.7086	1.0000	1.0000
March 2016	0.7791	0.2034	0.5140	0.0053	1.0000	1.0000	0.2783	0.4350	0.9231	0.7741
April 2015	0.9174	0.2015	0.8689	0.0374	0.3354	0.0945	0.2972	0.7269	0.3462	0.0044
$\cdots$	-	-	-	$\cdots$	-	-	-	$\cdots$		

Step 2. The position of forecasting point is determined by comparing the F_1 value of April 2016 with other months, i.e., between December 2015 and March 2016. Two neighboring values, respectively 0.7791 and 0.4726, are used as two virtual values for April 2016. Thus, the upper- and lower-bound MCP sequences are constructed. They are:

$$X_1^{up(0)} = \{0.0000, 0.0105, \cdots, 0.4726, 0.7791, 0.7791, \cdots, 1.0000\}$$

$$X_1^{low(0)} = \{0.0000, 0.0105, \cdots, 0.4726, 0.4726, 0.7791, \cdots, 1.0000\}$$

Step 3. Two improved GM(0, N) models are respectively built from the upper- and lower-bound MCP sequences combined with ranked factor sequences. Additionally, the model parameters are identified by the time-varying PSO as described in Section 3.4. To find out the best ranges of c_1 and c_2, numerical simulations are carried out. Based on the values used in [49,50,54–58], the test range of $c_1 + c_2$ is set to [3.0, 5.0]. The average and standard deviation of the optimum solutions for 100 trials of different values of $c_1 + c_2$ are shown in Table 5, and curves are presented in Figure 6 accordingly. From the result, we can see that when $c_1 + c_2 = 4.2$, the performance is best. Thus, in our study, with the increasing of iterations, c_1 is decreasing from 3.7 down to 0.5, and c_2 is increasing from 0.5–3.7. In this way, the resulting MCP forecasting interval is [0.4787, 0.7910]. The evolution process of best fitness when calculating upper- and lower-bound forecasting values by the time-varying PSO is shown below in Figure 7.

Table 5. The average optimum value (AOV) and standard deviation (SD) for 100 trials of different values of $c_1 + c_2$.

	Different Values and Ranges of c_1 and c_2				
Statistical Index	$c_1 + c_2 = 3.1$ $c_1 = 2.6\sim0.5$ $c_2 = 0.5\sim2.6$	$c_1 + c_2 = 3.2$ $c_1 = 2.7\sim0.5$ $c_2 = 0.5\sim2.7$	$c_1 + c_2 = 3.3$ $c_1 = 2.8\sim0.5$ $c_2 = 0.5\sim2.8$	$c_1 + c_2 = 3.4$ $c_1 = 2.9\sim0.5$ $c_2 = 0.5\sim2.9$	$c_1 + c_2 = 3.5$ $c_1 = 3.0\sim0.5$ $c_2 = 0.5\sim3.0$
AOV (%) SD ($\times10^{-2}$)	4.41 1.67	4.16 0.81	4.06 0.60	4.13 0.64	4.10 1.04
Statistical Index	$c_1 + c_2 = 3.6$ $c_1 = 3.1\sim0.5$ $c_2 = 0.5\sim3.1$	$c_1 + c_2 = 3.7$ $c_1 = 3.2\sim0.5$ $c_2 = 0.5\sim3.2$	$c_1 + c_2 = 3.8$ $c_1 = 3.3\sim0.5$ $c_2 = 0.5\sim3.3$	$c_1 + c_2 = 3.9$ $c_1 = 3.4\sim0.5$ $c_2 = 0.5\sim3.4$	$c_1 + c_2 = 4.0$ $c_1 = 3.5\sim0.5$ $c_2 = 0.5\sim3.5$
AOV (%) SD ($\times10^{-2}$)	4.12 0.76	4.26 1.46	3.99 0.48	3.76 0.73	3.74 0.49
Statistical Index	$c_1 + c_2 = 4.1$ $c_1 = 3.6\sim0.5$ $c_2 = 0.5\sim3.6$	$c_1 + c_2 = 4.2$ $c_1 = 3.7\sim0.5$ $c_2 = 0.5\sim3.7$	$c_1 + c_2 = 4.3$ $c_1 = 3.8\sim0.5$ $c_2 = 0.5\sim3.8$	$c_1 + c_2 = 4.4$ $c_1 = 3.9\sim0.5$ $c_2 = 0.5\sim3.9$	$c_1 + c_2 = 4.5$ $c_1 = 4.0\sim0.5$ $c_2 = 0.5\sim4.0$
AOV (%) SD ($\times10^{-2}$)	3.80 1.01	3.67 0.35	3.87 0.72	3.90 0.91	3.93 1.59
Statistical Index	$c_1 + c_2 = 4.6$ $c_1 = 4.1\sim0.5$ $c_2 = 0.5\sim4.1$	$c_1 + c_2 = 4.7$ $c_1 = 4.2\sim0.5$ $c_2 = 0.5\sim4.2$	$c_1 + c_2 = 4.8$ $c_1 = 4.3\sim0.5$ $c_2 = 0.5\sim4.3$	$c_1 + c_2 = 4.9$ $c_1 = 4.4\sim0.5$ $c_2 = 0.5\sim4.4$	$c_1 + c_2 = 5.0$ $c_1 = 4.5\sim0.5$ $c_2 = 0.5\sim4.5$
AOV (%) SD ($\times10^{-2}$)	4.19 0.69	4.17 0.57	4.16 0.71	4.10 0.56	4.28 0.74

Notes: $c_1 = 2.6 \sim 0.5$ means that c_1 is changing from 2.6 to 0.5, and similar to the others.

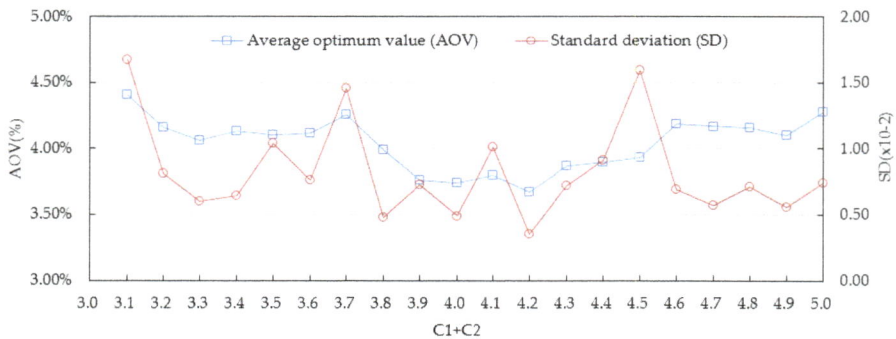

Figure 6. The curves of the average optimum value (AOV) and standard deviation (SD) for 100 trials of different values of $c_1 + c_2$.

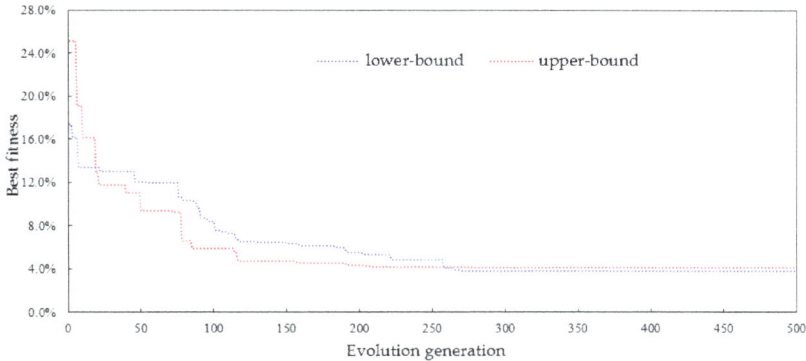

Figure 7. The evolution process of best fitness when calculating upper- and lower-bound forecasting values.

Step 4. As described in Section 3.3, the proportion coefficient α is calculated by Equation (21). Thus, the definite forecasting value of April 2016 is:

$$\hat{x}_1^{(0)} = \alpha \cdot \hat{x}_1^{low(0)} + (1 - \alpha) \cdot \hat{x}_1^{up(0)} = 0.4700 \times 0.4787 + (1 - 0.4700) \times 0.7910 = 0.6442$$

Therefore, the forecasting value of April 2016 is 0.2803 after inverse normalization transformation. As the observed value is 0.2893, the MAE and MAPE are respectively $|0.2893 - 0.2803| = 0.0090$ and $|0.2893 - 0.2803| \div 0.2893 \times 100\% = 3.10\%$. The forecasting result is acceptable. It is important to note that the forecasting result is based on the assumption that the input factor sequences of the forecasting point are accurately predicted.

5.2. Sensitivity Analysis of Position

To validate the applicability of the proposed interval GM(0, N) model, a sensitivity analysis of the position of the forecasting point is discussed. Here, also take the MCP forecasting of April 2016 as an example. The true position of April 2016 in the ranked MCP sequence is Position 1. Suppose that the position is somehow misjudged; define the left position as Position 2 and the right as Position 3, as shown in Figure 8.

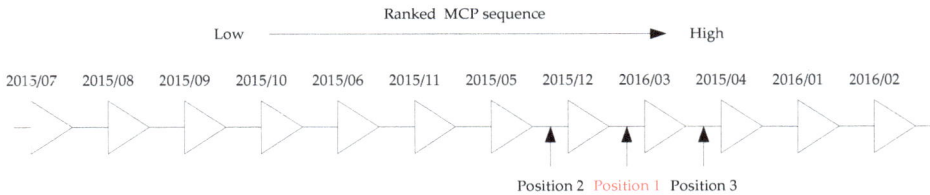

Figure 8. Different positions of April 2016 in the ranked market clearing price (MCP) sequence.

The proposed interval GM(0, N) model is applied to three different positions respectively, and the results are shown in Table 6. The MAPE for Position 1 is 3.10%, which is the best among the three positions. It obviously shows that the correct position estimation can provide a more accurate forecasting interval, which has a direct influence on the forecasting result. For Positions 2 and 3, the MAPEs are 8.28% and 4.79%, respectively. Compared to Position 1, although their forecasting performances are deteriorating, they are also acceptable in mid-term forecasting. The MAPE of the three positions is 5.39%. By taking into account the correlation between electricity MCP and influence factors, the coefficient α given by the developed novel whitenization method has a certain degree of correction to the forecasting interval. As shown in Figure 9, whether the position of the forecasting point is correct or not, the coefficient α will guide the forecasting interval towards the actual value, which also shows a better result than the equal proportion method (α = 0.5). For Positions 1 and 2, the forecasting interval is on the left of the observed value, so the direction of α is from left to right. On the other hand, the forecasting interval of Position 3 is on the right; the direction is accordingly from right to left. Although the position of the forecasting point is not exactly judged in some cases, if the deviation of the position is not too big, the proposed model can obtain an acceptable forecasting result.

Table 6. Forecasting results of April 2016 in Positions 1–3.

Position	Observed Value		Virtual Values		Forecasting Interval		Forecasting Value			MAPE (%)
	Nor.	*Inv.*	Lower	Upper	Lower	Upper	α	*Nor.*	*Inv.*	
Position 1			0.4726	0.7791	0.4787	0.7910	0.4700	0.6442	0.2803	3.10
Position 2	0.7330	0.2893	0.4128	0.4726	0.4186	0.4702	−0.5047	0.4962	0.2653	8.28
Position 3			0.7791	0.9174	0.7735	0.9098	2.3012	0.5961	0.2755	4.79
Average	0.7330	0.2893	0.5548	0.7230	0.5569	0.7237	0.7555	0.5789	0.2737	5.39

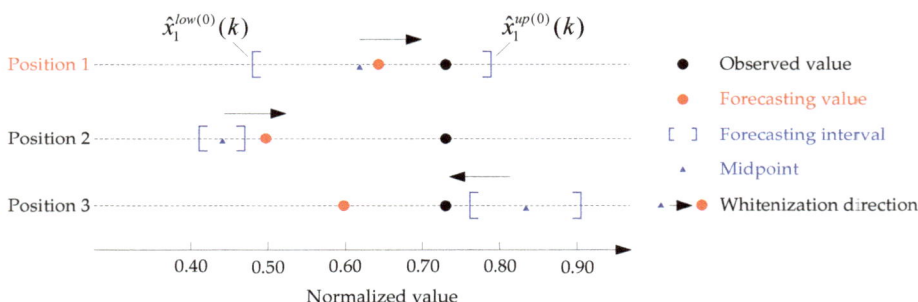

Figure 9. Definite forecasting values of Positions 1–3 determined by the novel whitenization method.

5.3. Comparison with Other Models

To further illustrate the forecasting performance of the proposed interval GM(0, N) model (Model 4), three other forecasting models are established for comparison, i.e., the multiple linear regression (MLR) model (Model 1), the traditional GM(0, N) model (Model 2) and the artificial neural network (ANN) model (Model 3). The MLR model is a typical static model for price forecasting by investigating the relationship between a dependent variable and several independent variables, which is also shaped like the proposed model. The used ANN model is a typical three-layered back propagation neural network model, including one input layer, one hidden layer and one output layer, which is usually preferred in practical engineering applications [23,59,60]. The sigmoid transfer function is used in both the neurons of the hidden and output layers. Nine inputs (influence factors) and one output (market price) are applied to the ANN model. At the training stage, the Levenberg–Marquardt algorithm is used to train the ANN model. Various numbers of neurons in the hidden layer are tested, and all statistical criteria of these network structures are recorded and compared. The best results are produced with 11 hidden units. The maximum iterations used in this paper is 1000. All four models are applied to forecast electricity MCP for each month in turn from April 2015–April 2016 under the same experimental conditions. For each month, the data of other months are used as input data for building these models respectively.

Forecasting results of these models are shown in Table 7 and Figure 10. We can see that the proposed interval GM(0, N) model shows a higher precision than the other compared models. The MAPE of the proposed model is 3.80%, and that of the MLR model is 22.06%; the traditional GM(0, N) model is 23.70%; the ANN model is 8.00%. Although the MAPE of the ANN model does not differ very greatly from that of the proposed model, the fluctuation of its results is much higher. The maximum absolute percentage error of the ANN model is 21.12%, which is not acceptable. The error variance of the proposed model is 0.0297, which is also much less than the MLR model (0.1581), the traditional GM(0, N) model (0.1801) and the ANN model (0.0550). For fitting precision checking, the *C* and *P* of the proposed model are 0.32 and 100.0%, respectively, showing a good fitting precision grade. While these of the traditional GM(0, N) model are 2.22 and 30.78%, respectively, the forecasting result is unqualified. This truly shows that our improvements on the traditional GM(0, N) model are effective in mid-term electricity MCP forecasting, and good forecasting results are obtained.

Table 7. Forecasting results of the proposed model compared with the other three models.

Month	Observed Value	Forecasting Value				Absolute Percentage Error (%)			
		Model 1	Model 2	Model 3	Model 4	Model 1	Model 2	Model 3	Model 4
April 2015	0.4471	0.3394	0.0000	0.3627	0.3929	13.08	54.32	10.25	6.58
May 2015	0.3107	0.3253	0.1370	0.2841	0.2778	2.14	25.30	3.87	4.79
June 2015	0.2124	0.3101	0.1957	0.2815	0.1920	16.62	2.83	11.75	3.46
July 2015	0.1990	0.1023	0.1901	0.1794	0.2334	16.81	1.55	3.42	5.98
August 2015	0.2018	0.2403	0.1369	0.2595	0.2005	6.65	11.24	9.98	0.23
September 2015	0.2070	0.0733	0.3286	0.1949	0.2056	22.94	20.85	2.09	0.25
October 2015	0.2072	0.1795	0.3583	0.2202	0.2033	4.75	25.90	2.23	0.68
November 2015	0.2448	0.3530	0.1696	0.3759	0.3111	17.43	12.11	21.12	10.68
December 2015	0.3268	0.1926	0.0917	0.3543	0.3069	19.10	33.45	3.91	2.83
January 2016	0.4494	0.6114	1.0000	0.3468	0.4981	19.62	66.71	12.43	5.89
February 2016	0.4695	0.0991	0.3321	0.3601	0.4307	43.80	16.24	12.93	4.59
March 2016	0.4097	0.9014	0.5564	0.4701	0.4125	62.58	18.66	7.68	0.35
April 2016	0.3972	0.1793	0.2502	0.4157	0.3732	28.18	19.01	2.39	3.10
MAPE	-	-	-	-	-	22.06	23.70	8.00	3.80

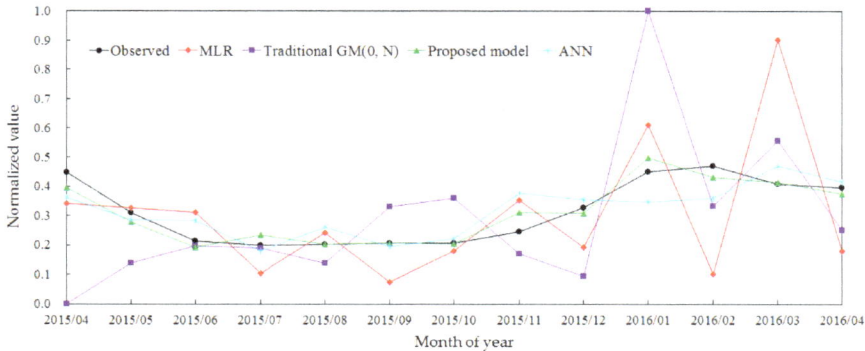

Figure 10. Forecasting results of different models from April 2015–April 2016.

5.4. Forecasting Evaluation Based on the Modified Diebold–Mariano (MDM) Test

In this part, the forecasting performance is compared via the MDM test. The MDM test is carried out respectively between our proposed model and the other three forecasting models. The evaluation results are summarized in Table 8.

73

Table 8. Forecasting evaluation results based on the modified Diebold–Mariano (MDM) test.

Indicators	MDM Test Results between Different Models		
	Models 1 and 4	Models 2 and 4	Models 3 and 4
MDM-MAE	3.2614 ***	3.5409 ***	3.4856 ***
MDM-MSE	2.0354 *	2.0850 *	2.8789 **

Notes: MDM-MAE denotes the MDM test based on the MAE loss function; MDM-MSE denotes the MDM test based on the MSE loss function; ***, ** and * denote significance at the 1%, 5% and 10% levels, respectively.

From Table 8, conclusions can be drawn by comparison of Model 1 and Model 4:

1. According to the MDM test based on the MAE loss function, the null hypothesis is rejected at the 1% level of significance. In other words, the observed differences are pretty significant, and the forecasting performance of Model 4 is better than Model 1.
2. According to the MDM test based on the MSE loss function, the null hypothesis is rejected at the 10% level of significance. That is to say, the observed differences are also significant, and the forecasting performance of Model 4 is better than Model 1.

Similarly, according to the comparisons of Models 2 and 4 and Models 3 and 4, both the MDM test by MAE and MSE loss functions evaluate that the forecasting performance of Model 4 is better than Models 2 and 3. Therefore, the superiority of the proposed model is further validated statistically.

6. Conclusions

Accurate mid-term electricity MCP forecasting is essential for all market participants. In this paper, a novel interval GM(0, N) model for mid-term electricity MCP forecasting is proposed, aiming to present a feasible method for the electricity market with few available data. In the proposed model, two improved GM(0, N) models are included to respectively estimate the upper- and lower-bound forecasting values. Firstly, all input sequences (not containing further values) including the MCP sequence and factor sequences are ranked in accordance with the ascending order of the MCP sequence. Then, one of the factor sequences is selected as the benchmark ranking sequence according to its ranking order and correlation with the MCP sequence. The position of forecasting point in the MCP sequence thus can be determined by the benchmark ranking sequence. Finally, the upper and lower points of the forecasting point in the MCP sequence are regarded as two virtual values to construct two new MCP sequences, which are respectively used as input for two improved GM(0, N) models, obtaining the forecasting interval. In the two improved GM(0, N) models, the input sequences used for modeling include not only known data, but also the virtual MCP and predicted factors of the forecasting point, and model parameters are identified by an improved PSO instead of LSM. Based on the forecasting interval, a novel whitenization method, taking correlation coefficient weighted gaps of factor sequences into consideration, is also established to determine the definite forecasting value.

The proposed model has been applied to the newly-reformed Yunnan electricity market. Being at the early stage of reform, this market shares many common features with restructuring practices in other countries, while simultaneously exhibiting many unique characteristics, as well. After careful analysis of its market conditions, the input variables are appropriately determined, among which the energy production of medium/large hydro power is selected as the benchmark ranking sequence. For April 2016, the absolute percentage error of the forecasting value is 3.10%. Additionally, the MAPE of the proposed model for MCP forecasting from April 2015–April 2016 is 3.80%, which shows a better forecasting performance compared with the MLR model, the traditional GM(0, N) model and the ANN model, and its superiority is also confirmed by the modified Diebold–Mariano test. The good forecasting results indicate that the proposed model is suitable for mid-term electricity MCP forecasting in a data-sparse electricity market.

Acknowledgments: This study is supported by the National Natural Science Foundation of China (No. 91547201) and the Major International Joint Research Project from the National Nature Science Foundation of China (51210014).

Author Contributions: Chuntian Cheng and Bin Luo developed and solved the proposed model and carried out this analysis. Bin Luo, Shumin Miao and Xinyu Wu wrote the manuscript and contributed to the revisions.

Conflicts of Interest: The authors declare no conflict of interest.

References

1. The Communist Party of China (CPC) Central Committee and State Council of PRC. Opinions on Further Deepening the Reforms of electricity power system. Available online: http://www.cec.org.cn/huanbao/xingyexinxi/fazhangaige/2015-03-25/135625.html (accessed on 25 March 2015).
2. Ma, C.; He, L. From state monopoly to renewable portfolio: Restructuring China's electric utility. *Energy Policy* **2008**, *36*, 1697–1711. [CrossRef]
3. Yan, X.; Chowdhury, N.A. Mid-term electricity market clearing price forecasting: A hybrid LSSVM and ARMAX approach. *Int. J. Electr. Power Energy Syst.* **2013**, *53*, 20–26. [CrossRef]
4. Weron, R. Electricity price forecasting: A review of the state-of-the-art with a look into the future. *Int. J. Forecast.* **2014**, *30*, 1030–1081. [CrossRef]
5. Aggarwal, S.K.; Saini, L.M.; Kumar, A. Electricity price forecasting in deregulated markets: A review and evaluation. *Int. J. Electr. Power Energy Syst.* **2009**, *31*, 13–22. [CrossRef]
6. Rubin, O.D.; Babcock, B.A. The impact of expansion of wind power capacity and pricing methods on the efficiency of deregulated electricity markets. *Energy* **2013**, *59*, 676–688. [CrossRef]
7. Sapio, S.; Wylomanska, A. The impact of forward trading on the spot power price volatility with Cournot competition. In Proceedings of the 5th international conference on the European electricity market, Lisboa, Portugal, 28–30 May 2008; pp. 105–110.
8. Ruibal, C.M.; Mazumdar, M. Forecasting the mean and the variance of electricity prices in deregulated markets. *IEEE Trans. Power Syst.* **2008**, *23*, 25–32. [CrossRef]
9. Wood, A.J.; Wollenberg, B.F. *Power Generation, Operation and Control*; Wiley: New York, NY, USA, 1996.
10. Batlle, C.; Barquin, J. A strategic production costing model for electricity market price analysis. *IEEE Trans. Power Syst.* **2005**, *20*, 67–74. [CrossRef]
11. Kian, A.; Keyhani, A. Stochastic price modeling of electricity in deregulated energy markets. In Proceedings of the 34th Annual Hawaii International Conference on System Sciences, Maui, HI, USA, 6 January 2001; pp. 1–7.
12. Robinson, T.A. Electricity pool prices: A case study in nonlinear time-series modelling. *Appl. Econ.* **2000**, *32*, 527–532. [CrossRef]
13. Carpio, J.; Juan, J.; López, D. Multivariate exponential smoothing and dynamic factor model applied to hourly electricity price analysis. *Technometrics* **2014**, *56*, 494–503. [CrossRef]
14. Nogales, F.J.; Contreras, J.; Conejo, A.J.; Espinola, R. Forecasting next-day electricity prices by time series models. *IEEE Trans. Power Syst.* **2002**, *17*, 342–348. [CrossRef]
15. Obradovic, Z.; Tomsovic, K. Time series methods for forecasting electricity market pricing. In Proceedings of the IEEE Power Engineering Society Summer Meeting, Edmonton, AB, Canada, 18–22 July 1999; pp. 1264–1265.
16. Cuaresma, J.C.; Hlouskova, J.; Kossmeier, S.; Obersteiner, M. Forecasting electricity spot-prices using linear univariate time-series models. *Appl. Energy* **2004**, *77*, 87–106. [CrossRef]
17. Conejo, A.J.; Plazas, M.A.; Espinola, R.; Molina, A.B. Day-ahead electricity price forecasting using the Wavelet Transform and ARIMA Models. *IEEE Trans. Power Syst.* **2005**, *20*, 1035–1042. [CrossRef]
18. Kim, C.; Yu, I.K.; Song, Y.H. Prediction of system marginal price of electricity using wavelet transform analysis. *Energy Convers. Manag.* **2002**, *43*, 1839–1851. [CrossRef]
19. Zhang, J.; Tan, Z. Day-ahead electricity price forecasting using WT, CLSSVM and EGARCH model. *Int. J. Electr. Power Energy Syst.* **2013**, *45*, 362–368. [CrossRef]
20. Contreras, J.; Espinola, R.; Nogales, F.J.; Conejo, A.J. ARIMA models to predict next-day electricity prices. *IEEE Trans. Power Syst.* **2003**, *18*, 1014–1020. [CrossRef]

21. Jakaša, T.; Andročec, I.; Sprčić, P. Electricity price forecasting - ARIMA model approach. In Proceedings of the 8th International Conference on the European Energy Market (EEM), Zagreb, Croatia, 25–27 May 2011; pp. 222–225.

22. Xie, M.; Sandels, C.; Zhu, K. A seasonal ARIMA model with exogenous variables for elspot electricity prices in Sweden. In Proceedings of the 10th International Conference on the European Energy Market (EEM), Stockholm, Sweden, 27–31 May 2013; pp. 1–4.

23. Szkuta, B.R.; Sanabria, L.A.; Dillon, T.S. Electricity price short-term forecasting using artificial neural networks. *IEEE. Trans. Power Syst.* **1999**, *14*, 851–857. [CrossRef]

24. Keles, D.; Scelle, J.; Paraschiv, F.; Fichtner, W. Extended forecast methods for day-ahead electricity spot prices applying artificial neural networks. *Appl. Energy* **2016**, *162*, 218–230. [CrossRef]

25. Shrivastava, N.A.; Khosravi, A.; Panigrahi, B.K. Prediction interval estimation of electricity prices using PSO-Tuned Support Vector Machines. *IEEE Trans. Ind. Inf.* **2015**, *11*, 322–331. [CrossRef]

26. Papadimitriou, T.; Gogas, P.; Stathakis, E. Forecasting energy markets using support vector machines. *Energy Econ.* **2014**, *44*, 135–142. [CrossRef]

27. Zhao, J.H.; Dong, Z.; Li, X.; Wong, K. A framework for electricity price spike analysis with advanced data mining methods. *IEEE Trans. Power Syst.* **2007**, *22*, 376–385. [CrossRef]

28. Huang, D.; Zareipour, H.; Rosehartet, W.D.; Amjady, N. Data mining for electricity price classification and the application to demand-side management. *IEEE Trans. Smart Grid* **2012**, *3*, 808–817. [CrossRef]

29. Tan, Z.F.; Zhang, J.L.; Wang, J.H.; Xu, J. Day-ahead electricity price forecasting using wavelet transform combined with ARIMA and GARCH models. *Appl. Energy* **2010**, *87*, 3606–3610. [CrossRef]

30. Gonzalez, V.; Contreras, J.; Bunn, D. Forecasting power prices using a hybrid Fundamental-Econometric model. *IEEE Trans. Power Syst.* **2012**, *27*, 363–372. [CrossRef]

31. Cerjan, M.; Matijaš, M.; Delimar, M. Dynamic hybrid model for short-term electricity price forecasting. *Energies* **2014**, *7*, 3304–3318. [CrossRef]

32. Torghaban, S.S.; Zareipour, H.; Tuan, L.A. Medium-term electricity market price forecasting: a data-driven approach. In Proceedings of the 2010 North American Power Symposium (NAPS 2010), Arlington, TX, USA, 26–28 September 2010; pp. 1–7.

33. Torghaban, S.S.; Motamedi, A.; Zareipour, H.; Tuan, L.A. Medium-term electricity price forecasting. In Proceedings of the 2012 North American Power Symposium (NAPS 2012), Champaign, IL, USA, 9–11 September 2012; pp. 1–8.

34. Bello, A.; Reneses, J.; Munoz, A. Medium-term probabilistic forecasting of extremely low prices in electricity markets: Application to the Spanish case. *Energies* **2016**, *9*, 193. [CrossRef]

35. Yan, X.; Chowdhury, N.A. Mid-term electricity market clearing price forecasting: A multiple SVM approach. *Int. J. Electr. Power Energy Syst.* **2014**, *58*, 206–214. [CrossRef]

36. Liu, S.; Forrest, J.; Yang, Y. A brief introduction to grey systems theory. *Grey Systems: Theory Appl.* **2012**, *2*, 89–104. [CrossRef]

37. Deng, J.L. The control problem of grey systems. *Syst. Control Lett.* **1982**, *1*, 288–294.

38. Qiu, B.J.; Zhang, J.H.; Qi, Y.T.; Liu, Y. Grey-theory-based optimization model of emergency logistics considering time uncertainty. *PLoS ONE* **2015**, *10*, e0139132. [CrossRef] [PubMed]

39. Wei, J.C.; Zhou, L.; Wang, F.; Wu, D.S. Work safety evaluation in Mainland China using grey theory. *Appl. Math. Model.* **2015**, *39*, 924–933. [CrossRef]

40. Zhou, H.; Wu, X.H.; Wang, W.; Chen, L.P. Forecast of next day clearing price in deregulated electricity market. In Proceedings of the IEEE International Conference on Systems, Man and Cybernetics, San Antonio, TX, USA, 11–14 October 2009; pp. 4397–4401.

41. Lei, M.; Feng, Z. A proposed grey model for short-term electricity price forecasting in competitive power markets. *Int. J. Electr. Power Energy Syst.* **2012**, *43*, 531–538. [CrossRef]

42. Ou, S.L. Forecasting agricultural output with an improved grey forecasting model based on the genetic algorithm. *Comput. Electron. Agr.* **2012**, *85*, 33–39. [CrossRef]

43. Wang, C.H.; Hsu, L.C. Using genetic algorithms grey theory to forecast high technology industrial output. *Appl. Math. Comput.* **2008**, *95*, 256–263. [CrossRef]

44. Harvey, D.; Leybourne, S.; Newbold, P. Testing the equality of prediction mean squared errors. *Int. J. Forecast.* **1997**, *13*, 281–291. [CrossRef]

45. Diebold, F.X.; Mariano, R.S. Comparing predictive accuracy. *J. Bus. Econ. Stat.* **1995**, *13*, 253–263.

46. Eberhart, R.; Kennedy, J. A new optimizer using particle swarm theory. In Proceedings of the 6th micro-machine and human, science, Nagoya, Japan, 4–6 October 1995; pp. 39–43.

47. Shi, Y.; Eberhart, R. Empirical study of particle swarm optimization. In Proceedings of the IEEE International Congress, Evolutionary Computation, Washington, DC, USA, 6–9 July 1999; pp. 101–106.

48. Chatterjee, A.; Siarry, P. Nonlinear inertia weight variation for dynamic adaptation in particle swarm optimization. *Comput. Oper. Res.* **2006**, *33*, 859–871. [CrossRef]

49. Ratnaweera, A.; Halgamuge, S.K.; Watson, H.C. Self-organizing hierarchical particle swarm optimizer with time-varying acceleration coefficients. *IEEE Trans. Evol. Comput.* **2004**, *8*, 240–255. [CrossRef]

50. Wang, Y.; Zhou, J.Z.; Zhou, C. An improved self-adaptive PSO technique for short-term hydrothermal scheduling. *Expert Syst. Appl.* **2012**, *39*, 2288–2295. [CrossRef]

51. Shi, Y.; Eberhart, R. A modified particle swarm optimizer. In Proceedings of the IEEE World Congress on Computational Intelligence, Anchorage, AK, USA, 4–9 May 1998; pp. 69–73.

52. National Development and Reform Commission. Notice to Accelerate Reform of Transmission and Distribution Price. Available online: http://jgs.ndrc.gov.cn/zcfg/201504/t20150416_688233.html (accessed on 13 April 2014).

53. Gao, F.; Guan, X.; Cao, X.; Papalexopoulos, A. Forecasting power market clearing price and quantity using a neural network method. In Proceedings of the Power Engineering Society Summer Meeting, Seattle, WA, USA, 16–20 July 2000; pp. 2183–2188.

54. Jiang, Y.; Liu, C.; Huang, C. Improved particle swarm algorithm for hydrological parameter optimization. *Appl. Math. Comput.* **2010**, *217*, 3207–3215. [CrossRef]

55. Montalvo, I.; Izquierdo, J.; Pérez, R. A diversity-enriched variant of discrete PSO applied to the design of water distribution networks. *Eng. Optimiz.* **2008**, *40*, 655–668. [CrossRef]

56. Gong, Y.J.; Zhang, J.; Chung, H.S. An efficient resource allocation scheme using particle swarm optimization. *IEEE Trans. Evol. Comput.* **2012**, *16*, 801–816. [CrossRef]

57. Clerc, M.; Kennedy, J. The particle swarm-explosion, stability, and convergence in a multidimensional complex space. *IEEE Trans. Evol. Comput.* **2002**, *6*, 58–73. [CrossRef]

58. Trelea, I.C. The particle swarm optimization algorithm: convergence analysis and parameter selection. *Inf. Process. Lett.* **2003**, *85*, 317–325. [CrossRef]

59. Zhang, J.; Cheng, C.T. Day-ahead electricity price forecasting using artificial intelligence. In Proceedings the of IEEE Electric Power and Energy Conference, Vancouver, BC, Canada, 6–7 October 2008; pp. 1–5.

60. Cheng, C.T.; Niu, W.J.; Feng, Z.K.; Shen, J.J.; Chau, K.W. Daily reservoir runoff forecasting method using artificial neural network based on quantum-behaved particle swarm optimization. *Water* **2015**, *7*, 4232–4246. [CrossRef]

Chapter 3:
Fuzzy Logic, Artificial Neural Networks and Hybrid Methods

energies

MDPI

Article

A Hybrid Multi-Step Model for Forecasting Day-Ahead Electricity Price Based on Optimization, Fuzzy Logic and Model Selection

Ping Jiang, Feng Liu * and Yiliao Song

School of Statistics, Dongbei University of Finance and Economics, Dalian 116025, China;
pjiang@dufe.edu.cn (P.J.); songyl13@lzu.edu.cn (Y.S.)
* Correspondence: liuf13@lzu.edu.cn or feng.liu.1990@ieee.org; Tel.: +86-139-1992-6679

Academic Editor: Javier Contreras
Received: 5 May 2016; Accepted: 27 July 2016; Published: 4 August 2016

Abstract: The day-ahead electricity market is closely related to other commodity markets such as the fuel and emission markets and is increasingly playing a significant role in human life. Thus, in the electricity markets, accurate electricity price forecasting plays significant role for power producers and consumers. Although many studies developing and proposing highly accurate forecasting models exist in the literature, there have been few investigations on improving the forecasting effectiveness of electricity price from the perspective of reducing the volatility of data with satisfactory accuracy. Based on reducing the volatility of the electricity price and the forecasting nature of the radial basis function network (RBFN), this paper successfully develops a two-stage model to forecast the day-ahead electricity price, of which the first stage is particle swarm optimization (PSO)-core mapping (CM) with self-organizing-map and fuzzy set (PCMwSF), and the second stage is selection rule (SR). The PCMwSF stage applies CM, fuzzy set and optimized weights to obtain the future price, and the SR stage is inspired by the forecasting nature of RBFN and effectively selects the best forecast during the test period. The proposed model, i.e., CM-PCMwSF-SR, not only overcomes the difficulty of reducing the high volatility of the electricity price but also leads to a superior forecasting effectiveness than benchmarks.

Keywords: selection rule (SR); reducing volatility; self-organizing-map; fuzzy logic; particle swarm optimization (PSO); forecasting

---

1. Introduction

Electricity is one of the most essential energy inputs to the industry and has increasingly significant influences on modern industry. Meanwhile, the management of operation process is more sensitive and vulnerable to the electricity supply fluctuations and its cost changes more than ever before. This demands more stable and reliable energy supply, cost management, as well as risk management. There is rising demand for more accurate analysis and forecasting of the electricity price movement [1]. To obtain accurate estimated electricity prices, modeling and prediction techniques are frequently applied to bid or hedge against the volatility of electricity prices [2,3]. Overall, it is not difficult to find that the electricity price is not only related to the interests of market participants but also affects many aspects of society and the economy. Thus, it is necessary to explore its nature in order to aid participants of the electricity market.

To show the significance of this paper better, some effective forecasting approaches for the electricity price from previous research investigations will be introduced here. One forecast strategy is a new two-stage feature selection (FS) algorithm, which is proposed by Keynia [4] and is based on the mutual information (MI) criterion; it selects representative features of the composite neural network (CNN) among feature candidates. Yan et al. [5,6] applied a multiple support vector machine (SVM) to forecast mid-term electricity price and developed a hybrid mid-term electricity price forecasting model by combining SVM and auto-regressive moving average with external input (ARMAX) modules. The Markov-switching generalized autoregressive conditional heteroskedasticity (MS-GARCH) model was developed to forecast low and high volatility electricity prices by Cifter [7]. Anbazhagan and Kumarappan proposed feed-forward neural network (FFNN) featured by one-dimensional discrete cosine transforms (DCT) and day-ahead electricity price classification using three-layered FFNN, cascade-forward neural network (CFNN) and generalized regression neural network (GRNN) [8–10]. A novel grey model was proposed using particle swarm optimization (PSO) algorithm by Lei and Feng [11]. Based on panel co-integration and particle filter (PCPF), Li et al. [12] investigated a two-stage hybrid model to achieve two main goals: (1) to expand the dimension of the dataset; and (2) to consider the model parameters as a time-varying process. Zhang and Tan [13,14] proposed new hybrid methods based on wavelet transform (WT), autoregressive integrated moving average (ARIMA) and least squares support vector machine (LSSVM) optimized by PSO and WT, chaotic least squares support vector machine (CLSSVM) and exponential generalized autoregressive conditional heteroskedastic (EGARCH) to predict electricity prices. Liu et al. [2] applied various autoregressive moving average (ARMA) models with generalized autoregressive conditional heteroskedasticity (GARCH) processes, namely ARMA-GARCH models, along with their modified forms, ARMA-GARCH-in-mean (ARMA-GARCH-M), to model and forecast hourly-ahead electricity prices. Najeh Chaâbane, based on the idea of choosing forecasting models, proposed a model that exploited the feature and strength of the auto-regressive fractionally integrated moving average (ARFIMA) model, as well as the feedforward neural networks model [15]. A new hybrid ARIMA-ANN model for the prediction of time series data based on the linear ARIMA and nonlinear artificial neural network (ANN) models was proposed by Babu et al. [16]. Shrivastava et al. [17] investigated the performance of extreme learning machine (ELM) in the price forecasting problem. Shayeghi et al. [18] proposed a new combination of the FS technique based on the MI technique and WT in. The delta and bootstrap methods were employed for the construction of prediction intervals (PIs) for uncertainty quantification by Khosravi et al. [19–21]. Bordignon et al. [22] studied combined versus individual forecasts for the prediction of British electricity prices. Grimes et al. [23] showed that simply optimizing price forecasts based on classical regression error metrics did not work well for scheduling. Nowotarski et al. [24] applied seven averaging and one selection scheme and performed backtesting analysis on day-ahead electricity prices in three major markets. From a dynamical system perspective, Sharma and Srinivasan [25] proposed a hybrid model that employed a synergistic combination of recurrent neural network (RNN) and coupled excitable system for electricity price forecasting. Dev and Martin [26] proposed an approach for the predictive capacity of neural networks and applied Australian National Electricity Market data to test their model. Wang et al. [27] proposed a forecasting model of electricity price using chaotic sequences for forecasting short-term electricity prices. The forecasting performances of four ARMAX-GARCH models for five MISO pricing hubs (Cinergy, First Energy, Illinois, Michigan, and Minnesota) were analyzed by Hickey et al. [28]. Christensen et al. [29] focused on the prediction of price spikes using a nonlinear variant of the autoregressive conditional hazard model. Amjady and Keynia [30] proposed a strategy that included a new closed-loop prediction mechanism composed of probabilistic neural network (PNN) and hybrid neuro-evolutionary system (HNES) forecast engines to forecast Pennsylvania–New Jersey–Maryland (PJM) electricity prices. Dudek [31] applied Multilayer perceptron for GEFCom2014 probabilistic electricity price forecasting. Panapakidis and Dagoumas [32] reviewed recent literature related to electricity price forecasting and applied ANN to predict future electricity prices. The K-support vector regression (K-SVR), a hybrid model to combine clustering algorithms, SVM, and SVR to forecast electricity price of PJM, is presented by Feijoo et al. [33]. Abedinia et al. [34] proposed a Combinatorial Neural Network-based forecasting engine to forecast the electricity price. The curvelet denoising-based approach was proposed to improve the forecasting effectiveness of the electricity price by He et al. [35]. Ziel et al. [36] gave an introduction of an econometric model for the hourly time series of electricity prices that incorporated specific features such as renewable energy. Hong et al. [37] applied a principal component analysis (PCA) network cascaded with a multi-layer feedforward (MLF) network for forecasting locational marginal prices (LMPs). By combining statistical techniques for pre-processing data and a multi-layer neural network, a dynamic hybrid model was proposed by Cerjan et al. [38] for forecasting electricity prices and price spike detection. Monteiro et al. [39] showed comparisons of forecasts, which led to the identification of the most important variables for forecasting purposes. By relying on simple models, forecasting approaches were derived and analyzed by Jónsson et al. [40]. Weron [41] reviewed literature related to electricity price forecasting and speculated on the directions electricity price forecasting should take in the next decade or so.

In this paper, based on reducing the volatility of the electricity price and the forecasting nature of the radial basis function network (RBFN), we successfully develop a two-stage model to forecast the day-ahead electricity price, of which the first stage is PSO-core mapping (CM) with self-organizing-map and fuzzy set (PCMwSF) and the second stage is selection rule (SR). The PCMwSF stage aims to apply CM, fuzzy set and optimized weights to obtain the future price, and the SR stage is inspired by the forecasting nature of RBFN and effectively selects the best forecast during the test period. The highlights of this paper are as follows:

➤ We successfully overcome the volatility of the electricity price through the CM method.
➤ Improvement from reducing the volatility is obvious during the test period.
➤ Self-organizing map (SOM) is assigned to divide the original data into three parts: low, medium and high.
➤ Divided price is weighted by the PSO algorithm and performs well during forecasting.
➤ SR is based on three new defined criteria and effectively selects the forecasting model.

2. Self-Organizing-Map

Figure 1 shows an application of SOM.

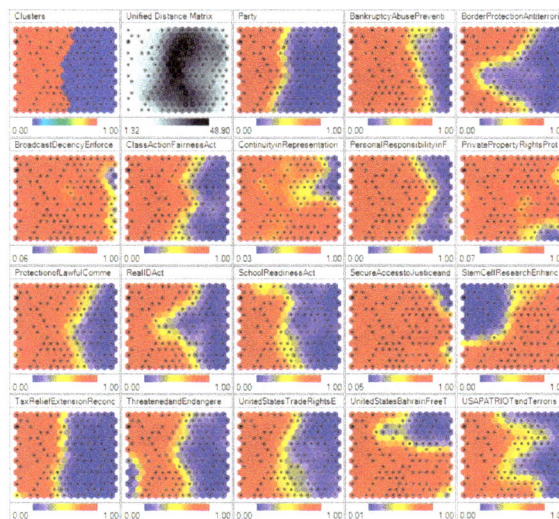

Figure 1. A self-organizing map showing U.S. Congress voting patterns visualized in Synapse. The first two boxes show clustering and distances, while the remaining ones show the component planes. Red means a yes vote, while blue means a no vote in the component planes (except the party component, where red is Republican and blue is Democratic) [42].

Because this paper focuses on the pre-process of forecasting, RBFN, i.e., the main forecasting tool, will not be introduced. Details of this method are described in [43], and the introduction of fuzzy logic and PSO can be found in [44–47]. As an ANN, SOM maps the training samples into low dimensional (typically two-dimensional), discretized representations in the input space using unsupervised learning. Unlike the other ANNs, SOM can preserve the topological properties of the input space by introducing a neighborhood function. Thus, SOM is able to visualize high-dimensional or multi-dimensional data as low-dimensional vectors. [48]. Besides, the ability of handling a high number of nodes makes SOM a powerful tool in clustering [49]. Details of the learning algorithm of SOM can be found in [50].

3. Core Mapping-Particle Swarm Optimization-Core Mapping with Self-Organizing-Map and Fuzzy Set-Selection Rule for Electricity Price Forecasting

To illustrate these approaches specifically, this section will give details of these models for forecasting electricity price.

3.1. Core Idea of This Paper

To demonstrate the core idea of this paper, the reason why high forecasting errors occur will be shown initially. In the process of forecasting, data firstly will be pre-processed to suit for model, which will be obtained by training through pre-processed data. Then, this trained model is utilized in the forecast. From research related to forecasting, it is apparent that the volatility of data has a huge effect on forecasting accuracy, which means that the volatility of data directly determines the accuracy level the model can reach. Thus, legitimately reducing volatility is an important problem in forecasting and is also the inspiration of this paper. However, from the above section, many researchers have concentrated on the promotion of algorithms, such as BP neuron network, LSSVM, ARIMA, GARCH and so on, rather than on the pre-processing of data or initial transformation of data. To improve this part of the entire forecasting process, mapping f is proposed in this paper:

$$f(x) = \int_0^x \ln(t+1)\, dt \tag{1}$$

Thus, for discrete data, Equation (1) can be expressed by:

$$f(price(x)) = \sum_{i=1}^x \ln(price(i)+1) \tag{2}$$

that means:

$$f : price(x) \rightarrow \sum_{i=1}^x \ln(price(i)+1) \tag{3}$$

This mapping is also called **CM** in this paper.

Furthermore, to reduce the volatility of the electricity price, it is divided into high price, low price and medium price by a SOM. Then, a fuzzy logic is established:

- **IF** *price(i)* **IS** High price, **THEN** *price(i)* equals *price(i)* × Highweight;
- **IF** *price(i)* **IS** Medium price, **THEN** *price(i)* equals *price(i)* × Mediumweight; and
- **IF** *price(i)* **IS** Low price, **THEN** *price(i)* equals *price(i)* × Lowweight.

Thus, the CM will be changed to:

$$f : price \rightarrow \sum_{High\_price} HIGHWeight \times \ln(price+1) + \sum_{Medium\_price} \ln(price+1) + \sum_{Low\_price} LOWWeight \times \ln(price+1) \tag{4}$$

Finally, PSO is used to optimize Highweight and Lowweight to make sure that a greater forecasting accuracy can be obtained. In post-processing, the formula of post-processing is as follows (where n is the length of forecasting series):

$$price_{forecast}(i) = e^{price_{forecast}^{pre-processed}(i)} - e^{price_{forecast}^{pre-processed}(i-1)} - 1, \ i = 2, ..., n \tag{5}$$

Thus, the CM method and PSO-CM with SOM and fuzzy logic (PCMwSF) method are proposed and used to pre-process price data in this paper. The pre-processed data will be given to RBFN to forecast the day-ahead electricity price. Mean absolute percentage error ($MAPE$), mean absolute error (MAE) and root mean square error ($RMSE$) obtained from the forecasting results demonstrate that the proposed model can efficiently forecast the price.

Furthermore, to obtain excellent forecasting accuracy of electricity prices, a rule of model selection is proposed to choose which model should be used. The final forecasting model, named CM-PCMwSF-SR, outperforms the others in each season of 2002 in the PJM power market, which is commonly recognized as one of the most successful markets in the US.

3.2. Basic Pre-Process

Before introducing proposed methods, simple pre-processes of data need to be defined first. In this paper, basic pre-processes can be expressed by:

$$price\,(i) = \begin{cases} \max\limits_{0 < i < N}(price\,(i)), & price\,(i) > 10 \times \frac{1}{N}\sum\limits_{i=1}^{N} price\,(i) \\ price\,(i), & \text{otherwise} \end{cases} \tag{6}$$

then:

$$price\,(i) = \begin{cases} \frac{price(i-1)+price(i+1)}{2}, & 0.8 < \frac{price(i)}{\frac{price(i-1)+price(i+1)}{2}} < 1 \\ \frac{price(i-1)+price(i+1)}{2}, & price\,(i) < 1 \\ price\,(i), & \text{otherwise} \end{cases} \tag{7}$$

where N is the length of the electricity price, which is prepared to train RBFN and $i = 1, 2, ..., N$. Equations (6) and (7) indicate that if the gap of $price(i)$ and mean of $price(i-1)$ and $price(i+1)$ are less than 20% or if $price(i)$ is too small, $price(i)$ will be changed to the mean value of $price(i-1)$ and $price(i+1)$. This can be observed in Figure 2. Obviously, the linearized line is smoother than the actual line.

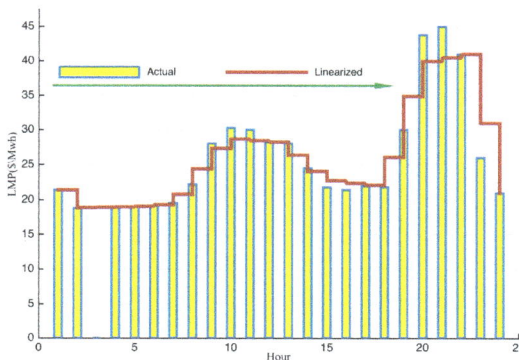

Figure 2. Actual and linearized one day Pennsylvania–New Jersey–Maryland (PJM) electricity price on 7 April 2002.

3.3. Core Mapping Method

In this section, the CM approach will be described using an actual example. Taking the electricity price of 26 June 2002 in the PJM electricity market as an example, the data are first linearized and then mapped by CM. The mapped data are shown in Figure 3.

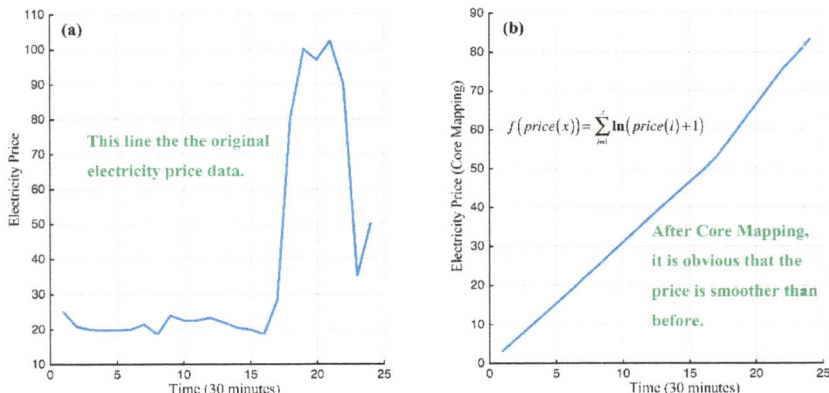

Figure 3. (**a**) Actual price; and (**b**) core-mapped price of 26 June 2001 in PJM electricity market.

It is obvious that the volatility of mapped data is smaller than the volatility of actual data. This means that the CM method can reduce the volatility of data and consequently makes the accuracy of the forecasted electricity price much higher than that of the original method, which is shown in the experiments in Section 4.

3.4. Swarm Optimization Algorithm-Core Mapping with Self-Organizing Map and Fuzzy (Particle Swarm Optimization-Core Mapping with Self-Organizing-Map and Fuzzy Set) Method

Although the CM method can reduce the volatility of the electricity price, there are always high prices or low prices, which increase this volatility of the electricity price. In this section, PCMwSF is proposed to address this problem.

3.4.1. Forecasting Rules

To evaluate the effectiveness of the methods, this paper uses three rules in the forecasting process:

(1)   A previous month's data are used to forecast the price of the target day.
(2)   There is only the historical electricity price considered in this paper (without data of demand or environmental data (for the environmental data, we do not find the corresponding dataset (24 h in one day))).
(3)   All forecasting results are day-ahead forecasting, and the forecasting mode is shown in Figure 4.

Remark 1. *In some literatures related to electricity price forecasting, electricity demand is regarded as a feature to predict the electricity price. However, adding electricity demand as one of the features cannot help to improve forecasting effectiveness after the experiment (the final experiment shows that the forecasting results with electricity demand is similar to the results without it, which means that electricity demand is not a key factor to influence the forecasting effectiveness). Thus, this paper does not select the electricity demand as one of features in our paper, which is the reason why there is only the historical electricity price considered in this paper.*

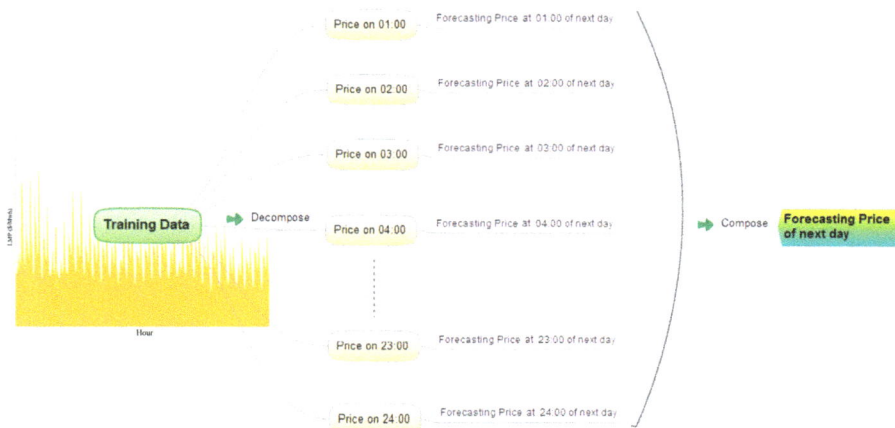

Figure 4. Forecasting mode.

3.4.2. Classification of Price with Self-Organizing Map and Fuzzy Logic

Before linearizing the price and applying CM, the PCMwSF method is used to divide the processed price into three categories: High price, Medium price and Low price by using SOM. The price mentioned above is the historical price prior to the price that needs to be forecasted. For example, if the price data on 26 April 2002 need to be forecasted, the PCMwSF method will divide the price data that are between 1 January and 25 April into three categories.

In the introduction section, a fuzzy logic was established to change CM to ensure good forecasting accuracy. When three classifications of the historical price are obtained, Highweight and Lowweight need to be determined to forecast the next spot price. How to determine both of them is a very important problem in the predication process, and the PSO algorithm, which is a powerful tool for optimizing parameters, is used to solve this problem.

3.4.3. Applying of Swarm Optimization Algorithm Algorithm

In the process of PSO, the fitness function is key to the optimization problem. Before identifying the fitness function, the index of measuring the degree of volatility needs to be established.

Definition 1. *The identity of volatility of price is defined as:*

$$vop\,(i) = var\left(\left[var\left(price\left(i,\,1:\tfrac{T}{4}\right)\right),\,var\left(price\left(i,\,\tfrac{T}{4}+1:\tfrac{2T}{4}\right)\right),\,\ldots,\,var\left(price\left(i,\,\tfrac{3T}{4}:T\right)\right)\right]\right) \qquad (8)$$

where vop(i) is the volatility of price of the ith day, T represents the number of points observed in one day, and ***var*** *refers to the variance of a specific series in the ith day.*

Then, another index to evaluate the forecasting accuracy is proposed for PSO algorithm.

Definition 2. *The index to evaluate the in-sample forecasting effectiveness can be expressed as following:*

$$aob\,(i) = \frac{1}{T}\sqrt{\sum_{t=1}^{T}\left(price_{\text{forecast}}\,(i-1,\,t) - price_{\text{actual}}\,(i-1,\,t)\right)^2} \qquad (9)$$

where T represents the number of points observed in one day, price_{forecast} represents the forecasting value at time point t of the ith day, and price_{actual} represents the observed value at time point t of the day.

This index is the forecasting accuracy of the previous day of the day the needs to be forecasted. Next, the fitness function of PSO is identified as follows.

Definition 3. *The fitness function* $\Phi(\cdot)$ *of PSO algorithm used in PCMwSF model is defined as:*

$$\Phi_i(\cdot) = aob(i) \times vop(i) \tag{10}$$

where i represents the ith day and this definition indicates that lower fitness values can represent lower values of vop and aob, indicating lower volatility and higher forecasting accuracy.

We assign **Ind** to represent the output values of $\Phi(\cdot)$. In the last step, Highweight and Lowweight are changed by the PSO algorithm to make sure **Ind** reaches a minimum. Then, the optimized HIGHWeight and Lowweight are used to forecast the next-day price with RBFN.

3.5. Selection Rule Based on Forecasting Nature of Radial Basis Function Network

The CM method and PCMwSF method have different merits when forecasting the electricity price. Thus, it is important to correctly select a method to pre-process the original data. To solve this problem, this paper studies the properties of the RBF network in forecasting.

- RBF Network in Forecasting

Initially, this paper applies the RBF network to forecast price with the CM method and compares results with the previous day's actual price. Then, it is observed that the forecasting values of RBFN have little changes compared with the former day's electricity prices (shown in Section 4.2). Thus, the index of changes of price (**ICP**) is proposed as a criterion to measure the magnitude of price changes.

Definition 4. *ICP is defined as follows:*

$$ICP(P_1, P_2; i, j) = \frac{1}{T} \sum_{t=1}^{T} \frac{|P_1(i, t) - P_2(j, t)|}{P_1(i, t)} \tag{11}$$

where $P_c(i, t)$ is the ith day's price (actual or forecasted) at t hour (c = 1, 2).

Based on Equation (11), we define a criterion to evaluate what extent the former day's electricity price changes.

Definition 5. *Index of changes of actual price (ICP-P) is defined as follows:*

$$ICP - P(i) = ICP(P_{actual}, P_{actual}; i - 1, i) = \frac{1}{T} \sum_{t=1}^{T} \frac{|P_{actual}(i - 1, t) - P_{actual}(i, t)|}{P_{actual}(i - 1, t)} \tag{12}$$

where $P_{actual}(i, t)$ is the ith day's actual price at t hour (c = 1, 2).

Additionally, if we obtain the forecasting values of the electricity price, we can define another criterion to evaluate to what extent the forecasting electricity price changes from the former day.

Definition 6. *Index of changes of forecasting price (ICP-F) is expressed as follows:*

$$ICP - F\left(i; P_{\text{forecast}}\right) = ICP\left(P_{\text{actual}}, P_{\text{forecast}}; i - 1, i\right) = \frac{1}{T} \sum_{t=1}^{T} \frac{\left|P_{\text{actual}}\left(i - 1, t\right) - P_{\text{forecast}}\left(i, t\right)\right|}{P_{\text{actual}}\left(i - 1, t\right)} \tag{13}$$

where $P_{\text{actual}}(i, t)$ is the ith day's actual price at t hour (c = 1, 2) and $P_{\text{forecast}}(i, t)$ is the ith day's forecasting price at t hour (c = 1, 2).

From Definition 4, it is obvious that different forecasting values have their own **ICP-F**, meaning that this new criterion can help us select the best forecasting models under the condition that we do not know the actual electricity price of the ith day. Thus, this paper proposes a **SR** to choose the best forecasting model based on **ICP-F**.

Definition 7. *When forecasting the electricity price of the ith day, the **SR** can be expressed as follows:*

$$\textbf{SR}\left(i\right) = \{m | \textbf{ICP} - \textbf{F}(i; P_{\text{forecast}}^{(m)}) = \max_{m=1, 2, \dots, M} (\textbf{ICP} - \textbf{F}(i; P_{\text{forecast}}^{(m)}))\} \tag{14}$$

where M is the number of forecasting models and $P_{\text{forecast}}^{(m)}$ is the forecasting values of the ith day obtained by the mth model.

It is obvious that the **SR** is an integer series. Thus, for the ith day, we should select $P_{\text{forecast}}^{(\text{SR}(i))}$ as the forecasting values of this day. Algorithm 1 demonstrates the Pseudo code of forecasting the electricity price of the ith day using the CM-PCMwSF-SR model.

Algorithm 1 Pseudo code of forecasting the electricity price of the ith day using the CM-PCMwSF-SR model.
P: The electricity price series
T: Number of time points in one-day electricity price series.
Iter: Number of iterations.
t = 1.
1   Assign Equations (6) and (7) to pre-process *P*
2   According to CM method, map *P* to P_{CM}
3   Divide P_{CM} into *T* subseries and denote them by $P_{\text{CM1}}, P_{\text{CM2}}, \dots, P_{\text{CMT}}$
4   According to CM method, map *P* to P_{PCM}
5   Divide P_{PCM} into *T* subseries and denote them by $P_{\text{PCM1}}, P_{\text{PCM2}}, \dots, P_{\text{PCMT}}$
6   **While** t < T + 1
7   Assign ***Pcm**_t* and RBFN, forecast the time *t* of electricity price of ith day and denote it by *pfcm(i, t)*.
8   Assign ***Ppcm**_t* and RBFN, forecast the time *t* of electricity price of ith day and denote it by *pfpcm(i, t)*.
9   t = t + 1
10   **End**
11   Calculate ***ICP-F**(i; pfcm)* and ***ICP-F**(i; pfpcm)*
12   **IF** ***ICP-F**(i; pfcm) > **ICP-F**(i; pfpcm)*
13   *Pf = pfcm*
14   **Else**
15   *Pf = pfpcm*
16   **End**
17   ***Return** Pf*

3.6. Forecasting Principle and Evaluation Criteria

Because the input of RBFN must be between 0 and 1, the processed data need to be changed by the following formula:

$$Price = \frac{Price - P_{min}}{P_{max} - P_{min}} \tag{15}$$

where P_{min} is the minimum value of the training data of RBFN and P_{max} is the maximum value of the training data of RBFN. To evaluate the accuracy of the forecast, the *MAPE*, *MAE* and *RMSE* are all used. The *MAPE*, *MAE*, and *RMSE* are defined as:

$$MAPE = \frac{1}{T}\sum_{t=1}^{T} \frac{|P_t^{actual} - P_t^{forecast}|}{P_t^{actual}} \tag{16}$$

$$MAE = \frac{1}{T}\sum_{t=1}^{T} |P_t^{actual} - P_t^{forecast}| \tag{17}$$

$$RMSE = \sqrt{\frac{1}{T}\sum_{t=1}^{T} |P_t^{actual} - P_t^{forecast}|^2} \tag{18}$$

where P_t^{actual} is the actual price at time t and $P_t^{forecast}$ is the forecasted price at time t. The range of Highweight is 0.9–1.05 and the range of Lowweight is 0.9–1.05 in the PSO algorithm.

4. Data Analyses and Numerical Results

PJM electricity price is selected to test the proposed methods. In Case 1, the forecasting results show that the PCMwSF method is better than the CM method. In Case 2, we illustrate the forecasting natures of RBFN, *ICP-P* and *ICP-F*, which lay a strong foundation for **SR**. In other cases, weeks in different seasons are selected to test models. The details of each case are shown in Table 1.

Table 1. Six cases to evaluate effectiveness of the forecasting models.

Case	Forecasted Data	Remarks
1	26 June 2002	Test data 1
2	28 June 2002	Test data 2
3	18–22 March 2002	Spring week
4	24–28 June 2002	Summer week
5	23–27 September 2002	Autumn week
6	23–27 December 2002	Winter week

4.1. Study of Case 1

Figure 5 shows the day-ahead price forecasting results of RBFN for Case 1. Figure 6 shows the day-ahead price forecasting (CM method) for Case 1. Figure 7 shows the day-ahead price forecasting (PCMwSF method) for Case 1. The forecasting results are compared with the actual LMP value.

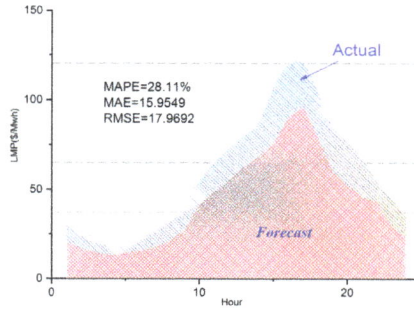

Figure 5. Actual PJM electricity price and forecasted values using radial basis function network (RBFN) in Case 1. *MAPE*: mean absolute percentage error; *MAE*: mean absolute error; and *RMSE*: root mean square error.

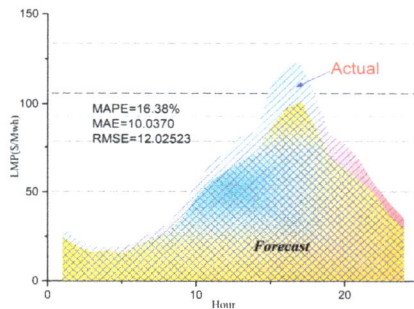

Figure 6. Actual PJM electricity price and forecasted values using CM in Case 1.

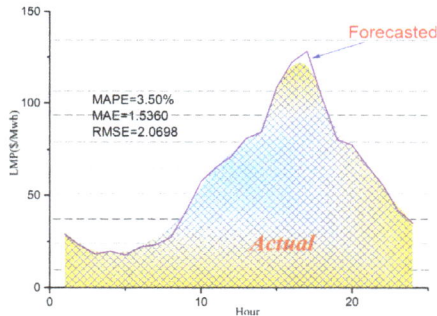

Figure 7. Actual PJM electricity price and forecasted values using particle swarm optimization (PSO)-core mapping (CM) with self-organizing-map and fuzzy set (PCMwSF) in Case 1.

Obviously, the forecasting result with the PCMwSF method is better than the others in Case 1. Details of the forecasting process are shown in Table 2. Table 2 collects data of the forecasting process with the PCMwSF method. The optimal Lowweight, optimal Highweight, optimal **Ind**, *vop*, accuracy of price forecasting on 25 June with optimized weight, actual price and forecasted price on 26 June, *MAPE* in the forecasting process and lower limit, upper limit of high price, medium price and low price are shown. It is obvious that **Ind** and *vop* are well optimized. The forecast on 25 June achieves desired results with the optimal Highweight and Lowweight, which means that the PCMwSF method has the ability to improve the forecasting effectiveness of the electricity price. The *MAPE* of the forecasted price in this model varies from a low of 0.01% at 20:00 to a high of 16% at 7:00. Figure 8 shows a flowchart of the PCMwSF method.

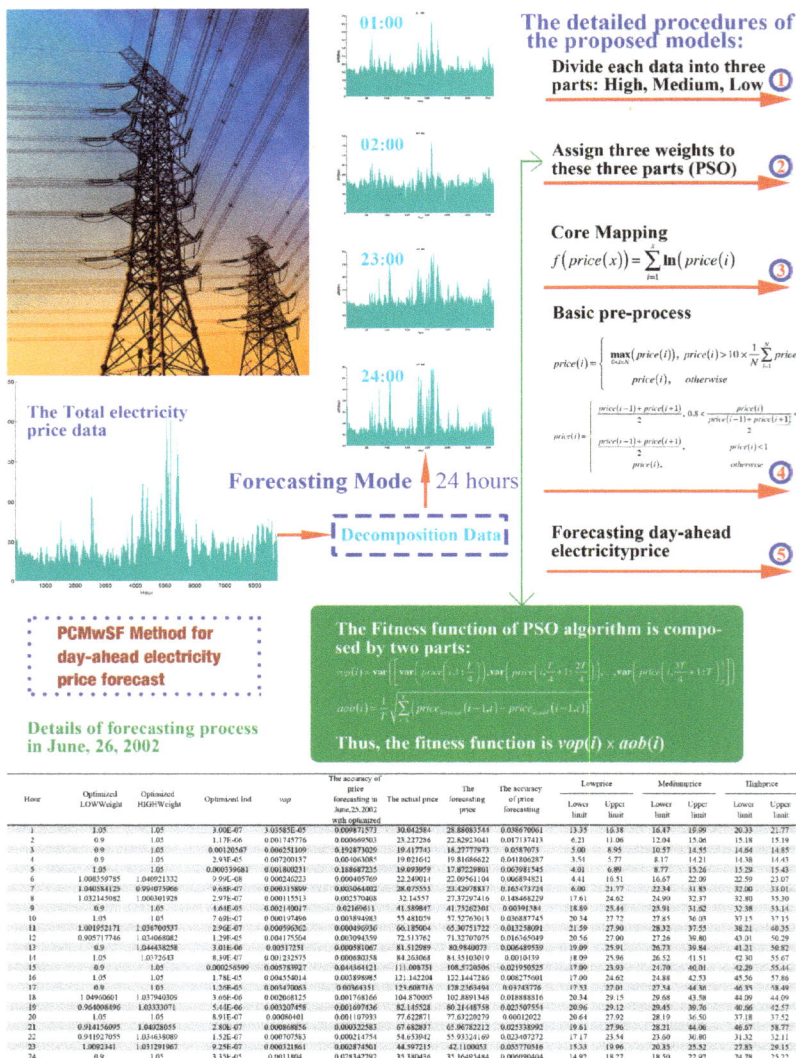

Figure 8. The flowchart of PCMwSF method. The "forecasting model" part illustrates how to predict 24-h eletricity prices for the next day. The "detailed procedures of the proposed models" demonstrates procedures of PCMwSF model and provides fitness function of PSO algorithm. The table in this figure demonstrates details of forecasting process on 26 June 2002.

Table 2. Details of forecasting process on 26 June 2002.

Hour	Optimized Lowweight	Optimized Highweight	Optimized Ind	vop	Accuracy of Price Forecasting in 25 June with Optimized Weight	Actual Price	The Forecasting Price	The MAPE in Forecasting	Lowprice Lower Limit	Lowprice Upper Limit	Mediumprice Lower Limit	Mediumprice Upper Limit	Highprice Lower Limit	Highprice Upper Limit
1	1.05	1.05	3.00×10^{-7}	3×10^{-5}	0.009871573	30.042584	28.88083544	0.038670061	13.35	16.38	16.47	19.99	20.33	21.77
2	0.9	1.05	1.17×10^{-6}	0.00175	0.000669503	23.227286	22.82923041	0.017137413	6.21	11.06	12.04	15.06	15.18	15.19
3	0.9	1.05	0.0012057	0.00625	0.192873029	19.417743	18.27777973	0.0587073	5.00	8.95	10.57	14.55	14.64	14.85
4	0.9	1.05	2.93×10^{-5}	0.0072	0.004063085	19.021642	19.81686622	0.041806287	3.54	5.77	8.17	14.21	14.38	14.43
5	1.05	1.05	0.0003397	0.0018	0.188687235	19.093959	17.87229801	0.063981545	4.01	6.89	8.77	15.26	15.29	15.43
6	1.00835979	1.04092133	9.99×10^{-8}	0.00025	0.000405769	22.24014	22.09561104	0.006894821	4.41	16.51	16.67	22.09	22.59	23.07
7	1.04058413	0.99407597	9.68×10^{-7}	0.00032	0.003064402	28.075555	23.42978837	0.165473724	6.00	21.77	22.34	31.85	32.00	33.01
8	1.03214508	1.00030193	2.97×10^{-7}	0.00012	0.002570408	32.14557	27.37297416	0.148468229	17.61	24.62	24.90	32.37	32.80	35.30
9	0.9	1.05	4.64×10^{-5}	0.00214	0.02169611	41.589847	41.75262301	0.00391384	18.89	25.44	25.91	31.62	32.38	33.14
10	1.05	1.05	7.69×10^{-7}	0.0002	0.003894983	55.481059	57.52763013	0.036887745	20.34	27.72	27.85	36.03	37.15	37.15
11	1.00195217	1.03870054	2.96×10^{-7}	0.0006	0.00496936	66.185004	65.30751722	0.013258091	21.59	27.90	28.32	37.55	38.21	40.35
12	0.90571775	1.03406808	1.29×10^{-5}	0.00418	0.003094359	72.513762	71.32707075	0.016365049	20.56	27.00	27.26	39.80	43.01	50.29
13	0.9	1.04443826	3.01×10^{-6}	0.00517	0.000581067	81.512989	80.9840073	0.006489539	19.09	25.91	26.73	39.84	41.21	50.82
14	1.05	1.0372643	8.39×10^{-7}	0.00123	0.000680358	84.263068	84.35103019	0.0010439	18.09	25.96	26.52	41.51	42.30	55.67
15	0.9	1.05	0.0002566	0.00578	0.04436121	111.008751	108.5720506	0.021950525	17.09	23.93	24.70	40.01	42.29	55.44
16	1.05	1.05	1.78×10^{-5}	0.00455	0.003898985	121.142204	122.1447286	0.008275601	17.00	24.62	24.88	42.53	45.56	57.86
17	0.9	1.05	1.26×10^{-5}	0.00347	0.00364351	123.608716	128.2363494	0.03743776	17.53	27.01	27.54	44.36	46.35	58.49
18	1.04960601	1.03794031	3.66×10^{-6}	0.00207	0.001768166	104.870005	102.8891348	0.018888816	20.34	29.15	29.68	43.58	44.09	44.09
19	0.9640085	1.03333071	5.44×10^{-6}	0.00321	0.001697436	82.145528	80.21448758	0.023507554	20.96	29.12	29.45	39.76	40.66	42.57
20	1.05	1.05	8.91×10^{-7}	0.0008	0.001107933	77.622871	77.63220279	0.00012022	20.64	27.92	28.19	36.50	37.18	37.52
21	0.91415609	1.04028055	2.80×10^{-7}	0.00087	0.000322583	67.682837	65.96782212	0.025338992	19.61	27.96	28.21	44.06	46.67	58.77
22	0.94192706	1.03463809	1.52×10^{-7}	0.00071	0.000214754	54.653942	55.93324169	0.023407272	17.17	23.54	23.60	30.80	31.32	32.11
23	1.0092341	1.03129197	9.25×10^{-7}	0.00032	0.002874501	44.597215	42.1100053	0.055770516	15.33	19.96	20.35	25.52	27.83	29.15
24	0.9	1.05	3.35×10^{-5}	0.00118	0.028342292	35.380436	35.16495484	0.006090404	14.92	18.27	18.50	22.92	24.78	25.22

4.2. Study of Case 2

In this case, we will illustrate *ICP-P*, *ICP-F* and the forecasting results of CM and PCMwSF and show that the **SR** is an effective tool to select the best model to forecast the next-day electricity price.

Figure 9 shows the day-ahead price forecasting from the CM method for Case 2. Figure 10 shows the day-ahead price forecasting from PCMwSF for Case 2. The forecasted results are compared with the actual LMP value, including the price on 27 June. It is obvious that the CM method is better than the PCMwSF method and that both methods are able to forecast the price changing trend. Thus, it is important to select a method of pre-processing correctly. Based on the **SR** defined in Section 3, the *ICP-F* of CM is more than that of PCMwSF; thus, CM is the selected model, indicating that the **SR** correctly selects the model with higher precision.

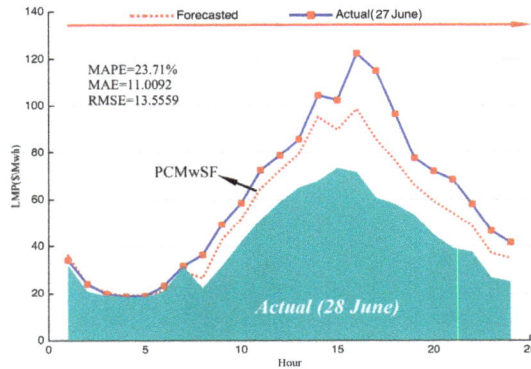

Figure 9. Actual PJM electricity price and forecasted values using CM in Case 2.

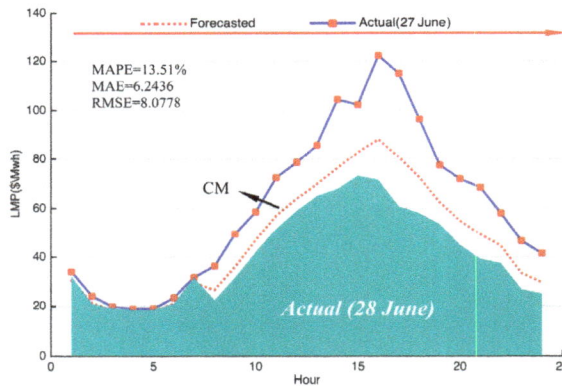

Figure 10. Actual PJM electricity price and forecasted values using PCMwSF in Case 2.

From Figure 11, it is obvious that the RBF network is conservative in the forecasting process. It makes little change in the forecasting process, and the change in price is observed to be relatively larger than that of the forecasted price in this figure.

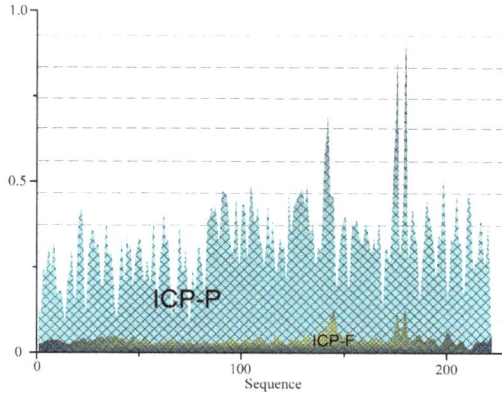

Figure 11. Illustration of *ICP-F* (index of changes of forecasting price) and *ICP-P* (index of changes of actual price).

4.3. Study of Case 3

In this section, the forecasting effectiveness of each model is highlighted. Figure 12 shows the day-ahead price forecasting for 18 March using both forecasting methods. It is apparent that the forecasted values of PCMwSF change more significantly than that of CM. Thus, PCMwSF is chosen to forecast the electricity price, and the *MAPE, MSE* and *RMSE* are 9.71%, 2.8821 and 3.6112, respectively. Figure 13 shows the day-ahead price forecasting for 19 March using both forecasting methods. The price from PCMwSF is selected as the final forecasted price because CM's forecasted price changes within a small range. The *MAPE, MSE* and *RMSE* are 6.25%, 1.6700 and 1.9214, respectively.

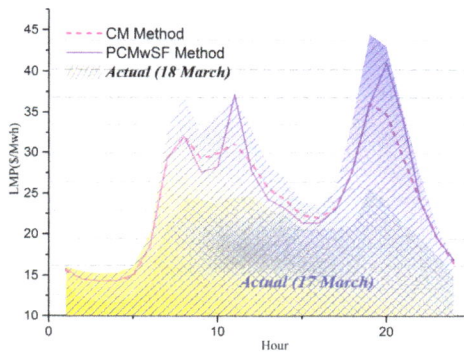

Figure 12. Actual PJM electricity price and forecasted values of 18 March.

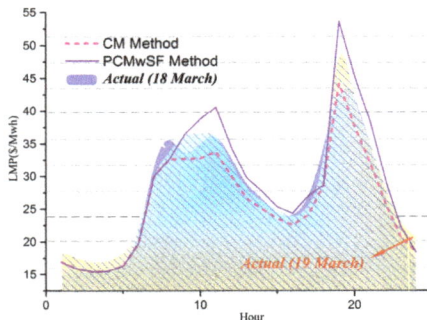

Figure 13. Actual PJM electricity price and forecasted values of 19 March.

In Figure 14, the forecasted price from the CM method is chosen because it changes more significantly than the forecasted price from the other method, and the *MAPE, MSE* and *RMSE* are 3.67%, 1.0212% and 1.1589%, respectively. The forecasted price from PCMwSF is selected as the final chosen price because the *ICP-F* of PCMwSF is larger than the index of the CM method. The *MAPE, MSE* and *RMSE* are 2.53%, 0.8096% and 1.1965%, respectively (Figure 15). By using the **SR** in Figure 16, the forecasted price from PCMwSF is regarded as the final result of forecasting, and the *MAPE, MSE* and *RMSE* are 13.40%, 4.9012% and 5.3977%, respectively.

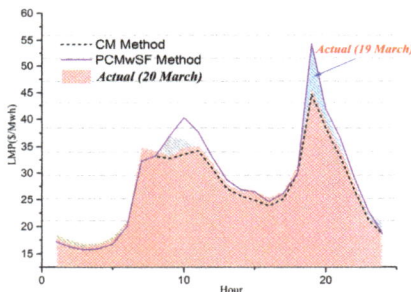

Figure 14. Actual PJM electricity price and forecasted values of 20 March. The blue area represents the actual electricity prices of 19 March.

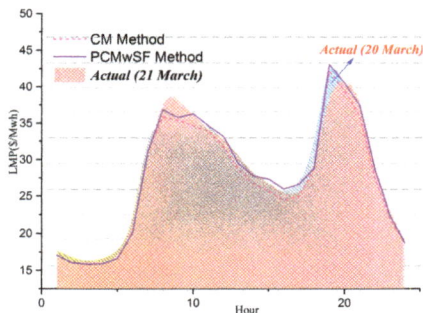

Figure 15. Actual PJM electricity price and forecasted values of 21 March. The blue area represents the actual electricity prices of 20 March.

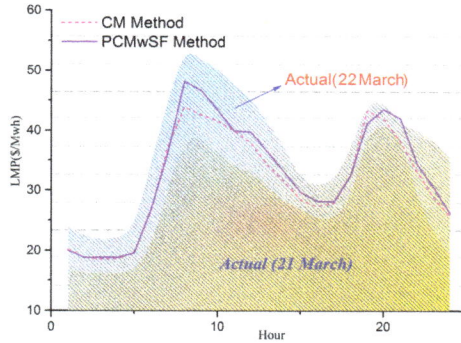

Figure 16. Actual PJM electricity price and forecasted values of 22 March. It is obvious that actual electricity prices of 21 March are less than those of 22 March.

The details of the forecasting results of Case 3 are shown in Table 3 and Figure 17 illustrates the forecasting results. In Table 3, the forecasting details from 18 March to 22 March are demonstrated. It is clearly seen that four days are forecasted using the PCMwSF method. For 18 March, the *MAPE* ranges from 1.6% at 1:00 to 19.7% at 10:00. The average *MAPE* is 9.70%. The *MAPE* of the PCMwSF forecasting model on 19 March varies from a low of 0.6% at 23:00 to a high of 11.2% at 24:00, and the average *MAPE* of this day is 6.25%. The *MAPE* varies from 0.2% at 1:00 to 7.5% at 9:00. The average *MAPE* on 21 March is 2.53%. Similarly, the lowest *MAPE* on 22 March is 2.7% at 21:00, and this day's highest *MAPE* is 26.1% at 24:00. The average *MAPE* of this day is 13.40%. The CM method is chosen to forecast the price on 20 March, and the *MAPE* of this day varies from a low of 1.2% at 9:00 to a high of 8.0% at 7:00. The average *MAPE* on 20 March is 3.67%. Thus, the lowest *MAPE* of Case 3 is 0.2% at 2:00 on 21 March, and the highest *MAPE* of this case is 26.1% at 24:00 on 22 March.

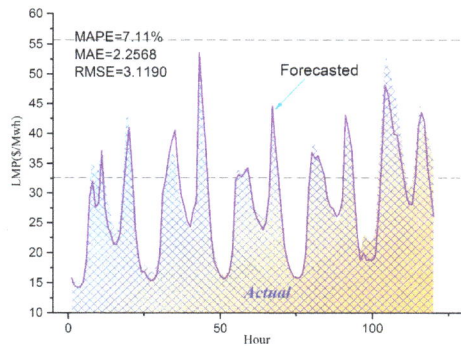

Figure 17. Actual PJM electricity price and forecasted values of Case 3. In this figure, the area represents the actual electricity prices of this week and the line is the forecasted values.

Table 3. Details of forecasting process for Case 3.

Hour	18 March (PCMwSF)			19 March (PCMwSF)			20 March (CM)			21 March (PCMwSF)			22 March (PCMwSF)		
	Actual	Forecasted	MAPE	Actual	Forecasted	MAPE	Actual	Forecasted	MAPE	Actual	Forecasted	MAPE	Actual	Forecasted	MAPE
1	16.03	15.774	0.016	18.51	16.898	0.087	17.72	17.227	0.028	17	17.038	0.002	24	19.967	0.168
2	15.45	14.524	0.06	17.51	15.878	0.093	16.72	16.222	0.03	16.01	16.046	0.002	22.33	18.79	0.159
3	15.207	14.256	0.063	16.79	15.403	0.083	16.15	15.703	0.028	16.16	15.861	0.018	21.9	18.856	0.139
4	15.312	14.254	0.069	16.94	15.472	0.087	16.58	15.946	0.038	16	15.904	0.006	22	18.852	0.143
5	16	15.082	0.057	18.01	16.295	0.095	17.43	16.778	0.037	16.58	16.606	0.002	23.5	19.555	0.168
6	20.27	18.624	0.081	21.01	19.687	0.063	20.51	20.001	0.025	20.356	20.084	0.013	35	26.786	0.235
7	33	29.261	0.113	31	29.958	0.034	35	32.211	0.08	30.177	31.02	0.028	45	36.19	0.196
8	37	32.009	0.135	33.51	32.574	0.028	34.164	33.182	0.029	39.109	36.824	0.058	54.255	48.145	0.113
9	32.42	27.554	0.15	36.917	36.381	0.015	33.144	32.734	0.012	38.668	35.768	0.075	51.7	46.633	0.098
10	35.146	28.222	0.197	36.345	38.725	0.065	34.75	33.562	0.034	36.114	36.264	0.004	50.447	43.354	0.141
11	36.94	37.145	0.006	37.09	40.525	0.093	35.1	34.246	0.024	33.785	34.628	0.025	47.568	39.927	0.161
12	33.005	27.715	0.16	31.724	34.423	0.085	31.889	30.771	0.035	33.101	33.054	0.001	45.617	39.734	0.129
13	28.522	24.293	0.148	28	29.776	0.063	27.981	27.152	0.03	30.465	29.432	0.034	41.289	36.414	0.118
14	26.891	23.162	0.139	26.033	27.753	0.066	26.677	25.695	0.037	28.066	27.746	0.011	37.516	33.067	0.119
15	23.95	21.386	0.107	25.03	25.276	0.01	26.565	24.861	0.064	26.811	27.337	0.02	31.726	29.668	0.065
16	23.685	21.376	0.097	23.233	24.273	0.045	25.415	23.799	0.064	25.212	26.002	0.031	30.707	28.166	0.083
17	26.071	23.105	0.114	25.011	26.981	0.079	26.54	25.115	0.054	24.998	26.637	0.066	31.55	28.129	0.108
18	34.44	29.583	0.141	29.573	28.589	0.033	31.429	29.819	0.051	28.076	28.791	0.025	37	32.474	0.122
19	44.535	36.006	0.192	54.635	53.567	0.02	45.681	44.629	0.023	40.057	43.06	0.075	45	41.186	0.085
20	43.06	40.996	0.048	41	45.297	0.105	39.436	38.278	0.029	41.064	40.447	0.015	45	43.544	0.032
21	34.96	33.898	0.03	35.51	38.372	0.081	34.421	33.04	0.04	39.486	37.466	0.051	40.9	41.984	0.027
22	27.41	24.028	0.123	28	29.499	0.054	28	26.745	0.045	28.537	28.673	0.005	40.242	34.35	0.146
23	21	19.499	0.071	22.44	22.312	0.006	21.65	21.129	0.024	23	22.493	0.022	38.26	30.501	0.203
24	17.03	16.82	0.012	20.7	18.375	0.112	19	18.599	0.021	19	18.713	0.015	35.31	26.084	0.261

4.4. Study of Cases 4–6

By applying two forecasting models and **SR**, Cases 4–6 can be solved. Figures 18–20 separately illustrate the forecasting results of CM-PCMwSF-SR in the three cases.

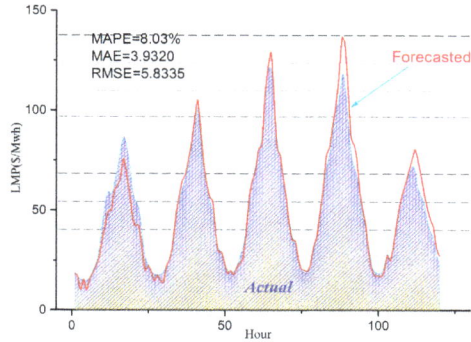

Figure 18. Actual PJM electricity price and forecasted values of Case 4. In this figure, the area represents the actual electricity prices of this week and the line is the forecasted values.

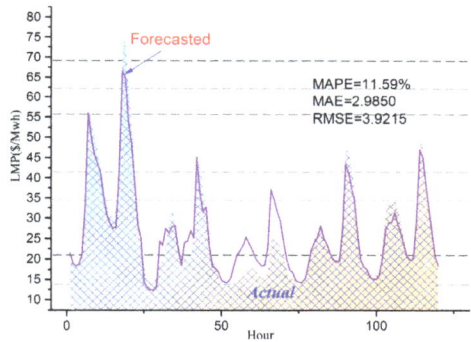

Figure 19. Actual PJM electricity price and forecasted values of Case 5. In this figure, the area represents the actual electricity prices of this week and the line is the forecasted values.

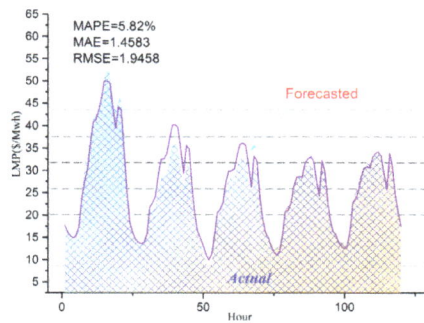

Figure 20. Actual PJM electricity price and forecasted values of Case 6. In this figure, the area represents the actual electricity prices of this week and the line is the forecasted values.

Details of the forecasting results of Cases 4–6 are shown in Tables 4–6. Table 4 shows the forecasting details from 24 June to 28 June. It is clearly seen that three days are forecasted using the PCMwSF method. For 25 June, the *MAPE* ranges from 0.1% at 2:00 to 30.6% at 6:00. The average *MAPE* is 6.0%. The *MAPE* of the PCMwSF forecasting model on 26 June varies from a low of 0.012% at 20:00 to a high of 19.8% at 7:00, and the average *MAPE* of this day is 3.78%. Similarly, the lowest *MAPE* on 27 June is 0.8% at 1:00, and this day's highest *MAPE* is 13.4% at 20:00. The average *MAPE* of this day is 6.32%. The CM method is chosen to forecast the prices on 24 June and 28 June. The *MAPE* of the previous day varies from a low of 1.5% at 2:00 to a high of 22.80% at 23:00. The average *MAPE* on 24 June is 15.30%. The other day's *MAPE* ranges from 0.1% at 12:00 to 18.10% at 17:00, and the average *MAPE* is 8.48%. Thus, the lowest *MAPE* of Case 4 is 0.012% at 20:00 on 26 June, and the highest *MAPE* of this case is 30.6% at 6:00 on 25 June.

The forecasting details from 23 September to 27 September are shown in Table 5. It is obvious that three days are forecasted using the CM method. For 24 September, the *MAPE* ranges from 0.019% at 6:00 to 20.9% at 19:00. The average *MAPE* is 8.74%. The *MAPE* of the CM forecasting model on 25 September varies from a low of 0.6% at 24:00 to a high of 35.7% at 4:00, and the average *MAPE* of this day is 8.30%. The *MAPE* varies from 0.009% at 15:00 to 8.4% at 4:00. The average *MAPE* on 27 September is 3.10%. The PCMwSF method is chosen to forecast the prices on 23 September and 26 September. The *MAPE* of 23 September varies from a low of 0.016% at 13:00 to a high of 12.4% at 22:00. The average *MAPE* is 4.61%. The *MAPE* of 26 September ranges from 0.011% at 12:00 to 13.1% at 4:00, and the average *MAPE* of this day is 4.34%. Thus, the lowest *MAPE* of Case 5 is 0.009% at 15:00 on 27 September, and the highest *MAPE* of this case is 35.7% at 4:00 on 25 September.

Table 6 lists the details of the forecasting result from 23 December to 27 December. It is easy to see that three days are forecasted using the CM method. For 24 December, the *MAPE* ranges from 2.6% at 24:00 to 37.1% at 16:00. The average *MAPE* is 14.04%. The *MAPE* on 25 December varies from a low of 6.0% at 2:00 to a high of 47.1% at 7:00, and the average *MAPE* of this day is 24.43%. Similarly, the lowest *MAPE* on 26 December is 0.3% at 24:00, and this day's highest *MAPE* is 17.3% at 7:00. The average *MAPE* of this day is 4.94%. 23 December and 27 December use the PCMwSF method as their forecasting method. The *MAPE* of 23 December varies from a low of 0.5% at 3:00 to a high of 14.9% at 20:00. This day's *MAPE* is 7.60%. The last column of this table shows that the *MAPE* of 27 December ranges from 1.7% at 22:00 to 17.3% at 7:00, and the average *MAPE* is 6.98%. Thus, the lowest *MAPE* of Case 6 is 0.3% at 24:00 on 26 December, and the highest *MAPE* of this case is 47.1% at 7:00 on 25 December.

Table 4. Details of forecasting process for Case 4.

Hour	24 June (CM)			25 June (PCMwSF)			26 June (PCMwSF)			27 June (PCMwSF)			28 June (CM)		
	Actual	Forecast	MAPE	Actual	Forecast	MAPE	Actual	Forecast	MAPE	Actual	Forecast	MAPE	Actual	Forecast	MAPE
1	18.901	18.338	0.03	23.573	22.374	0.054	30.043	28.881	0.04	33.965	34.238	0.008	32.087	28.339	0.132
2	16.62	16.365	0.015	19.062	19.075	0.001	23.227	22.829	0.017	24.03	25.41	0.054	21.035	19.954	0.054
3	15.51	10.128	0.347	18.238	14.721	0.239	19.418	18.278	0.062	19.814	20.565	0.037	19.06	17.323	0.1
4	14.8	15.295	0.033	17.64	17.48	0.009	19.022	19.817	0.04	18.8	19.973	0.059	18.43	17.126	0.076
5	14.92	9.917	0.335	17.651	14.32	0.233	19.094	17.872	0.068	18.96	19.873	0.046	18.43	16.809	0.096
6	16.17	16.076	0.006	19.268	14.759	0.306	22.249	22.096	0.007	23.35	24.628	0.052	21.14	20.013	0.056
7	19.966	18.865	0.055	23.186	22.684	0.022	28.076	23.43	0.198	31.55	32.574	0.031	31.41	27.63	0.137
8	23.219	22.491	0.031	25.918	25.693	0.009	32.146	27.373	0.174	36.25	37.572	0.035	22.376	24.931	0.102
9	29.326	26.367	0.101	34.028	34.768	0.021	41.59	41.753	0.004	49.29	49.998	0.014	31.538	33.637	0.062
10	44.447	35.531	0.201	49.044	48.853	0.004	55.481	57.528	0.036	58.442	61.623	0.052	41.975	43.397	0.033
11	55.216	43.013	0.221	54.909	57.439	0.044	66.185	65.308	0.013	72.358	77.943	0.072	51.399	52.636	0.023
12	61.1	49.62	0.188	60.652	64.9	0.065	72.514	71.327	0.017	78.544	81.822	0.04	58.915	58.952	0.001
13	56.153	50.403	0.102	66.413	67.902	0.022	81.513	80.984	0.007	85.559	91.998	0.07	64.995	64.462	0.008
14	64.393	59.737	0.072	71.894	75.777	0.051	84.263	84.351	0.001	104.294	102.081	0.022	67.832	70.257	0.035
15	71.692	62.255	0.132	80.659	84.246	0.043	111.009	108.572	0.022	102.227	118.042	0.134	73.239	75.898	0.035
16	81.962	67.378	0.178	97.686	97.306	0.004	121.142	122.145	0.008	122.228	136.956	0.108	71.587	80.658	0.112
17	89.208	75.863	0.15	104.657	105.039	0.004	123.609	128.236	0.036	114.878	130.954	0.123	60.741	74.127	0.181
18	82.122	66.567	0.189	84.824	89.533	0.053	104.87	102.889	0.019	96.25	110.481	0.129	57.986	66.78	0.132
19	72.609	56.682	0.219	67.768	72.747	0.068	82.146	80.214	0.024	77.466	86.297	0.102	53.425	57.771	0.075
20	57.966	44.994	0.224	62.682	62.609	0.001	77.623	77.632	1.16E	71.847	82.991	0.134	44.856	50.84	0.118
21	53.88	42.471	0.212	54.379	56.513	0.038	67.683	65.968	0.026	68.396	74.936	0.087	39.225	45.684	0.141
22	52.403	41.59	0.206	49.969	53.021	0.058	54.654	55.933	0.023	57.934	58.991	0.018	37.543	41.408	0.093
23	38.681	29.851	0.228	35.46	37.805	0.062	44.597	42.11	0.059	46.488	50.42	0.078	26.801	31.185	0.141
24	24.805	19.889	0.198	30.043	29.191	0.029	35.38	35.165	0.006	41.45	41.919	0.011	25.081	27.661	0.093

Table 5. Details of forecasting process for Case 5.

Hour	23 September (PCMwSF)			24 September (CM)			25 September (CM)			26 September (PCMwSF)			27 September (CM)		
	Actual	Forecast	MAPE	Actual	Forecast	MAPE	Actual	Forecast	MAPE	Actual	Forecast	MAPE	Actual	Forecast	MAPE
1	17.93	17.77	0.01	16.05	16.15	0.01	14.12	14.93	0.06	15.59	14.97	0.04	15.72	15.59	0.01
2	16.39	16.26	0.01	14.41	14.64	0.02	12.06	13.15	0.09	13.52	13.09	0.03	14.17	13.84	0.02
3	15.50	15.29	0.01	13.79	13.85	0.01	9.16	11.19	0.22	13.08	11.88	0.09	13.84	13.03	0.06
4	15.00	14.90	0.01	13.55	13.56	0.00	7.33	9.94	0.36	12.71	11.04	0.13	13.59	12.44	0.08
5	15.73	15.33	0.03	13.80	13.89	0.01	9.00	11.11	0.24	13.14	11.87	0.10	13.74	12.97	0.06
6	17.83	17.34	0.03	15.90	15.91	0.00	15.11	15.31	0.01	16.33	15.51	0.05	16.96	16.49	0.03
7	25.75	23.80	0.08	21.76	21.66	0.00	19.87	20.50	0.03	22.13	20.86	0.06	24.37	22.95	0.06
8	30.90	27.53	0.11	20.81	22.77	0.09	20.47	21.30	0.04	21.68	21.03	0.03	25.58	23.64	0.08
9	32.88	29.91	0.09	21.59	24.14	0.12	22.23	22.86	0.03	23.61	22.72	0.04	27.51	25.49	0.07
10	37.70	35.46	0.06	26.34	28.96	0.10	26.47	27.29	0.03	26.60	26.34	0.01	29.85	28.60	0.04
11	42.22	40.73	0.04	28.96	32.51	0.12	28.20	29.85	0.06	28.48	28.50	0.00	31.27	30.46	0.03
12	42.24	41.36	0.02	28.54	32.55	0.14	28.32	29.93	0.06	28.59	28.59	0.00	31.98	30.85	0.04
13	43.55	43.56	0.00	28.91	33.66	0.16	28.16	30.36	0.08	28.15	28.59	0.02	31.48	30.59	0.03
14	46.80	46.07	0.02	33.04	37.01	0.12	31.19	33.48	0.07	31.32	31.65	0.01	33.16	33.04	0.00
15	51.12	49.93	0.02	35.99	40.15	0.12	32.83	35.79	0.09	31.01	32.57	0.05	33.89	33.89	0.00
16	52.72	50.01	0.05	36.09	40.21	0.11	33.49	36.18	0.08	31.71	33.11	0.04	33.96	34.21	0.01
17	52.62	49.39	0.06	35.58	39.70	0.12	33.07	35.72	0.08	30.09	32.06	0.07	33.27	33.31	0.00
18	45.78	43.83	0.04	30.70	34.84	0.14	29.15	31.43	0.08	26.01	27.99	0.08	29.86	29.46	0.01
19	42.10	39.39	0.07	24.37	29.45	0.21	23.98	26.24	0.09	23.69	24.41	0.03	26.12	25.72	0.02
20	46.69	44.22	0.05	32.30	35.68	0.10	31.72	33.17	0.05	32.94	32.27	0.02	34.17	33.90	0.01
21	46.91	43.38	0.08	31.12	34.75	0.12	30.93	32.31	0.05	28.67	29.75	0.04	30.96	30.97	0.00
22	38.92	34.11	0.12	23.18	26.72	0.15	23.02	24.47	0.06	22.74	23.09	0.02	25.28	24.62	0.03
23	28.09	25.11	0.11	18.24	20.44	0.12	18.54	19.20	0.04	20.07	19.23	0.04	21.57	20.74	0.04
24	18.48	18.33	0.01	16.19	16.47	0.02	16.24	16.15	0.01	17.38	16.42	0.06	18.27	17.62	0.04

Table 6. Details of forecasting process for Case 6.

Hour	23 December (PCMwSF)			24 December (CM)			25 December (CM)			26 December (CM)			27 December (PCMwSF)		
	Actual	Forecast	MAPE	Actual	Forecast	MAPE	Actual	Forecast	MAPE	Actual	Forecast	MAPE	Actual	Forecast	MAPE
1	20.129	21.535	0.07	17.91	14.333	0.2	15.723	16.953	0.078	15.8	16.669	0.055	18.516	17.551	0.052
2	18.492	18.993	0.027	15.245	13.179	0.136	14.128	14.972	0.06	14.117	14.8	0.048	16.394	15.562	0.051
3	18.295	18.391	0.005	14.528	12.476	0.141	13.517	14.35	0.062	13.859	14.358	0.036	15.983	15.135	0.053
4	18.99	18.707	0.015	14.426	12.426	0.139	13.41	14.325	0.068	13.858	14.348	0.035	16.076	15.174	0.056
5	20.425	20.885	0.023	15.632	13.304	0.149	13.884	15.284	0.101	14.789	15.314	0.035	16.932	16.087	0.05
6	33.275	32.466	0.024	19.035	24.49	0.287	14.386	18.253	0.269	18.242	18.603	0.02	22.397	20.394	0.089
7	61.526	56.038	0.089	19.331	23.353	0.208	14.234	20.942	0.471	22.959	22.372	0.026	32.67	27.019	0.173
8	53.876	50.023	0.072	23.866	27.451	0.15	15.085	22.029	0.46	25.22	24.063	0.046	34.544	28.811	0.166
9	51.101	45.891	0.102	28.265	26.419	0.065	16.427	23.424	0.426	26.332	25.36	0.037	33.155	28.97	0.126
10	47.329	43.639	0.078	34.867	27.883	0.2	17.952	25.459	0.418	29.655	28.061	0.054	35.156	31.376	0.108
11	43.324	41.221	0.049	29.488	28.192	0.044	17.395	23.729	0.364	25.718	25.205	0.02	32.18	28.453	0.116
12	38.111	35.248	0.075	23.901	22.755	0.048	18.277	22.203	0.215	24.337	23.696	0.026	29.215	26.286	0.1
13	32.935	30.741	0.067	21.015	18.606	0.115	17.323	20.269	0.17	22.202	21.599	0.027	24.851	23.144	0.069
14	30.412	28.817	0.052	20.133	24.026	0.193	16.185	19.104	0.18	20.307	20.038	0.013	20.983	20.483	0.024
15	30.398	27.43	0.098	19.868	24.496	0.233	15.47	18.407	0.19	19.698	19.362	0.017	20.201	19.756	0.022
16	30.43	27.764	0.088	19.642	26.931	0.371	15.573	18.471	0.186	20.017	19.552	0.023	20.769	20.13	0.031
17	47.471	41.881	0.118	27.353	25.298	0.075	16.743	23.011	0.374	31.276	27.336	0.126	32.265	29.666	0.081
18	76.514	66.692	0.128	47.185	45.225	0.042	26.72	37.133	0.39	48.81	43.524	0.108	50.867	46.992	0.076
19	74.707	64.847	0.132	43.588	36.176	0.17	24.514	34.632	0.413	47.498	41.472	0.127	49.224	45.126	0.083
20	63.244	53.816	0.149	37.947	31.452	0.171	24.05	31.629	0.315	42.577	37.508	0.119	39.31	38.351	0.024
21	55.39	48.473	0.125	31.732	32.804	0.034	23.802	29.33	0.232	38.793	34.46	0.112	30.677	32.477	0.059
22	43.07	38.821	0.099	25.028	22.954	0.083	20.914	24.556	0.174	26.971	26.261	0.026	27.121	26.657	0.017
23	31.186	28.374	0.09	20.14	18.346	0.089	18.203	20.089	0.104	19.084	19.963	0.046	21.076	20.49	0.028
24	26.393	25.116	0.048	18.361	17.881	0.026	15.329	17.503	0.142	18.057	18.112	0.003	18.844	18.455	0.021

4.5. Comparison Study

In this section, a comparison study will be provided to present the forecasting effectiveness of the proposed model. In detail, genetic algorithm (GA) will be applied to optimize weights of low and high prices, and backward propagation neural network (BPNN), elman neural network (ENN) and GRNN are selected as benchmarks. For the GA-based method, we use CM-GCMwSF-SR to present it in Table 7, which shows the forecasting results of models in Cases 3–6. In addition, we provide an experiment to show the forecasting effectiveness when electricity demand is considered as one of features, which is represented as CM-PCMwSF-SR (with demand) in Table 7.

Table 7. Comparison with other algorithms in Cases 3–6. SR: selection rule; BPNN: backward propagation neural network; ENN: elman neural network; GRNN: generalized regression neural network. MAE: mean absolute error; RMSE: root mean square error.

Season	Criteria	CM-PCMwSF-SR	CM-GCMwSF-SR	CM-PCMwSF-SR (with Demand)	PCMwSF	CM	BPNN	ENN	GRNN
Spring	MAPE	7.11%	7.01%	7.21%	7.51%	10.08%	20.90%	21.75%	21.90%
	MAE	2.2568	2.1948	2.3761	2.5682	3.5268	6.2178	6.4994	6.3687
	RMSE	3.119	3.098	3.202	3.9865	5.1268	7.8962	7.9463	8.2122
Summer	MAPE	8.03%	9.28%	8.01%	11.21%	15.58%	26.60%	27.81%	26.79%
	MAE	3.932	4.8329	3.917	6.1025	8.0256	14.8875	15.3582	15.2014
	RMSE	5.8335	6.9726	5.8017	8.1564	10.1526	19.9902	20.0877	20.9055
Autumn	MAPE	5.82%	7.20%	5.72%	10.54%	8.25%	16.30%	16.53%	16.95%
	MAE	1.4583	1.9872	1.2918	2.8658	2.2139	4.2477	4.3638	4.4515
	RMSE	1.9458	2.8977	1.3681	4.0213	2.9684	5.8511	6.1312	6.0429
Winter	MAPE	11.59%	12.29%	12.33%	15.68%	12.86%	29.16%	30.57%	29.21%
	MAE	2.985	3.6298	3.7288	4.2681	3.0254	7.29	7.3474	7.5995
	RMSE	3.9215	4.7892	4.4025	5.9812	4.1285	9.6061	10.0723	10.0547
Average	MAPE	8.14%	8.95%	8.32%	11.24%	11.69%	23.24%	24.16%	23.71%
	MAE	2.66	3.16	2.83	3.95	4.20	8.16	8.39	8.41
	RMSE	3.70	4.44	3.69	5.54	5.59	10.84	11.06	11.30

In Table 7, it is obvious that the proposed model has better performance than benchmarks and the following conclusions can be made:

(a) PSO is a better selection to optimize the weights of low and high electricity prices than GA because CM-PCMwSF-SR has better overall forecasting effectiveness than CM-GCMwSF-SR.
(b) PCMwSF and CM have ability to improve the forecasting accuracy.
(c) The electricity price of autumn can be predicted more precisely.
(d) Although some literature regard the electricity demand as features to predict electricity price, adding the electricity demand data as a feature cannot help to improve forecasting effectiveness of prices in this paper (forecasting results are similar in Table 7).

As a demonstration of (d), regarding electricity demand as a feature cannot improve the forecasting effectiveness, which is different with some electricity price forecasting methods. The main reasons are demonstrated as following:

(1) The proposed model mostly concentrates on reducing the volatility of electricity price for a higher accuracy, which means the electricity demand is not important compared to the pre-processed electricity price.
(2) Model performance under specific conditions should be analyzed and understood and incremental improvements made based on knowledge gained. Moghram and Rahman review five short-term load forecasting methods:

(i)   multiple linear regression;
(ii)   time series;
(iii) general exponential smoothing;
(iv)   state space and Kallman filter; and
(v)   knowledge-based approach.

The forecasting results show that no one method was determined to be superior. The transfer function approach was the second worst predictor over the winter months but was the best method over the summer months. The authors conclude that because of its strong dependency on historical data, the transfer function approach did not respond well to abrupt changes as did the knowledge based approaches. The conclusion reached is that there is no one best approach, which means that it is possible that regarding electricity demand as a feature cannot improve the forecasting effectiveness [51].

Thus, the proposed method combining PCMwSF method, CM method and SR is better than traditional approaches according to the numerical calculating results. Concretely, the CM method is helpful to reduce the volatility of the electricity price and, consequently, to improve the forecasting effectiveness. For other techniques presented in this paper, PSO aims to obtain the best weights of high and low electricity prices, and SOM and Fuzzy logic are effective tools to confirm three levels of electricity prices (high, medium and low), and the purpose of SR is to select the best model for each day based on the nature of RBFN.

5. Conclusions

Forecasting electricity is a key problem for generators and consumers in a deregulated electricity market, and the difficulty of an accurate forecast is due to the high volatility of the electricity price. The reduction of this volatility is the key to improving prediction accuracy. In this paper, based on SOM, Fuzzy logic, PSO and the forecasting nature of the RBF network, the PCMwSF method, CM method and SR were developed to reduce the volatility of the electric price and to improve the accuracy of the forecast. The final model, CM-PCMwSF-SR, successfully reduced the volatility of the electricity price and was able to obtain a higher accuracy compared to other benchmarks. In the numerical simulation of four seasons, the proposed model exhibited the best performance, where the *MAPEs* are 7.11%, 8.03%, 5.82%, and 11.59% for each season (spring, summer, autumn and winter respectively). The PCMwSF method and CM method were the best models (except when using the SR approach) for two different seasons. The BP network, i.e., a classical neuron network method for forecasting the electricity price, did not have a good performance compared to the other models in these four seasons. The experimental results showed that reducing the volatility and effectively selecting forecasting models not only improve the forecasting effectiveness of the electricity price but also obtained a satisfactory forecasting accuracy.

Acknowledgments: The work was supported by the National Natural Science Foundation of China (Grant No. 71573034) and National Social Science Post-Funded Projects (Grant No. 15FTJ002).

Author Contributions: Feng Liu and Yiliao Song conceived and designed the experiments; Feng Liu performed the experiments; Ping Jiang and Feng Liu analyzed the data; and Feng Liu wrote the paper.

Conflicts of Interest: The authors declare no conflict of interest.

Abbreviations

RBFN	Radial basis function network
PCMwSF	Particle swarm optimization-core mapping with self-organizing-map and fuzzy set
CM	Core mapping
SR	Selection rule
MI	Mutual information
CNN	Composite neural network
SVM	Support vector machine

ARMAX	Auto-regressive moving average with external input
MS-GARCH	Markov-switching generalized autoregressive conditional heteroskedasticity
DCT	Discrete cosine transforms
FFNN	Feed-forward neural network
CFNN	Cascade-forward neural network
GRNN	Generalized regression neural network
PSO	Particle swarm optimization
PCPF	Panel cointegration and particle filter
WT	Wavelet transform
ARIMA	Autoregressive integrated moving average
LSSVM	Least squares support vector machine
CLSSVM	Chaotic least squares support vector machine
EGARCH	Exponential generalized autoregressive conditional heteroskedastic
ARMA	Autoregressive moving average
GARCH	Generalized autoregressive conditional heteroskedasticity
ARMA-GARCH-M	ARMA-GARCH-in-mean
ARFIMA	Auto-regressive fractionally integrated moving average
ANN	Artificial neural network
ELM	Extreme learning machine
PIs	Prediction intervals
RNN	Recurrent neural network
PNN	Probabilistic neural network
HNES	Hybrid neuro-evolutionary system
PCA	Principal component analysis
MLF	Multi-layer feedforward
BPNN	Backward propagation neural network
ENN	Elman neural network
GA	Genetic algorithm

References

1. Sáez, Á.E. Modeling electricity prices: International evidence. In Proceedings of the European Financial Management Association (EFMA), London, UK, 26–29 June 2002; pp. 2–34.
2. Liu, H.; Shi, J. Applying ARMA–GARCH approaches to forecasting short-term electricity prices. *Energy Econ.* **2013**, *37*, 152–166. [CrossRef]
3. García-Martos, C.; Rodríguez, J.; Sánchez, M.J. Forecasting electricity prices and their volatilities using Unobserved Components. *Energy Econ.* **2011**, *33*, 1227–1239. [CrossRef]
4. Keynia, F. A new feature selection algorithm and composite neural network for electricity price forecasting. *Eng. Appl. Artif. Intell.* **2012**, *25*, 1687–1697. [CrossRef]
5. Yan, X.; Chowdhury, N.A. Mid-term electricity market clearing price forecasting: A hybrid LSSVM and ARMAX approach. *Int. J. Electr. Power Energy Syst.* **2013**, *53*, 20–26. [CrossRef]
6. Yan, X.; Chowdhury, N.A. Mid-term electricity market clearing price forecasting utilizing hybrid support vector machine and auto-regressive moving average with external input. *Int. J. Electr. Power Energy Syst.* **2014**, *63*, 64–70. [CrossRef]
7. Cifter, A. Forecasting electricity price volatility with the Markov-switching GARCH model: Evidence from the Nordic electric power market. *Electr. Power Syst. Res.* **2013**, *102*, 61–67. [CrossRef]
8. Anbazhagan, S.; Kumarappan, N. Day-ahead deregulated electricity market price classification using neural network input featured by DCT. *Int. J. Electr. Power Energy Syst.* **2012**, *37*, 103–109. [CrossRef]
9. Anbazhagan, S.; Kumarappan, N. A neural network approach to day-ahead deregulated electricity market prices classification. *Electr. Power Syst. Res.* **2012**, *86*, 140–150. [CrossRef]
10. Anbazhagan, S.; Kumarappan, N. Day-ahead deregulated electricity market price forecasting using neural network input featured by DCT. *Energy Convers. Manag.* **2014**, *78*, 711–719. [CrossRef]
11. Lei, M.; Feng, Z. A proposed grey model for short-term electricity price forecasting in competitive power markets. *Int. J. Electr. Power Energy Syst.* **2012**, *43*, 531–538. [CrossRef]

12. Li, X.R.; Yu, C.W.; Ren, S.Y.; Chiu, C.H.; Meng, K. Day-ahead electricity price forecasting based on panel cointegration and particle filter. *Electr. Power Syst. Res.* **2013**, *95*, 66–76. [CrossRef]

13. Zhang, J.; Tan, Z.; Yang, S. Day-ahead electricity price forecasting by a new hybrid method. *Comput. Ind. Eng.* **2012**, *63*, 695–701. [CrossRef]

14. Zhang, J.; Tan, Z. Day-ahead electricity price forecasting using WT, CLSSVM and EGARCH model. *Int. J. Electr. Power Energy Syst.* **2013**, *45*, 362–368. [CrossRef]

15. Chaâbane, N. A hybrid ARFIMA and neural network model for electricity price prediction. *Int. J. Electr. Power Energy Syst.* **2014**, *55*, 187–194. [CrossRef]

16. Babu, C.N.; Reddy, B.E. A moving-average filter based hybrid ARIMA–ANN model for forecasting time series data. *Appl. Soft Comput.* **2014**, *23*, 27–38. [CrossRef]

17. Shrivastava, N.A.; Panigrahi, B.K. A hybrid wavelet-ELM based short term price forecasting for electricity markets. *Int. J. Electr. Power Energy Syst.* **2014**, *55*, 41–50. [CrossRef]

18. Shayeghi, H.; Ghasemi, A. Day-ahead electricity prices forecasting by a modified CGSA technique and hybrid WT in LSSVM based scheme. *Energy Convers. Manag.* **2013**, *74*, 482–491. [CrossRef]

19. Khosravi, A.; Nahavandi, S.; Creighton, D. A neural network-GARCH-based method for construction of Prediction Intervals. *Electr. Power Syst. Res.* **2013**, *96*, 185–193. [CrossRef]

20. Khosravi, A.; Nahavandi, S.; Creighton, D. Quantifying uncertainties of neural network-based electricity price forecasts. *Appl. Energy* **2013**, *112*, 120–129. [CrossRef]

21. Khosravi, A.; Nahavandi, S. Effects of type reduction algorithms on forecasting accuracy of IT2FLS models. *Appl. Soft Comput.* **2014**, *17*, 32–38. [CrossRef]

22. Bordignon, S.; Bunn, D.W.; Lisi, F.; Nan, F. Combining day-ahead forecasts for British electricity prices. *Energy Econ.* **2013**, *35*, 88–103. [CrossRef]

23. Grimes, D.; Ifrim, G.; O'Sullivan, B.; Simonis, H. Analyzing the impact of electricity price forecasting on energy cost-aware scheduling. *Sustain. Comput. Inform. Syst.* **2014**, *4*, 276–291. [CrossRef]

24. Nowotarski, J.; Raviv, E.; Trück, S.; Weron, R. An empirical comparison of alternative schemes for combining electricity spot price forecasts. *Energy Econ.* **2014**, *46*, 395–412. [CrossRef]

25. Sharma, V.; Srinivasan, D. A hybrid intelligent model based on recurrent neural networks and excitable dynamics for price prediction in deregulated electricity market. *Eng. Appl. Artif. Intell.* **2013**, *26*, 1562–1574. [CrossRef]

26. Dev, P.; Martin, M.A. Using neural networks and extreme value distributions to model electricity pool prices: Evidence from the Australian National Electricity Market 1998–2013. *Energy Convers. Manag.* **2014**, *84*, 122–132. [CrossRef]

27. Wang, J.; Zhang, W.; Li, Y.; Wang, J.; Dang, Z. Forecasting wind speed using empirical mode decomposition and Elman neural network. *Appl. Soft Comput.* **2014**, *23*, 452–459. [CrossRef]

28. Hickey, E.; Loomis, D.G.; Mohammadi, H. Forecasting hourly electricity prices using ARMAX-GARCH models: An application to MISO hubs. *Energy Econ.* **2012**, *34*, 307–315. [CrossRef]

29. Christensen, T.M.; Hurn, A.S.; Lindsay, K.A. Forecasting spikes in electricity prices. *Int. J. Forecast.* **2012**, *28*, 400–411. [CrossRef]

30. Amjady, N.; Keynia, F. A new prediction strategy for price spike forecasting of day-ahead electricity markets. *Appl. Soft Comput. J.* **2011**, *11*, 4246–4256. [CrossRef]

31. Dudek, G. Multilayer perceptron for GEFCom2014 probabilistic electricity price forecasting. *Int. J. Forecast.* **2016**, *32*, 1057–1060. [CrossRef]

32. Panapakidis, I.P.; Dagoumas, A.S. Day-ahead electricity price forecasting via the application of artificial neural network based models. *Appl. Energy* **2016**, *172*, 132–151. [CrossRef]

33. Feijoo, F.; Silva, W.; Das, T.K. A computationally efficient electricity price forecasting model for real time energy markets. *Energy Convers. Manag.* **2016**, *113*, 27–35. [CrossRef]

34. Abedinia, O.; Amjady, N.; Shafie-Khah, M.; Catalão, J.P.S. Electricity price forecast using Combinatorial Neural Network trained by a new stochastic search method. *Energy Convers. Manag.* **2015**, *105*, 642–654. [CrossRef]

35. He, K.; Xu, Y.; Zou, Y.; Tang, L. Electricity price forecasts using a Curvelet denoising based approach. *Phys. A Stat. Mech. Its Appl.* **2015**, *425*, 1–9. [CrossRef]

36. Ziel, F.; Steinert, R.; Husmann, S. Efficient modeling and forecasting of electricity spot prices. *Energy Econ.* **2015**, *47*, 98–111. [CrossRef]

37. Hong, Y.Y.; Wu, C.P. Day-ahead electricity price forecasting using a hybrid principal component analysis network. *Energies* **2012**, *5*, 4711–4725. [CrossRef]
38. Cerjan, M.; Matijaš, M.; Delimar, M. Dynamic hybrid model for short-term electricity price forecasting. *Energies* **2014**, *7*, 3304–3318. [CrossRef]
39. Monteiro, C.; Fernandez-Jimenez, L.A.; Ramirez-Rosado, I.J. Explanatory information analysis for day-ahead price forecasting in the Iberian electricity market. *Energies* **2015**, *8*, 10464–10486. [CrossRef]
40. Jónsson, T.; Pinson, P.; Nielsen, H.A.; Madsen, H. Exponential smoothing approaches for prediction in real-time electricity markets. *Energies* **2014**, *7*, 3710–3732. [CrossRef]
41. Weron, R. Electricity price forecasting: A review of the state-of-the-art with a look into the future. *Int. J. Forecast.* **2014**, *30*, 1030–1081. [CrossRef]
42. Wikipedia File: Synapse Self-Organizing Map. Available online: http://en.wikipedia.org/wiki/File: Synapse_Self-Orga (accessed on 1 August 2016).
43. Shafie-Khah, M.; Moghaddam, M.P.; Sheikh-El-Eslami, M.K. Price forecasting of day-ahead electricity markets using a hybrid forecast method. *Energy Convers. Manag.* **2011**, *52*, 2165–2169. [CrossRef]
44. Zadeh, L.A. Fuzzy logic. *Computer* **1988**, *21*, 83–93. [CrossRef]
45. Zadeh, L.A. Fuzzy sets. *Inf. Control* **1965**, *8*, 338–353. [CrossRef]
46. Zhao, J.; Guo, Z.H.; Su, Z.Y.; Zhao, Z.Y.; Xiao, X.; Liu, F. An improved multi-step forecasting model based on WRF ensembles and creative fuzzy systems for wind speed. *Appl. Energy* **2016**, *162*, 808–826. [CrossRef]
47. Wang, Z.; Liu, F.; Wu, J.; Wang, J. A hybrid forecasting model based on bivariate division and a backpropagation artificial neural network optimized by chaos particle swarm optimization for day-ahead electricity price. *Abstr. Appl. Anal.* **2014**, *2014*. [CrossRef]
48. Kohonen, T. Self-organized formation of topologically correct feature maps. *Biol. Cybern.* **1982**, *43*, 59–69. [CrossRef]
49. Ultsch, A. Emergence in Self Organizing Feature Maps. In Proceedings of the 6th International Workshop on Self-Organizing Maps, Bielefeld, Germany, 3–6 September 2007.
50. Kohonen, T. *Self-Organizing Maps*; Springer-Verlag New York, Inc.: New York, NY, USA, 1997.
51. Moghram, I.S.; Rahman, S. Analysis and evaluation of five short-term load forecasting techniques. *IEEE Trans. Power Syst.* **1989**, *4*, 1484–1491. [CrossRef]

energies

MDPI

Article

Enhanced Forecasting Approach for Electricity Market Prices and Wind Power Data Series in the Short-Term

Gerardo J. Osório [1], Jorge N. D. L. Gonçalves [2], Juan M. Lujano-Rojas [1,3] and João P. S. Catalão [1,2,3,]*

[1] C-MAST, University of Beira Interior, Covilhã 6201-001, Portugal;
    gjosilva@gmail.com (G.J.O.); lujano.juan@gmail.com (J.M.L.-R.)
[2] INESC TEC and the Faculty of Engineering of the University of Porto, Porto 4200-465, Portugal;
    ee10191@fe.up.pt
[3] INESC-ID, Instituto Superior Técnico, University of Lisbon, Lisbon 1049-001, Portugal
*   Correspondence: catalao@fe.up.pt; Tel.: +351-220-413-295

Academic Editor: Frede Blaabjerg
Received: 31 July 2016; Accepted: 25 August 2016; Published: 31 August 2016

Abstract: The uncertainty and variability in electricity market price (EMP) signals and players' behavior, as well as in renewable power generation, especially wind power, pose considerable challenges. Hence, enhancement of forecasting approaches is required for all electricity market players to deal with the non-stationary and stochastic nature of such time series, making it possible to accurately support their decisions in a competitive environment with lower forecasting error and with an acceptable computational time. As previously published methodologies have shown, hybrid approaches are good candidates to overcome most of the previous concerns about time-series forecasting. In this sense, this paper proposes an enhanced hybrid approach composed of an innovative combination of wavelet transform (WT), differential evolutionary particle swarm optimization (DEEPSO), and an adaptive neuro-fuzzy inference system (ANFIS) to forecast EMP signals in different electricity markets and wind power in Portugal, in the short-term, considering only historical data. Test results are provided by comparing with other reported studies, demonstrating the proficiency of the proposed hybrid approach in a real environment.

Keywords: adaptive neuro-fuzzy inference system (ANFIS); differential evolutionary particle swarm optimization (DEEPSO); electricity market prices (EMP); forecasting; short-term; time series; wavelet transform (WT); wind power

1. Introduction

In competitive and deregulated electricity markets, potential integration of renewables, especially wind power, which naturally introduces its stochastic, volatile, and uncertain behaviour, is totally reflected in the market players' strategies and presents more difficulties for a sustainable and robust management of the power framework. Even more when the renewable potential is introduced very widely, yielding higher production costs, inflexibility, and unnecessary penalties due to wrong strategies by players or an increment in emissions caused by conventional producers filling the gaps, especially when the renewable resources suddenly fail or do not cover the required demand [1]. Moreover, with the growing need for smart grids, for example to meet the growing interest in electric vehicles and their integration, the above concerns may be even more pronounced without the use of innovative tools or mechanisms to ensure the quality, safety, and robustness of the electrical system [2].

One of the approaches discussed nowadays in the scientific domain to mitigate some of the problems described above and to achieve a profitable and sustainable management of the electrical framework involves the integration of energy storage systems, which makes the electrical system more flexible due to the increased exploitation of potential usage of renewables, especially under peak loads, reducing the operational cost or curtailment events; however their implementation is still highly costly and in experimental phases in some cases [3].

An alternative way to tackle the aforementioned concerns in power systems and in competitive electricity markets, which are by nature more economical and useful for all agent players, is through the use of innovative forecasting tools to determine the future behaviour of the renewable potential or electricity market price (EMP) signals; making the creation of sets of possible market strategies suitable, considering other important indicators such as social behaviour, environmental factors, electrical constraints, and the behaviours of other electricity agents; in other words, the forecasting tools may be used as a first stage of defense for all market players [4]. In the last years, massive efforts, supported by the scientific community, have been made to propose more viable and reliable solutions, allowing mitigation of the countless concerns regarding power systems, which are reflected in widespread techniques and forecasting approaches for EMPs or wind power behaviour, considering statistical or physical models in soft or hard computing, as shown for instance in [5–7], considering very short-, short-, and long-term horizon forecasting [8,9].

Regarding EMP forecasting tools, since 2005, models such as autoregressive integrated moving average (ARIMA) combined with wavelet transform (WT) [10] can be found. This model belongs to the family of hard computing tools, which require a large amount of physical information and an exact modelling of the system, resulting in high computational complexity, and in this sense, will not be considered in this review of the state of the art. However, soft computing models, such as fuzzy neural network (FNN) [11] or hybrid intelligent system (HIS) [12], are among the soft computing models, which require the usage of any auto learning process from historical sets to identify future patterns and therefore require less computational complexity or information to model the problem. In this regard, several examples can be found such as neural network (NN) models [13], adaptive wavelet NN (AWNN) [14], cascaded neuro-evolutionary algorithm (CNEA) [15], cascaded NN (CNN) [16], the hybrid neuro-evolutionary system (HNES) [17], and some hybrid forecasting models, such as those presented in [18], or a combination of WT with particle swarm optimization (PSO) and the adaptive neuro-fuzzy inference system (ANFIS) (WPA) [19] and other hybrids [20], the hybrid fundamental-econometric model [21], or two-stage approaches such as those reported in [22,23]. Furthermore, more approaches considering singular spectrum analysis [24], informative vector machine [25], or even new genetic algorithms such as Levenberg-Marquardt and cuckoo search algorithms [26] and genetic regression of relevance vector machines [27] can be found for different EMP prices analyses, considering the Spanish, Pennsylvania-New Jersey-Maryland (PJM), Australian National Electricity Market (ANEM), and other liberalized electricity markets around the world as real case studies.

In wind power forecasting, widespread use of forecasting models for the very short and short term can be found in specialized literature considering soft computing and statistical models. In this sense, several examples are usually found, such as an evolutionary algorithm using an artificial intelligence model [28], NN [29,30], ridgelet NN [31], hybrid approaches composed of WT and a neuro-fuzzy network (NF) [32], WT with NN [33], WT with ANFIS (WNF) [34], or WPA [35]. Also, wind power forecasting can be tackled by considering a combination of WT with support vector machine (SVM) and statistical analysis [36], adaptive WT combined with feed-forward NN (AWNN) [37], WT combined with ARTMAP [38], and optimized SVM using a genetic algorithm [39]. More recently some proposals have considered a principal component analysis algorithm [40], hybrid WT, PSO, and NN [41], multi-layer artificial NN improved with simplified swarm optimization [42], and WT combined with NN, trained by an improved clonal selection algorithm [43]. All of the aforementioned models were run considering real cases with data from wind farms or historical data collected from the public domain in different locations around the world.

In this paper, in accordance with the features demonstrated by hybrid forecasting models briefly presented above, a new approach to forecast the EMP or wind power performance in the short term (from a few to 168 h ahead) is proposed.

Specifically, in the case of EMP forecasting, the proposed approach will perform a forecast for the next 168 h ahead with a time step of 1 h, considering only historical data available from the public domain, without considering the inclusion of exogenous data such as load and other energy prices, among others, to allow a fair and clean comparison with other already published methodologies. In the case of wind power forecasting, the proposed forecasting approach will perform the forecasting for a range of 3 h ahead with a time-step of 15 min, refreshing the system (input data and forecast results) until completion of the forecasting results for 24 h ahead. As in the previous case study, in wind power forecasting the proposed approach does not consider the inclusion of exogenous data such as wind profile and atmospheric data, among others, in order to make a fair and clean comparison with previously published approaches.

Furthermore, the proposed approach is composed of an innovative combination of WT as the pre-processing tool, which provides a smoothing effect of all inputs, providing more flexibility and more convergence to forecast the future behaviour, differential evolutionary particle swarm optimization (DEEPSO), which is itself a hybrid method and will be responsible for augmenting the performance of ANFIS (which is by nature a hybrid tool) by tuning the ANFIS membership functions to attain a lower forecasting error. Finally, the inverse WT will be used to introduce again the smoothing information collected at the beginning, providing the final forecasting signal. In this sense, hereafter the proposed approach will be called the hybrid WT+DEEPSO+ANFIS (HWDA) approach. In all case studies, the real historical data used will be comparable to those data used in reported and published models [44,45]. The remainder of the manuscript is organized as follows: Section 2 describes the concepts used to create the HWDA approach, the algorithm used for EMP or wind power forecasting, and the criteria used to validate and compare the capabilities of the proposed HWDA approach with previous and published methodologies. Section 3 describes the historical data used to carry out the forecasting considering the EMP or wind power, the detailed results, and the comparison carried out; finally, Section 4 presents the main conclusions drawn in this paper.

2. Proposed Approach

The HWDA approach results from the successful combination of WT, DEEPSO, and ANFIS. The WT is employed as a pre-processing step to decompose the historical sets of EMP or wind power into new constitutive sets with better behaviour. Then, the forthcoming values of those constitutive sets are the feeding sets of ANFIS responsible for creating the forecast results. DEEPSO augments the performance of ANFIS by tuning the ANFIS membership functions, resulting in lower forecasting error. In comparison with its ancestor, evolutionary particle swarm optimization (EPSO), the underlying evolutionary and differential concepts make real differences in terms of robustness, convergence, and computational time. So the combination of DEEPSO features with the adaptive characteristics of ANFIS means that they complement each other in positive way. Finally, the inverse WT is used to reconstruct the forecasting signal, and thus the final forecasting results are obtained.

2.1. Wavelet Transform

As reported in most of the previously described works on the state of the art, the application of WT in forecasting approaches is important for overcoming the limitations of non-stationary time series such as EMP or wind power; however, it may be applied in other engineering fields, since it enables the analysis of time series in their natural state. WT is used as a pre-processing tool for understanding non-stationary or time varying data [46], with sensibility to the irregularities of input data. In this sense, WT is especially useful for showing different aspects that constitute the data without losing the real signal content [47]. Despite the problems related with continuous WT (CWT) analysis, discrete WT (DWT) was created to give, in an effective way, a description relative to CWT, which is widely used to decompose the time series under study:

$$DWT\left(m_{wt}, n_{wt}\right) = 2^{-\left(m_{wt}/2\right)} \sum_{h=0}^{H} p\left(t_{wt}\right) \varphi\left(\frac{t_{wt} - b}{a}\right) \tag{1}$$

where H represents the length $p\left(t_{wt}\right)$, and the parameters of scaling (a) and translation (b) are changed to integer variables $a_{wt} = 2^{m_{wt}}$ and $b_{wt} = n_{wt}\,2^{m_{wt}}$ respectively, with a time-step t_{wt}, i.e.,:

$$DWT\left(m_{wt}, n_{wt}\right) = 2^{-\left(m_{wt}/2\right)} \sum_{h=0}^{H} p\left(t_{wt}\right) \varphi\left(\frac{t_{wt} - n_{wt}2^{m_{wt}}}{2^{m_{wt}}}\right) \tag{2}$$

The DWT is performed by multi-resolution analysis, where a "father wavelet", responsible for the low-frequency series, is used with a complementary "mother wavelet", which is responsible for the high-frequency series components [38]. In this paper and following the description cited in [44,45] the Daubechies of fourth order, or Db4, was used as the mother-wavelet function. The Db4 has asymmetrical and continuous proprieties, where a higher order level will create a higher level oscillation, which is desirable in forecasting [38,47]. The coefficients of approximations A_n and details D_n are expressed as:

$$A_n = \sum_{n} DWT\left(m_{wt}, n_{wt}\right) \varphi_{mn}\left(t\right) \tag{3}$$

$$D_n = \sum_{n} DWT\left(m_{wt}, n_{wt}\right) \psi_{mn}\left(t\right) \tag{4}$$

where $\varphi_{mn}\left(t_{wt}\right)$ is the father-wavelet and $\psi_{mn}\left(t_{wt}\right)$ is the mother-wavelet, and $DWT\left(m_{wt}, n_{wt}\right)$ are the coefficients obtained from Equation (2) [33]. Furthermore, the Db4 is chosen as the mother-wavelet function due to a better trade-off between smoothness and length [19]. Also, the DWT used in this paper was created on four filters divided into two groups: the decomposition group, composed of low-pass and high-pass filters, and the reconstruction group, composed of low-pass and high-pass filters as described in [44,45]. Figure 1 shows a general decomposition model of WT, where approximation steps A_n are able to analyse the universal information of original sets; that is, the low-frequency representation and description of the high-frequency component and the detailed steps D_n are able to describe the difference between the successive approximations.

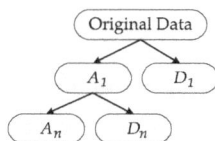

Figure 1. Universal n level decomposition model of WT.

2.2. Differential Evolutionary Particle Swarm Optimization

DEEPSO is a successful hybrid combination of the EPSO model [44,45], which is itself a hybrid combination of its ancestor model, namely PSO, where weight factors have self-adaptive features, with evolutionary programming, which brings self-adaptive operators [48], and a differential evolution algorithm, which provides a new solution from the current particle of the swarm by adding a fraction difference between two other points found from the previously evaluated swarm [49]. The DEEPSO schema is similar to EPSO [50]; however, the movement rule Equation (6) has new notation:

$$X_i^{new} = X_i + V_i^{new} \tag{5}$$

$$V_i^{new} = w_{i0}^* V_i + w_{i1}^* \left(X_{r1}^i - X_{r2}^i\right) + P\, w_{i2}^* \left(b_g^* - X_i\right) \tag{6}$$

where the weights w_{in-1}^* (inertia, memory and cooperation) are defined as:

$$w_{ik}^* = w_{ik} + \tau N\,(0,1) \tag{7}$$

and the global position is defined as:

$$b_g^* = b_g\,(1 + w_g N\,(0,1)) \tag{8}$$

From Equation (6), components X_r^i should be any pair of different particle already tested from the swarm, and ordered to minimize at the end of respective iteration, i.e.,:

$$f\left(X_{r1}^i\right) < f\left(X_{r2}^i\right) \tag{9}$$

From Equations (5)–(9), X_i^{new} is the new position of the particle, V_i^{new} is the new velocity found, P is a diagonal binary matrix with a value of 1 when the probability is p and 0 when the probability is $\{1 - p\}$, w_{ik}^* are the mutated weights of inertia, memory, and cooperation of the swarm, given by a learning parameter τ (fixed or mutated), and $N\,(0,1)$ is a random Gaussian variable with 0 mean and variance 1.

Also, b_g^* is the global position provided by the new weight w_g, which is collected from a diagonal matrix, having a self-adaptive feature, and in this sense, it is a mutated element [48,49]. Components X_{r1}^i and X_{r2}^i guarantee that a suitable extraction really happens, considering macro-gradient points in a descending direction depending on the structured comparison of $f\left(X_{r1}^i\right)$ and $f\left(X_{r2}^i\right)$. In this sense, component X_{r2}^i is assumed to be as $X_{r2}^i = X_i$, and component X_{r1}^i is sampled from the set of best ancestors from the swarm of n particles, that is, $S_{bA} = \{b_1, b_2, \ldots, b_n\}$ [50–52]. The main idea underlying DEEPSO movement is briefly illustrated in Figure 2.

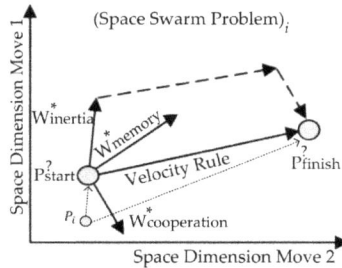

Figure 2. Brief illustration of DEEPSO (differential evolutionary particle swarm optimization) particle movement rule.

2.3. Adaptive Neuro-Fuzzy Inference System

ANFIS is a well-known hybrid combination of NN and fuzzy algorithms combining useful features such as low computational requirements, the possibility of dealing with a large number of data, and high response features. Furthermore, it has self-learning capabilities provided by the NN, which help it to self-adjust its parameters due to fuzzy capabilities [19,45]. The general ANFIS structure is based on several layers, which provide the fuzzification, rules, normalization data, desfuzzification, and data reconstruction process as described in [35,44]. Figure 3 briefly describes the multi-layer feed-forward network ANFIS structure. Mathematically, each of the five layers l_{n_k} used is:

$$\begin{cases} l_{1_k} = \mu A_i\,(x)\,, & k = 1,2 \\ l_{1_k} = \mu B_{i-2}\,(y)\,, & k = 3,4 \end{cases} \tag{10}$$

$$\mu A_k\left(x\right) = \cfrac{1}{1 + \left|\frac{x - r_k}{p_{ik}}\right|^{2q_k}} \tag{11}$$

$$l_{2_k} = w_k = \mu A_k\left(x\right)\mu B_k\left(y\right), \quad k = 1, 2 \tag{12}$$

$$l_{3_k} = \overline{w}_i = \frac{w_k}{w_1 + w_2}, \quad k = 1, 2 \tag{13}$$

$$l_{4_k} = \overline{w}_k z_k = \overline{w}_k\left(a_k x + b_k y + c_k\right), \quad k = 1, 2 \tag{14}$$

$$l_{5_k} = \sum_k \overline{w}_k z_k = \frac{\sum_k w_k z_k}{\sum_k w_k} \tag{15}$$

From Equation (10), all nodes k are adaptive nodes with node function l_{1_k}, where x and y are the input of the k_{th} node and A_k and B_{k-2} are the membership function, also called the linguistic label, associated with these nodes. In this paper, a triangular membership function is normally used [44,45], where $\{p_k, q_k, r_k\}$ are parameter sets, because it is a continuous and piecewise differentiable function, described in Equation (11), which represents the first layer. In Equation (12), all output nodes represent the firing strength of the rule w_k, where each node signal is multiplied by the previous inputs signals, representing the second layer. In Equation (13), the third layer, every node computes the ratio of firing strength rules k_{th} to the sum of all firing strength rules. Equation (14) represents the computation of all nodes' contribution to k_{th} rule with global output, where $\{a_k, b_k, c_k\}$ are parameter sets, and $\overline{w}_k$ is the layer output (fourth layer). Finally, Equation (15) defines the ANFIS output node, that is, the fifth layer where the summation Σ is made. As reported in [19,35], in this paper, the ANFIS structure follows the least-squares and back-propagation gradient descent method, considering the Takagi-Sugeno approach.

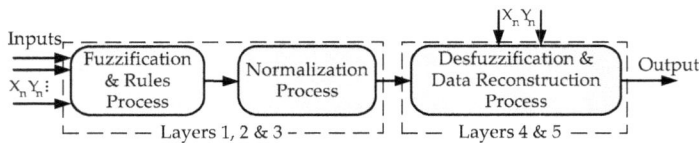

Figure 3. Brief illustration of ANFIS (adaptive neuro-fuzzy inference system) structure.

2.4. Hybrid Proposed Approach

As stated before, the HWDA approach results from a combination of WT, DEEPSO, and ANFIS. The WT is employed as a pre-processing step to decompose the historical sets. The DEEPSO augments the ANFIS performance by tuning the ANFIS membership functions. Finally, the inverse WT is used to reconstruct the forecasting signal, and then the final forecasting results are obtained. Figure 4 shows the HWDA flowchart. In detail, HWDA follows the following steps:

- Step 1: Initialize the HWDA approach with a historical data matrix of EMP or wind power, respectively, considering the forecasting time-scale of each forecast field;
- Step 2: Choose a set of historical data of the previous step to run the pre-processing process carried out by the WT tool. This step is performed by a backtracking process, in order to attain a smaller error at the end by choosing the best set of candidates. Also, the approach considered in this paper uses A_3, D_3, and D_1 steps as inputs for the next step;
- Step 3: Train the ANFIS tool with the previous sets of constitutive historical data obtained from WT. The optimization process of the ANFIS membership function parameters will be achieved with the DEEPSO method. All parameters considered from all methods are summarized in Table 1.

As in [44,45], the ANFIS inference rules are obtained by considering the automatic ANFIS mode, due to the nature of the data, which requires a large number of inference rules, and thus additional improvement is achieved.

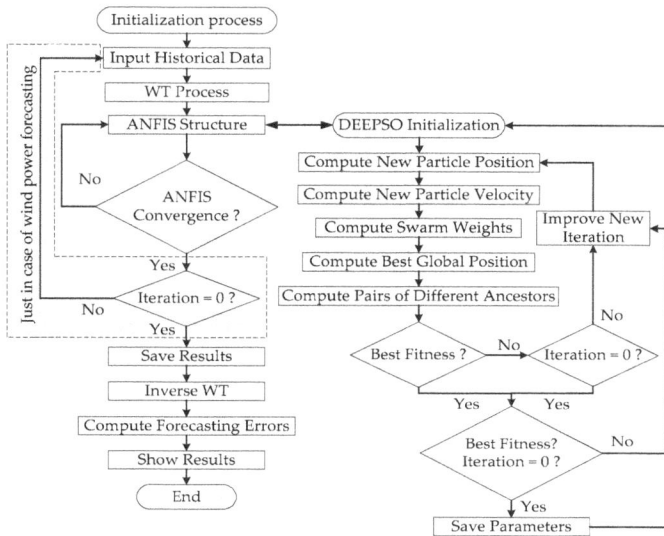

Figure 4. HWDA (hybrid WT+DEEPSO+ANFIS) forecasting approach flowchart.

Table 1. DEEPSO (differential evolutionary particle swarm optimization) and ANFIS (adaptive neuro-fuzzy inference system) parameters used for EMP (electricity market price) and wind power forecasting.

Methods	Parameters	Type or Size
WT	Decomposition Direction	Row
	Level of Decomposition	3
	Mother-Wavelet Function	Db4
	Denoising Methods	"sqtwolog"–"minimaxi"
	Multiplicative Thresholds Rescaling	"one"–"sln"
DEEPSO	Communication Probability	0.10
	Final Inertia Wight	0.01–0.15
	Initial Inertia Weight	0.50–0.90
	Initial Population Size	100
	Initial Sharing Acceleration	0.50–2.00
	Initial Swarm Learning Process	1.00–2.00
	Initial Swarm Sharing Process	2.00
	Learning Parameter	1
	Maximum Value of New Position	Set of Max. Inputs
	Minimum Value of New Position	Set of Min. Inputs
	Necessary iterations	100–1000
ANFIS	Structure Type	Takagi-Sugeno
	Style of Membership Function	Triangular
	Number of Inference Rules	Automatic
	Membership Functions	2–15
	Number of Epochs	2–50
	Number of Nodes	3–9
	Number of Inputs/Outputs	2–5/1

- Step 4: until the best results are obtained or convergence is reached:

  - Step 4.1: Jump to Step 4 in the case of EMP if convergence is not reached;
  - Step 4.2: Jump to Step 2 in the case of wind power forecasting, refreshing the historical data matrix.

When the best result is found or convergence is reached, the wind power data are forecasted for the next 3 h until the forecast for the next 24 h ahead is complete.

- Step 5: Apply the inverse WT. The output of the proposed HWDA approach is attained; that is, the forecasted EMP or wind power results are ready to be presented;
- Step 6: Compute the forecasting errors of EMP or wind power results with different criteria to validate the proposed HWDA approach and show the results.

2.5. Forecasting Error Evaluation

To compare the proposed approach with other methodologies for EMP or wind power forecasting previously published in the specialized literature, the mean absolute percentage error (MAPE) criterion is used. This criterion is given as [44,45]:

$$MAPE = \frac{100}{N} \sum_{n=1}^{N} \frac{|\hat{p}_n - p_n|}{\overline{p}} \tag{16}$$

$$\overline{p} = \frac{1}{N} \sum_{n=1}^{N} p_n \tag{17}$$

where $\hat{p}_n$ is the data forecasted at hour n, p_n is the real data at hour n, $\overline{p}$ is the average value for the forecasting time horizon, and N has the length value of observed points. Following the same concept from the MAPE criterion, the uncertainty of the HWDA model is evaluated using the error variance, described as [19,35]:

$$\sigma_{e,n}^2 = \frac{1}{N} \sum_{n=1}^{N} \left(\frac{|\hat{p}_n - p_n|}{\overline{p}} - e_n \right)^2 \tag{18}$$

$$e_n = \frac{1}{N} \sum_{n=1}^{N} \frac{|\hat{p}_n - p_n|}{\overline{p}} \tag{19}$$

Moreover, for wind power forecasting, the normalized mean absolute error (NMAE) criterion is used [35,45]:

$$NMAE = \frac{100}{N} \sum_{n=1}^{N} \frac{|\hat{p}_n - p_n|}{P_{installed}} \tag{20}$$

where $P_{installed} = 2700\,MW$, which corresponds to the total wind power capacity installed in accordingly to [53]. Furthermore, the normalized root mean square error (NRMSE) is also used and is described as [45]:

$$NRMSE = \sqrt{\frac{1}{N} \sum_{n=1}^{N} \left(\frac{\hat{p}_n - p_n}{P_{intalled}} \right)^2} \times 100 \tag{21}$$

3. Case Studies and Results

3.1. Electricity Market Prices Forecasting

As briefly stated before, the HWDA approach is used first to forecast EMP for the next 168 and 24 h considering the historical data from the Spanish market available in [54].

As mentioned in [10,21], this market has features that are difficult to forecast due to influences from dominant players, which are reflected in historical data. The EMP historical data used for the Spanish market date back to the year 2002, allowing a clear and fair comparison with the already published results from other proposed methodologies, considering the same four test weeks of the year 2002, which are consistent with the four seasons. As stated before, only EMP historical data sets were used, for the reasons stated above, otherwise a correct comparative study would not be possible.

The HWDA approach forecasts the next 168 h of EMP considering the previous 1008 h (six weeks), which are used as input sets. In order to avoid over-training during the learning process, very large training sets are not used. The output of the HWDA approach results in a set of 168 points representing the forecasting horizon. For day-ahead forecasting, the same idea may be followed; that is, the HWDA approach has as its input the previous six days, considering the historical data from the same market for the year 2006, which were analysed by the case studies reported in [44].

Furthermore, the HWDA approach is tested for the PJM market, forecasting the EMP for the next 24 and 168 h ahead. The historical data of electricity prices are available in [55]. Similarly to the Spanish market, no exogenous data were considered for the same reason as described above.

3.1.1. Spanish Market Results

The results obtained with the HWDA approach are provided in Figures 5–8 for the four test weeks (168 h ahead) of 2002, where the solid and dash-dot black lines represent the actual and forecasted EMP, respectively, while the blue line at the bottom of each figure represents the resulting errors as absolute values. Tables 2 and 3 shows the comparative MAPE criterion and weekly error variance criterion results, respectively, between the HWDA approach and ten previous published methodologies, namely NN [13], FNN [11], AWNN [14], HIS [12], CNEA [15], CNN [16], WPA [19], mutual information with composite NN (MI+CNN) [22], and hybrid evolutionary algorithm (HEA) [44], indicating the enhancements as the percentage evolution between the HWDA approach and the respective comparative methodology under analysis.

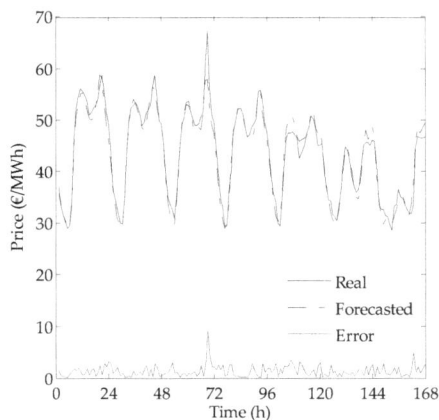

Figure 5. Winter week 2002 results for the Spanish market.

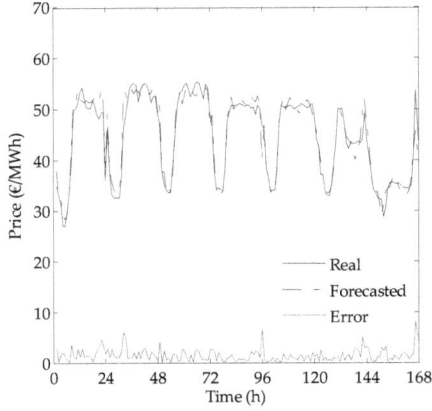

Figure 6. Spring week 2002 results for the Spanish market.

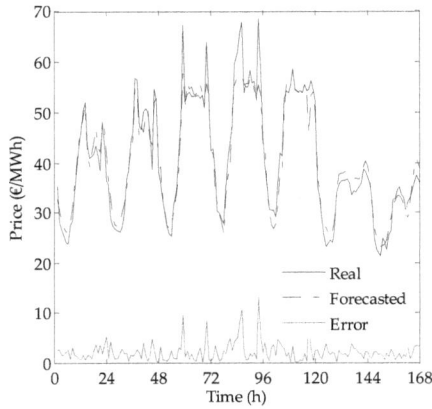

Figure 7. Summer week 2002 results for the Spanish market.

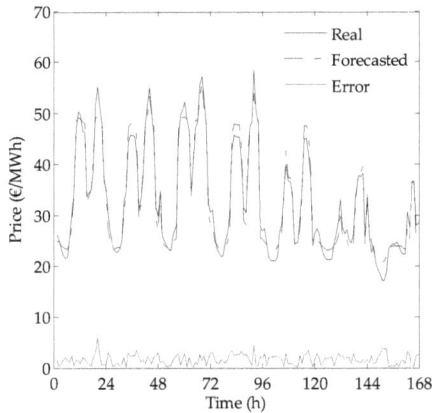

Figure 8. Autumn week 2002 results for the Spanish market.

Table 2. MAPE (the mean absolute percentage error) comparison considering the year 2002 Spanish market case study for 168 h ahead.

Methods	Winter	Spring	Summer	Fall	Average	Enhancement
NN [13], 2007	5.23	5.36	11.40	13.65	8.91	54.66%
FNN [11], 2006	4.62	5.30	9.84	10.32	7.52	46.28%
HIS [12], 2009	6.06	7.07	7.47	7.30	6.97	42.04%
AWNN [14], 2008	3.43	4.67	9.64	9.29	6.75	40.15%
CNEA [15], 2009	4.88	4.65	5.79	5.96	5.32	24.06%
CNN [16], 2009	4.21	4.76	6.01	5.88	5.22	22.61%
HNES [17], 2010	4.28	4.39	6.53	5.37	5.14	21.40%
MI+CNN [22], 2012	4.51	4.28	6.47	5.27	5.13	21.25%
WPA [19], 2011	3.37	3.91	6.50	6.51	5.07	20.32%
HEA [44], 2014	3.04	3.33	5.38	4.97	4.18	3.35%
HWDA	3.00	3.16	5.23	4.76	4.04	-

Table 3. Weekly error variance comparison considering the year 2002 Spanish market case study for 168 h ahead.

Methods	Winter	Spring	Summer	Fall	Average	Enhancement
NN [13], 2007	0.0017	0.0018	0.0109	0.0136	0.0070	82.86%
FNN [11], 2006	0.0018	0.0019	0.0092	0.0088	0.0054	77.78%
AWNN [14], 2008	0.0012	0.0031	0.0074	0.0075	0.0048	75.00%
HIS [12], 2009	0.0034	0.0049	0.0029	0.0031	0.0036	66.67%
CNEA [15], 2009	0.0036	0.0027	0.0043	0.0039	0.0036	66.67%
CNN [16], 2009	0.0014	0.0033	0.0045	0.0048	0.0035	65.71%
WPA [19], 2011	0.0008	0.0013	0.0056	0.0033	0.0027	55.56%
MI+CNN [22], 2012	0.0014	0.0014	0.0033	0.0022	0.0021	42.86%
HNES [17], 2010	0.0013	0.0015	0.0033	0.0022	0.0021	42.86%
HEA [44], 2014	0.0008	0.0011	0.0026	0.0014	0.0015	20.00%
HWDA	0.0007	0.0008	0.0022	0.0010	0.0012	-

When the HWDA approach was used, the MAPE criterion reached an average value of 4.04%, which is significant, even when it is compared for each week independently or considering the improvements over all comparative methodologies. The weekly error variance criterion results obtained using the HWDA approach reached an average value of 0.0012, showing a notable accuracy compared with the other methodologies described and reported, even when its improvements are analysed independently.

3.1.2. PJM (Pennsylvania-New Jersey-Mary) Land Market Results

The HWDA approach was also used to forecast the EMP considering the historical data from the PJM market, available in [55], providing results for the next 24 and 168 h ahead. As in the previous case study, no exogenous data are taken into account. Figures 9–11 illustrate some results for some days and weeks tested considering the historical data of 2006 for the PJM market, and the same condition as described in [44] is applied to give a clear and fair comparison with other published methodologies. Moreover, in all figures, the solid and dash-dot black lines represent the actual and forecasted EMP, respectively, while the blue line at the bottom of each figure represents the resulting errors as absolute values.

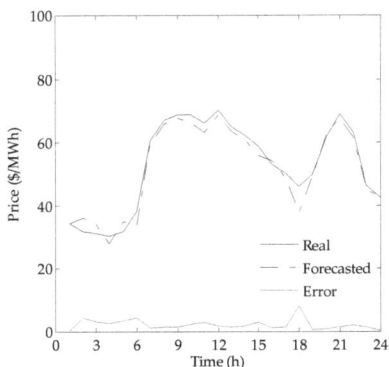

Figure 9. 7 April 2006 results for the PJM (Pennsylvania-New Jersey-Mary) market.

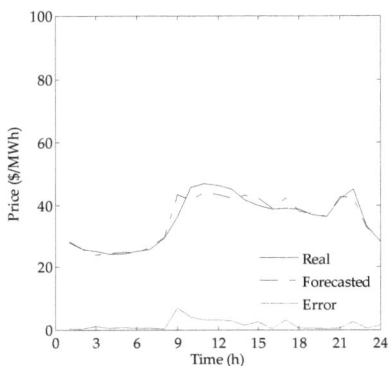

Figure 10. 13 May 2006 results for the PJM (Pennsylvania-New Jersey-Mary) land market.

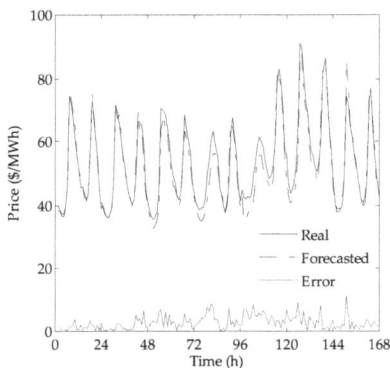

Figure 11. 22–28 February 2006 results for the PJM (Pennsylvania-New Jersey-Mary) land market.

Tables 4 and 5 shows the MAPE and error variance results, respectively, for the HWDA approach and four other methodologies. When using the HWDA approach, the MAPE criterion reached an average value of 3.16% and the error variance reached an average of 0.0011, which is notable for this competitive market.

Table 4. MAPE (the mean absolute percentage error) comparison considering the year 2006 PJM (Pennsylvania-New Jersey-Mary) land market case study for 24/168 h ahead.

	HNES [17], 2010	Hybrid [44], 2010	CNEA [15], 2009	HEA [44], 2014	HWDA
Jan. 20	4.98	3.71	4.73	3.29	3.22
Feb. 10	4.10	2.85	4.50	2.80	2.71
Mar. 5	4.45	5.48	4.92	3.32	3.27
Apr. 7	4.67	4.17	4.22	3.55	3.42
May 13	4.05	4.06	3.96	3.43	3.40
Feb. 1–7	4.62	5.27	4.02	3.11	3.09
Feb. 22–28	4.66	5.01	4.13	3.08	3.02
Average	4.50	4.36	4.35	3.23	3.16
Enhancement	29.78%	27.52%	27.36%	2.17%	-

Table 5. Error variance comparison considering the year 2006 PJM (Pennsylvania-New Jersey-Mary) land market case study for 24/168 h ahead.

	CNEA [15], 2009	Hybrid [44], 2010	HNES [17], 2010	HEA [44], 2013	HWDA
Jan. 20	0.0031	0.0010	0.0020	0.0010	0.0010
Feb. 10	0.0036	0.0015	0.0012	0.0009	0.0008
Mar. 5	0.0042	0.0033	0.0015	0.0011	0.0010
Apr. 7	0.0022	0.0013	0.0018	0.0011	0.0011
May 13	0.0027	0.0015	0.0013	0.0012	0.0012
Feb. 1–7	0.0044	0.0037	0.0016	0.0012	0.0011
Feb. 22–28	0.0035	0.0025	0.0017	0.0017	0.0016
Average	0.0034	0.0021	0.0016	0.0012	0.0011
Enhancement	67.65%	47.62%	45.45%	8.33%	-

3.2. Wind Power Forecasting

The HWDA approach was used to forecast the wind power for 3 h ahead with a time-step of 15 min until the forecast for the whole 24 h ahead was complete, considering the historical data of wind power in Portugal between 2007 and 2008 as described in [45,53] and considering the different seasons of the year. Also, as in the previous case studies, to allow a fair and clean comparison, only historical wind power data are considered, for the same reason as described above. Figures 12–15 show the numerical wind power results for winter, spring, summer, and autumn days, respectively, where solid and dash-dot black lines represent the actual and forecasted wind power, respectively, while the blue line in the bottom figures represents the errors as absolute values. For all results, it is possible to observe how the HWDA approach correctly forecasts the unexpected and abrupt changes of the wind power profile, that is, its uncertainty behaviour during the whole day of forecasting.

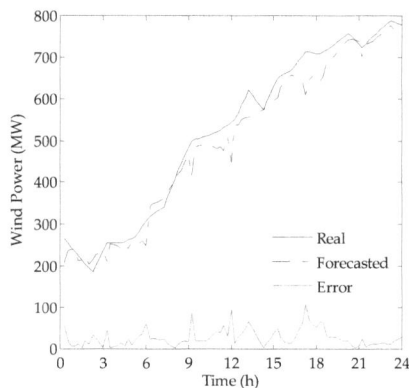

Figure 12. Real and forecasted wind power results (15 min intervals) for the Winter day.

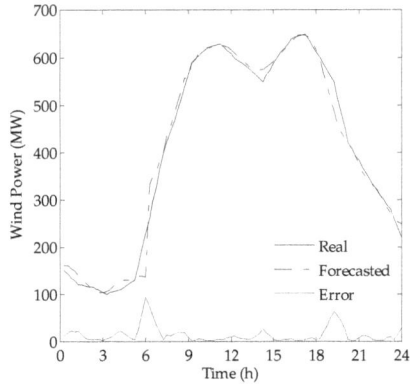

Figure 13. Real and forecasted wind power results (15-min intervals) for the Spring day.

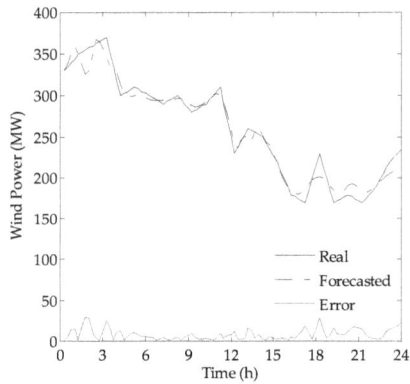

Figure 14. Real and forecasted wind power results (15-min intervals) for the Summer day.

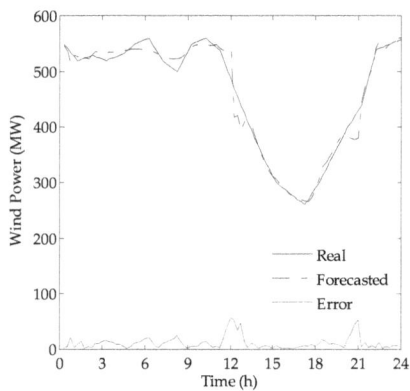

Figure 15. Real and forecasted wind power results (15-min intervals) for the Autumn day.

Tables 6 and 7 provide a comparative study between the HWDA approach using MAPE and the daily error variance criterion and five other previously published methodologies, namely NN [29],

NF [32], WNF [34], WPA [35], and HEA [45], respectively. When the HWDA approach is used, the MAPE criterion has an average value of 3.37%, representing an enhancement of 11.28% compared to the HEA methodology, which is again significant.

Table 6. MAPE (the mean absolute percentage error) comparison for wind power forecasting.

	Winter	Spring	Summer	Fall	Average	Enhancement
NN [29]	9.51	9.92	6.34	3.26	7.26	53.58%
NF [32]	8.85	8.96	5.63	3.11	6.64	49.25%
WNF [34]	8.34	7.71	4.81	3.08	5.99	43.74%
WPA [35]	6.47	6.08	4.31	3.07	4.98	32.33%
HEA [45]	5.74	3.49	3.13	2.62	3.75	11.28%
HWDA	5.08	3.19	2.96	2.27	3.37	-

Table 7. Daily error variance comparison for wind power forecasting.

	Winter	Spring	Summer	Fall	Average	Enhancement
NN [29]	0.0044	0.0106	0.0043	0.0010	0.0051	76.47%
NF [32]	0.0041	0.0086	0.0038	0.0008	0.0043	72.09%
WNF [34]	0.0046	0.0051	0.0021	0.0011	0.0032	62.50%
WPA [35]	0.0021	0.0035	0.0016	0.0011	0.0021	42.86%
HEA [45]	0.0019	0.0015	0.0010	0.0008	0.0013	7.69%
HWDA	0.0017	0.0016	0.0007	0.0006	0.0012	-

Furthermore, the daily error variance obtained using the HWDA approach has an average value of 0.0013%, presenting lower uncertainty in the forecasts done, and again, in all results the HWDA approach shows better accuracy in comparison with analyses of the same real case by all other previously published methodologies.

Finally, Tables 8 and 9 show a comparison of the results obtained with the HWDA approach according to the NMAE and NRMSE criteria, respectively. In all cases analysed, it is possible to observe that the HWDA approach gave better results than the other published methodologies considering the same cases studies. The proposed HWDA approach was performed on a standard PC equipped with an Intel Core i7-3537U, 2 GHz CPU and 4 GB of RAM with Windows 10 and the MATLAB®2016a platform. The authors used the ANFIS and WT structure functions available in MATLAB toolboxes, while DEEPSO was programmed from scratch in MATLAB considering the information available in [49–52].

Table 8. NMAE (the normalized root mean square error) comparison for wind power forecasting.

	Winter	Spring	Summer	Fall	Average	Enhancement
NN [29]	5.22	3.72	2.35	2.15	3.36	84.23%
NF [32]	4.86	3.36	2.09	2.05	3.09	82.85%
WNF [34]	4.58	2.89	1.78	2.03	2.82	81.21%
WPA [35]	3.56	2.28	1.60	2.02	2.37	77.64%
HEA [45]	2.73	1.48	0.74	1.10	1.51	64.90%
HWDA	0.94	0.49	0.28	0.39	0.53	-

Table 9. NRMSE (the normalized root mean square error) comparison for wind power forecasting.

	Winter	Spring	Summer	Fall	Average	Enhancement
HEA [45]	3.60	3.18	1.78	2.07	2.66	39.47%
HWDA	2.19	1.27	1.81	1.18	1.61	-

4. Conclusions

An enhanced HWDA approach was proposed in this paper for short-term EMP and wind power forecasting considering real cases studies, specifically the analyses from Spanish and PJM markets, as well as the wind power behavior in Portugal. The innovative and successful combination of WT, DEEPSO and ANFIS provided interesting and valuable results. The main findings resulting from this study are related to the lower forecasting errors attained while providing an acceptable computational time. The MAPE criterion reached an average value of 4.04% for the Spanish Market, surpassing all other methodologies, and for the PJM market reached an average value of 3.16%. Regarding the wind power forecasting results, the MAPE criterion had an average value of 3.37%. Lower error variances were also obtained in all cases. Moreover, the computational time required for HWDA approach was less than two min, on average, for the EMP results, and for wind power forecasting took less than one min per iteration. Hence, the overall results obtained with the HWDA approach provided an excellent trade-off between computational time and accuracy, which is crucial for real-life and real-time applications.

Acknowledgments: João Catalão and Juan Lujano-Rojas thank the EU Seventh Framework Programme FP7/2007–2013 under grant agreement No. 309048, FEDER through COMPETE and FCT, under FCOMP-01-0124-FEDER-020282 (Ref. PTDC/EEA-EEL/118519/2010), UID/CEC/50021/2013 and SFRH/BPD/103079/2014.

Author Contributions: All authors have worked on this manuscript together and all authors have read and approved the final manuscript.

Conflicts of Interest: The authors declare no conflict of interest.

Nomenclature

a	WT scaling integer variable
A_k	ANFIS linguistic label
a_k	ANFIS contribution parameter set
A_n	WT approximation coefficient
b	WT translation integer variable
b_g	DEEPSO actual global position
b_g^*	DEEPSO global position provided by a new weight w_g
B_k	ANFIS linguistic label
b_k	ANFIS contribution parameter set
c_k	ANFIS contribution parameter set
D_n	WT detail coefficient
DWT	Discrete wavelet transform set
e_n	Error at hour n
φ_{mn}	WT father-wavelet function
H	WT length of set $p\,(t_{wt})$
i	DEEPSO integer time-step from global search space
k	ANFIS number of nodes
k_{th}	ANFIS output node
l_{n_k}	ANFIS layer
$MAPE$	Mean absolute percentage error
m_{wt}	WT integer scaling parameter
N	Length of observed values points
$N\,(0,1)$	DEEPSO random Gaussian variable with 0 mean and variance 1
$NMAE$	Normalized mean absolute error
$NRMSE$	Normalized root mean square error
n_{wt}	WT integer translation parameter

$\bar{p}$	Average value for the forecasting horizon
P	DEEPSO probabilistic diagonal binary matrix
$\hat{p}_n$	Data forecasted at hour n
$P_{installed}$	Total wind power capacity installed
p_k	ANFIS parameter set of membership function
p_n	Real data at hour n
ψ_{mn}	WT mother-wavelet function
$p\,(t_{wt})$	WT signal input
q_k	ANFIS parameter set of membership function
r_k	ANFIS parameter set of membership function
$\sigma_{e,n}^2$	Error variance from the forecasting horizon
τ	DEEPSO learning parameter
t_{wt}	WT time-step
V_i	DEEPSO actual velocity
V_i^{new}	DEEPSO new velocity of the particle
w_g	DEEPSO new weight with self-adaptive features
w_{ik}^*	DEEPSO mutated weights of inertia, memory and cooperation
w_k	ANFIS firing strength
$\bar{w}_k$	ANFIS output firing strength
x	ANFIS input data
X_i	DEEPSO actual position
X_i^{new}	DEEPSO new position of the particle
X_{r1}^i	DEEPSO set of best ancestors from the swarm
X_{r2}^i	DEEPSO set of recorded positions of the swarm
y	ANFIS input data
z_k	ANFIS defuzzification parameters data

References

1. Dufo-López, R.; Bernal-Agustín, J.L.; Monteiro, C. New methodology for the optimization of the management of wind farms, including energy storage. *Appl. Mech. Mater.* **2013**, *330*, 183–187. [CrossRef]
2. Catalão, J.P.S. *Smart and Sustainable Power Systems: Operations, Planning, and Economics of Insular Electricity Grids*, 1st ed.; CRC Press, Taylor and Francis Group: Boca Raton, FL, USA, 2015.
3. Rodrigues, E.M.G.; Osório, G.J.; Godina, R.; Bizuayehu, A.W.; Lujano-Rojas, J.M.; Matias, J.C.O.; Catalão, J.P.S. Modelling and sizing of NaS (sodium sulfur) battery energy storage system for extending wind power performance in Crete Island. *Energy* **2015**, *90*, 1606–1617. [CrossRef]
4. Li, L.; Wang, J. Sustainable energy development scenario forecasting and energy saving policy analysis of China. *Renew. Sust. Energy Rev.* **2016**, *58*, 718–724.
5. Weron, R. Electricity price forecasting: A review of the state-of-the-art with a look into the future. *Int. J. Forecas.* **2014**, *30*, 1030–1081. [CrossRef]
6. Chang, W.Y. A literature review of wind forecasting methods. *J. Power Energy Eng.* **2014**, *2*, 161–168. [CrossRef]
7. Ren, Y.; Suganthan, P.N.; Srikanth, N. Ensemble methods for wind and solar power forecasting—A state of the art review. *Renew. Sust. Energy Rev.* **2015**, *50*, 82–91. [CrossRef]
8. Okumus, I.; Dinler, A. Current status of wind energy forecasting and a hybrid method for hourly predictions. *Energy Conv. Manag.* **2016**, *123*, 362–371. [CrossRef]
9. Wang, X.; Guo, P.; Huang, X. A review of wind power forecasting models. *Energy Proc.* **2011**, *12*, 770–778. [CrossRef]
10. Conejo, A.J.; Plazas, M.A.; Espínola, R.; Molina, A.B. Day-ahead electricity price forecasting using wavelet transform and ARIMA models. *IEEE Trans. Power Syst.* **2005**, *20*, 1035–1042. [CrossRef]
11. Amjady, N. Day-ahead price forecasting of electricity markets by a new fuzzy neural network. *IEEE Trans. Power Syst.* **2006**, *21*, 887–896. [CrossRef]

12. Amjady, N.; Hemmati, H. Day-ahead price forecasting of electricity markets by a hybrid intelligent system. *Euro Trans. Electron. Power* **2006**, *19*, 89–102. [CrossRef]

13. Catalão, J.P.S.; Mariano, S.J.P.S.; Mendes, V.M.F.; Ferreira, L.A.F.M. Short-term electricity prices forecasting in a competitive market: A neural network approach. *Electron Power Syst. Res.* **2007**, *77*, 1297–1304. [CrossRef]

14. Pindoriya, N.M.; Singh, S.N.; Singh, S.K. An adaptive wavelet neural network-based energy price forecasting, in electricity market. *IEEE Trans. Power Syst.* **2008**, *23*, 1423–1432. [CrossRef]

15. Amjady, N.; Keynia, F. Day-ahead price forecasting of electricity markets by mutual information technique and cascaded neuro-evolutionary algorithm. *IEEE Trans. Power Syst.* **2009**, *24*, 306–318. [CrossRef]

16. Amjady, N.; Daraeepour, A. Design of input vector for day-ahead price forecasting of electricity markets. *Exp. Syst. Appl.* **2009**, *36*, 12281–12294. [CrossRef]

17. Amjady, N.; Keynia, F. Application of a new hybrid neuro-evolutionary system for day-ahead price forecasting of electricity markets. *Appl. Soft Comput.* **2010**, *10*, 784–792. [CrossRef]

18. Wu, L.; Shahidehpour, M. A hybrid model for day-ahead price forecasting. *IEEE Trans. Power Syst.* **2010**, *25*, 1519–1530.

19. Catalão, J.P.S.; Pousinho, H.M.I.; Mendes, V.M.F. Hybrid wavelet-PSO-ANFIS approach for short-term electricity prices forecasting. *IEEE Trans. Power Syst.* **2011**, *26*, 137–144. [CrossRef]

20. Shafie-khah, M.; Moghaddam, M.P.; Sheikh-El-Eslami, M.K. Price forecasting of day-ahead electricity markets using a hybrid forecast method. *Energy Convers. Manag.* **2011**, *52*, 2165–2169. [CrossRef]

21. González, V.; Contreras, J.; Bunn, D.W. Forecasting power prices using a hybrid fundamental-econometric model. *IEEE Trans. Power Syst.* **2012**, *27*, 363–372. [CrossRef]

22. Keynia, F. A new feature selection algorithm and composite neural network for electricity price forecasting. *Eng. Appl. Artif. Intell.* **2012**, *25*, 1687–1697. [CrossRef]

23. Shayeghi, H.; Ghasemi, A. Day-ahead electricity prices forecasting by a modified CGSA technique and hybrid WT in LSSVM based scheme. *Energy Convers. Manag.* **2013**, *74*, 482–491. [CrossRef]

24. Miranian, A.; Abdollahzade, M.; Hassani, H. Day-ahead electricity price analysis and forecasting by singular spectrum analysis. *IET Gener. Trans. Distrib.* **2013**, *7*, 337–346. [CrossRef]

25. Elattar, E.; Shebin, E.K. Day-ahead price forecasting of electricity markets based on local informative vector machine. *IET Gener. Transm. Distrib.* **2013**, *7*, 1063–1071. [CrossRef]

26. Kim, M.K. Short-term price forecasting of Nordic power market by combination Levenberg-Marquardt and cuckoo search algorithms. *IET Gen. Trans. Distrib.* **2015**, *9*, 1553–1563. [CrossRef]

27. Alamaniotis, M.; Bargiotas, D.; Bourbakis, N.G.; Tsoulalas, L.H. Genetic optimal regression of relevance vector machines for electricity pricing signal forecasting in smart grids. *IEEE Trans. Smart Grid* **2015**, *6*, 2997–3005. [CrossRef]

28. Jursa, R.; Rohrig, K. Short-term wind power forecasting using evolutionary algorithms for the automated specification of artificial intelligence models. *Int. J. Forecast.* **2008**, *24*, 694–709. [CrossRef]

29. Catalao, J.P.S.; Pousinho, H.M.I.; Mendes, V.M.F. An artificial neural network approach for short-term wind power forecasting in Portugal. *Eng. Intell. Syst. Electron. Eng. Commun.* **2009**, *17*, 5–11.

30. Rosado, I.J.R.; Jimenez, L.A.F.; Monteiro, C.; Sousa, J.; Bessa, R. Comparison of two new short-term wind-power forecasting systems. *Renew. Energy* **2009**, *34*, 1848–1854. [CrossRef]

31. Amjady, N.; Keynia, F.; Zareipour, H. Short-term wind power forecasting using ridgelet neural network. *Electr. Power Syst. Res.* **2011**, *81*, 2099–2107. [CrossRef]

32. Catalão, J.P.S.; Pousinho, H.M.I.; Mendes, V.M.F. Hybrid intelligent approach for short-term wind power forecasting in Portugal. *IET Renew. Power Gener.* **2011**, *5*, 251–257. [CrossRef]

33. Catalão, J.P.S.; Pousinho, H.M.I.; Mendes, V.M.F. Short-term wind power forecasting in Portugal by neural network and wavelet transform. *Renew. Energy* **2011**, *36*, 1245–1251. [CrossRef]

34. Pousinho, H.M.I.; Mendes, V.M.F.; Catalão, J.P.S. Application of adaptive neuro-fuzzy inference for wind power short-term forecasting. *IEEJ Trans. Electr. Electron. Eng.* **2011**, *6*, 571–576. [CrossRef]

35. Catalão, J.P.S.; Pousinho, H.M.I.; Mendes, V.M.F. Hybrid wavelet-PSO-ANFIS approach for short-term wind power forecasting in Portugal. *IEEE Trans. Sustain. Energy* **2011**, *2*, 50–59.

36. Liu, Y.; Shi, J.; Yang, Y.; Lee, W.J. Short-term wind-power prediction based on wavelet transform-support vector machine and statistic-characteristics analysis. *IEEE Trans. Ind. Appl.* **2012**, *48*, 1136–1141. [CrossRef]

37. Bhaskar, K.; Singh, S. AWNN-assisted wind power forecasting using feedforward neural network. *IEEE Trans. Sustain. Energy* **2012**, *3*, 306–315. [CrossRef]

38. Haque, A.U.; Mandal, P.; Meng, J.; Srivastava, A.K.; Tseng, T.L.; Senjyu, T. A novel hybrid approach based on wavelet transform and fuzzy ARTMAP networks for predicting wind farm power production. *IEEE Trans. Ind. Appl.* **2013**, *49*, 2253–2261. [CrossRef]
39. Liu, D.; Niu, D.; Wang, H.; Fan, L. Short-term wind speed forecasting using wavelet transform and support vector machines optimized by genetic algorithm. *Renew. Energy* **2013**, *62*, 592–597. [CrossRef]
40. Skittides, C.; Früh, W.G. Wind forecasting using principal component analysis. *Renew. Energy* **2014**, *69*, 365–374. [CrossRef]
41. Mandal, P.; Zareipour, H.; Rosehart, W.D. Forecasting aggregated wind power production of multiple wind farms using hybrid wavelet-PSO-NNs. *Int. J. Energy Res.* **2014**, *38*, 1654–1666. [CrossRef]
42. Yeh, W.C.; Yeh, Y.M.; Chang, P.C.; Ke, Y.C. Forecasting wind power in the Mai Liao wind farm based on the multi-layer perceptron artificial neural network model with improved simplified swarm optimization. *Elect. Power Energy Syst.* **2014**, *55*, 741–748. [CrossRef]
43. Chitsaz, H.; Amjady, N.; Zareipour, H. Wind power forecast using wavelet neural network trained by improved clonal selection algorithm. *Energy Conv. Manag.* **2015**, *89*, 588–598. [CrossRef]
44. Osório, G.J.; Matias, J.C.O.; Catalão, J.P.S. Electricity prices forecasting by a hybrid evolutionary-adaptive methodology. *Energy Conv. Mang.* **2014**, *80*, 363–373. [CrossRef]
45. Osório, G.J.; Matias, J.C.O.; Catalão, J.P.S. Short-term wind power forecasting using adaptive neuro-fuzzy inference system combined with evolutionary particle swarm optimization, wavelet transform and mutual information. *Renew. Energy* **2015**, *75*, 301–307. [CrossRef]
46. Eynard, J.; Grieu, S.; Polit, M. Wavelet-based multi-resolution analysis and artificial neural networks, for forecasting temperature and thermal power consumption. *Eng. App. Art. Intell.* **2011**, *24*, 501–516. [CrossRef]
47. Amjady, N.; Keynia, F. Short-term loads forecasting of power systems by combining wavelet transform and neuro-evolutionary algorithm. *Energy* **2009**, *34*, 46–57. [CrossRef]
48. Miranda, V.; Carvalho, L.M.; Rosa, M.A.; Silva, A.M.L.; Singh, C. Improving power system reliability calculation efficiency with EPSO variants. *IEEE Trans. Power Syst.* **2009**, *24*, 1772–1779. [CrossRef]
49. Miranda, V.; Alves, R. Differential evolutionary particle swarm optimization (DEEPSO): A successful hybrid. In Proceedings of the 1st BRICS Congress on Computational Intelligence and 11th Brazilian Congress on Computational Intelligence (BRICS-CCI and CBIC), Recife, Brazil, 8–11 September 2013.
50. Carvalho, L.M.; Loureiro, F.; Sumaili, J.; Keko, H.; Miranda, V.; Marcelino, C.G.; Wanner, E.F. Statistical tuning of DEEPSO soft constraints in the security constrained optimal power flow problem. In Proceedings of the 2015 18th International Conference on Intelligent System Application to Power Systems (ISAP), Porto, Portugal, 11–17 September 2015.
51. Pinto, P.; Carvalho, L.M.; Sumaili, J.; Pinto, M.S.S.; Miranda, V. Coping with wind power uncertainty in unit commitment: A robust approach using the new hybrid metaheuristic DEEPSO. In Proceedings of the Towards Future Power Systems and Emerging Technologies, Powertech Eindhoven, Eindhoven, The Netherlands, 29 June–2 July 2015.
52. Differential Evolutionary Particle Swarm Optimization (DEEPSO). Available online: http://epso.inescporto.pt/deepso/deepso-basics (accessed on 10 February 2016).
53. Portuguese Transmission System Operator—REN. Available online: http://www.centrodeinformacao.ren.pt/ (accessed on 13 June 2016).
54. Electricity Market Operator—OMEL. Available online: http://www.omelholding.es/omel-holding/ (accessed on 10 February 2016).
55. Pennsylvania-New Jersey-Maryland (PJM) Electricity Markets. Available online: http://www.pjm.com (accessed on 20 June 2016).

energies

MDPI

Article

Short-Term Price Forecasting Models Based on Artificial Neural Networks for Intraday Sessions in the Iberian Electricity Market

Claudio Monteiro [1], Ignacio J. Ramirez-Rosado [2], L. Alfredo Fernandez-Jimenez [3,*] and Pedro Conde [1]

[1] Department of Electrical and Computer Engineering, Faculty of Engineering of the University of Porto (FEUP), Porto 4200-465, Portugal; cdm@fe.up.pt (C.M.); ee10153@fe.up.pt (P.C.)
[2] Electrical Engineering Department, University of Zaragoza, Zaragoza 50018, Spain; ijramire@unizar.es
[3] Electrical Engineering Department, University of La Rioja, Logroño 26004, Spain
* Correspondence: luisalfredo.fernandez@unirioja.es; Tel.: +34-941-299-473

Academic Editor: Javier Contreras
Received: 7 June 2016; Accepted: 26 August 2016; Published: 7 September 2016

Abstract: This paper presents novel intraday session models for price forecasts (ISMPF models) for hourly price forecasting in the six intraday sessions of the Iberian electricity market (MIBEL) and the analysis of mean absolute percentage errors (*MAPEs*) obtained with suitable combinations of their input variables in order to find the best ISMPF models. Comparisons of errors from different ISMPF models identified the most important variables for forecasting purposes. Similar analyses were applied to determine the best daily session models for price forecasts (DSMPF models) for the day-ahead price forecasting in the daily session of the MIBEL, considering as input variables extensive hourly time series records of recent prices, power demands and power generations in the previous day, forecasts of demand, wind power generation and weather for the day-ahead, and chronological variables. ISMPF models include the input variables of DSMPF models as well as the daily session prices and prices of preceding intraday sessions. The best ISMPF models achieved lower *MAPEs* for most of the intraday sessions compared to the error of the best DSMPF model; furthermore, such DSMPF error was very close to the lowest limit error for the daily session. The best ISMPF models can be useful for MIBEL agents of the electricity intraday market and the electric energy industry.

Keywords: short-term forecasting; electricity market prices; Iberian electricity market (MIBEL); daily session prices; intraday session prices

1. Introduction

The Iberian electricity market (MIBEL) was created in 2004 as a joint initiative from the governments of Portugal and Spain, involving the integration of their respective electric power systems and their previous electricity markets. The MIBEL allows any consumer in the Iberian region (mainland of Portugal and Spain) to purchase electrical energy under a free competition regime from any producer or retailer acting in that region. It represents a regional electricity market with a remarkable growth of renewable energy production that frequently pushes the most expensive thermal power stations outside the generation scheduling of the wholesale market [1]. The MIBEL consists of the forward markets, managed by the company Iberian Energy Market Operator–Portuguese Division (OMIP) [2], and the daily and intraday markets, both managed by the company Iberian Energy Market Operator–Spanish Division (OMIE) [3]. The daily and intraday markets are organized in a daily session, where next-day sale and electricity purchase transactions are carried out, and in six intraday sessions that consider energy offer and demand, which may arise in the hours following the daily viability schedule fixed after the daily session.

Short-term electricity price forecasting (STEPF) has attracted the attention of many researchers in the last years. The previous knowledge on the forecasted hourly prices that will be settled in the pool constitutes very valuable information for any agent involved in the electricity markets. A considerable amount of research has been dedicated to bidding procedures, trading strategies or electricity market offerings—especially for wind farms [4–6], price-makers [7,8], and wind farms enhanced with storage capability [9]—and even for bidding in micro grids with renewable generation [10]. Consequently, accurate price forecasts are of significant interest for electric power plants. Thus, according to pool price forecasts, mainly electric energy producers and also distribution utilities and large customers from the demand-side, can change their bidding policy in order to obtain the maximum profit. However, pool prices are hard to forecast due to some characteristics such as non-stationary mean and variance, multiple seasonality, calendar effect, high volatility and high percentage of outliers [11].

Additionally, the price forecasting can also influence the consumers' demand response [12–14]. On the one hand, effective demand response is related to demand forecasting (as well as renewable power forecasting) and, on the other hand, it is associated with price forecasting for the consumers as well as price forecasting of the pool market. Such demand response has to consider interactions among prices, consumer demands and renewable power generation. Research works are starting to develop advanced STEPF models and several other short-term forecasting models for other technical magnitudes, in diverse spatio-temporal scales, in order to support complex systems for research on demand response [15–17]. In this sense, suitable STEPF models for intraday sessions are expected that also will play a key role in related research developments.

The immediate application of STEPF models in bidding strategies has propelled the development of this kind of forecasting model. Most of the STEPF models reported are focused on the application to the daily market. The techniques used include those of traditional time series such as auto-regressive integrated moving average (ARIMA) [18–21] or other artificial intelligence-based techniques such as artificial neural networks (ANNs) [19,22–25] and fuzzy inference systems (FIS) [26]. Some authors propose hybrid approaches combining two or more techniques in the forecasting model [27–32]. In general, most of the published articles are focused on the description of the forecasting techniques and its application to the daily market. The analysis of the price explanatory variables used to build STEPF models is barely studied [33], although this analysis has been pointed out as the focus on STEPF models for the next years [34].

Only a few published works deal with the development of STEPF models in applications to the intraday prices of an electricity market [35,36], although intraday prices are of prime importance in day-to-day market operations, in particular for applications in trading of power plant productions [37,38], or for applications in implementing effective demand response as mentioned above [12–17,39]. In [35], a strategic energy bidding for a wind power farm is presented including a very brief description of a STEPF model used for intraday session prices forecasting based on classic time series, but their performance in each of the six sessions of the MIBEL is not indicated. Another research work [36] is focused on maximizing the profit of a wind power producer placed in Holland by using day-ahead and cross-border intraday markets; it seems to utilize a classic seasonal autoregressive integrated moving average (SARIMA) modeling, which is not described, for one-month price forecasts in the intraday German market. The hourly prices settled in intraday sessions in the MIBEL have been studied [40,41]. Their correlations [40] or realized volatility [41] have been highlighted, but no STEPF model for intraday session prices in the MIBEL has been presented in scientific literature describing the best selection of input variables among an extensive set of intraday price explanatory variables, by obtaining mean absolute percentage errors (*MAPEs*) of forecasts for each of the six intraday session prices and also by comparing them with respect to error of the day-ahead price forecast in the MIBEL.

In general, as indicated above, most of the published papers described the forecasting technique: there are advanced versatile techniques with similar accuracy when they are applied to a specific

STEPF case using the same variables and time period. Sometimes authors compare the results obtained with their models with respect to those reported in other works, using the same data and the same period. This paper is not concentrated on forecasting techniques, but it deals with the "forecasting modelling", that is, on the analysis of extensive sets of explanatory variables and their influence in price forecasts. Furthermore, the forecasting modelling of this paper permits to determine input data and structures of processing appropriate for application to the MIBEL.

This paper presents novel STEPF models developed to be applied to the six intraday sessions of the MIBEL, which are called intraday session models for price forecasts (ISMPF model). The best ISMPF model for each session is described in the paper as well as the selected combination of their input variables. The paper also analyses the forecasting errors achieved using different combinations of input (price explanatory) variables in order to determine the best model which uses the proper combination.

The process for analyzing combinations of explanatory variables is initially used for daily session models for price forecasts (DSMPF models) and for reference models for price estimation (RMPE models) in the daily session of the MIBEL. This process is applied to ISMPF models afterwards in the intraday sessions of the MIBEL. The main characteristics of these models are described in the following paragraphs:

- DSMPF models, developed for the day-ahead hourly price forecasting in the daily session of the MIBEL, consider an extensive set of explanatory variables which include recent prices, regional aggregation of power demands and power generations, hourly time series records of power demand forecasts, wind power generation forecasts and weather forecasts as well as chronological information.
- RMPE models, developed for the estimation of the hourly prices in the daily session, use actual power generations and actual power demands of the day-ahead instead of these variables in the previous day, and the same forecast variables, price variables and chronological information of DSMPF models. They allow the calculation of the lowest limit of error values achievable with the utilized explanatory variables.
- ISMPF models, developed for the hourly price forecasting in the six intraday sessions of the MIBEL, consider the input variables included in DSMPF models as well as the hourly prices of the daily session and hourly prices of previous intraday sessions.

The MIBEL was used to test the models of this paper. On one hand, the best ISMPF models achieved very satisfactory *MAPEs* for the intraday sessions of the MIBEL, which were lower errors for most of the intraday sessions than the error of the best DSMPF model. On the other hand, the *MAPE* of the best DSMPF model was very close to the lowest limit error of the best RPME model for the daily session. The best ISMPF model, its performance in the MIBEL and its input variables can constitute valuable information for MIBEL agents of the electricity intraday market and the electric energy industry.

The structure of this paper is the following: Section 2 contains a description of time frameworks for the forecasting models and reference models as well as the characteristics of data corresponding to the MIBEL for hourly price forecast purposes; Section 3 describes RMPE models (reference models) for the hourly prices estimation; Section 4 presents DSMPF models for day-ahead hourly price forecasts; Section 5 describes ISMPF models for hourly price forecasts of the six intraday sessions of the MIBEL; lastly, the conclusions of this paper are presented in Section 6.

2. Time Frameworks and Data Characteristics for Forecasting Models and Reference Models

Time frameworks for day-ahead and intraday MIBEL price forecasting models as well as for reference models are described in Section 2.1. Afterwards, Section 2.2 shows data characteristics corresponding to the MIBEL for the hourly price forecasting.

2.1. Time Frameworks

The description of time frameworks corresponding to DSMPF models and reference models is presented in Section 2.1.1; Section 2.1.2 describes a time framework for ISMPF models.

2.1.1. Time Frameworks for Daily Session Models for Price Forecasts and Reference Models

Bidding offers to the day-head electricity market and the implementation of other power system operation functions are mainly prepared based on short-term forecasting models that provide forecasted hourly prices of the day-ahead.

DSMPF models use as input variables (price explanatory variables) recorded time series of hourly prices in previous days, regional-aggregated hourly power demands and hourly power generations of most of the types of electricity production in the previous day, forecasts of demand, wind power generation and weather (hourly wind speed, temperature and irradiation) for the day-ahead in the region, and chronological variables.

The time framework of DSMPF models is shown in Figure 1. The price forecast $\hat{p}^d_{D+1,h|D,t}$ is obtained at hour t of the day D for each hour h of the 24 h in day $D + 1$. The delivery of the price forecast is assumed in hour t of day D which can be any instant prior to the opening of the daily market session and after the moment in which the forecasted variables corresponding to demand and wind power generation for the day $D + 1$ are known.

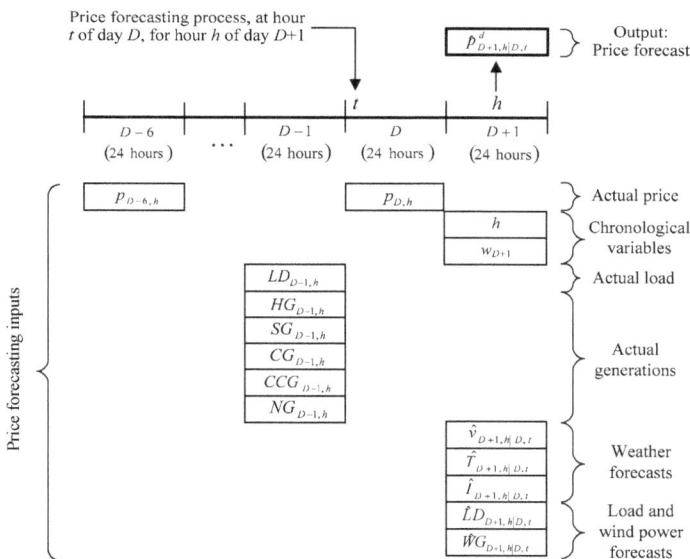

Figure 1. Time framework of daily session models for price forecasts (DSMPF models).

The price for hour h of the day D, $p_{D,h}$, and the price $p_{D-6,h}$ for hour h of day $D - 6$, are inputs to forecast the price for the hour h of day $D + 1$. Other inputs are the week day w_{D+1} and hour h of day $D + 1$, and the weather forecasts obtained at the first hours of day D for the geographical region corresponding to the electricity market and for hour h of day $D + 1$, that is, the regional weighted forecasted hourly wind speeds $\hat{v}_{D+1,h|D,t}$, regional weighted forecasted hourly temperatures $\hat{T}_{D+1,h|D,t}$ and regional weighted forecasted hourly irradiations $\hat{I}_{D+1,h|D,t}$. These last inputs are similar to those used by authors in [33].

Diverse input variables were included in Figure 1: power demand $LD_{D-1,h}$, hydropower generation $HG_{D-1,h}$, solar power generation and power cogeneration $SG_{D-1,h}$, coal power generation

$CG_{D-1,h}$, combined cycle power generation $CCG_{D-1,h}$, and nuclear power generation $NG_{D-1,h}$ at hour h of day $D - 1$. Two additional input variables of forecasting available before the opening of the daily session in day D were considered: power demand forecast $\hat{LD}_{D+1,h|D,t}$ and wind power generation forecast $\hat{WG}_{D+1,h|D,t}$ for hour h of day $D + 1$.

Figure 2 illustrates the time framework of the RMPE models. A part of the input variables of the RMPE models, mainly actual power generations and power demands of day $D + 1$, as well as the output variable price estimation $\tilde{p}_{D+1,h}$, are different from the input variables of DSMPF models. Note that DSMPF models mainly use actual power generations and power demands of day $D - 1$ as input variables, and the output variable corresponds to the price forecast $\hat{p}^d_{D+1,h|D,t}$. Thus, RMPE models are not forecasting models, but models for hourly price estimation.

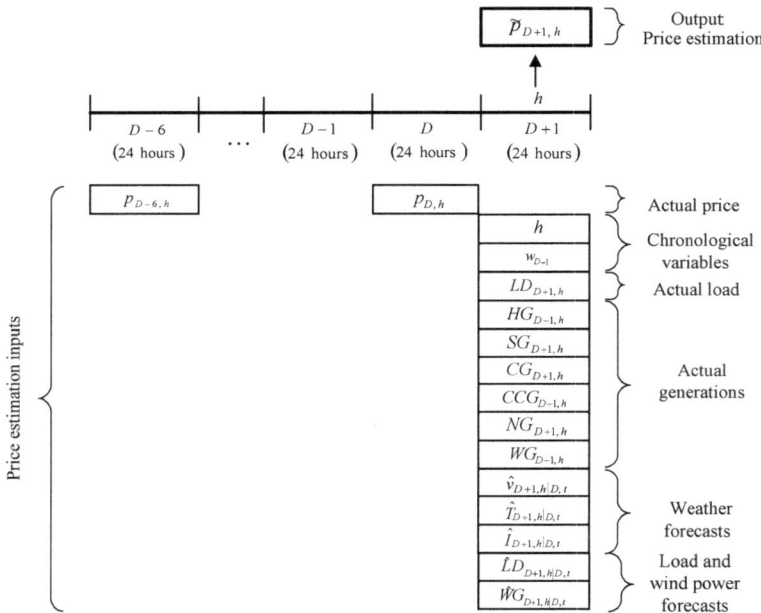

Figure 2. Time framework of reference models for price estimation (RMPE models).

2.1.2. Time Framework of Intraday Session Models for Price Forecasts

The MIBEL market is organized in a daily session whose closing time takes place at 12:00 a.m. (Spanish official hour) of day D and six intraday sessions whose structure is given in Table 1. As it is shown, the first intraday session covers the last 3 h of the current day (D) and the 24 h of the following day ($D + 1$), that is, a total of 27 h. The other sessions comprise only a shorter time period of day $D + 1$. The time period covered by each session is reduced session after session with a minimum of 9 h in the sixth intraday session.

The time framework for the ISMPF models is illustrated in Figure 3. The input variables of these ISMPF models for a given intraday session can include the hourly prices of previous intraday sessions and the hourly prices of the daily session of the MIBEL, as well as the set of input variables used by DSMPF models. Figure 3 shows in orange the periods in which the forecast can be carried out, from the moment when the prices of the previous session are known (around 45 min after its closing hour) to the closing hour of the corresponding intraday session. Figure 3 also partially shows in blue the period covered by each market session.

Table 1. Structure of the Iberian electricity market's (MIBEL's) intraday sessions.

Session Number	Session Opening Hour (Spanish Hour)	Session Closing Hour (Spanish Hour)	Time Period	Hours in Time Period
1	17:00	18:45	21 (D)–23 $(D + 1)$	27
2	21:00	21:45	00 $(D + 1)$–23 $(D + 1)$	24
3	01:00	01:45	04 $(D + 1)$–23 $(D + 1)$	20
4	04:00	04:45	07 $(D + 1)$–23 $(D + 1)$	17
5	08:00	08:45	11 $(D + 1)$–23 $(D + 1)$	13
6	12:00	12:45	15 $(D + 1)$–23 $(D + 1)$	9

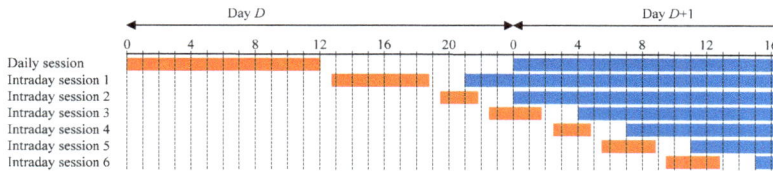

Figure 3. Time framework for outputs of intraday session models for price forecasts (ISMPF models).

2.2. Data Characteristics

For the development of the proposed hourly price forecasting models different kinds of variables have been considered, which are the following:

a   Actual hourly data prices for the day-ahead and intraday markets available from the market operator OMIE [3].

b   Actual hourly data of the power system: load demand, wind power generation, hydropower generation, cogeneration and solar power generation, nuclear power generation, coal power generation, combined cycle power generation and power exchanged with France. These data were obtained by aggregating a very large amount of information from the websites of Redes Energéticas Nacionais (REN), the Portuguese transmission system operator (TSO) [42] and Red Eléctrica de España (REE), the Spanish TSO [43].

c   Hourly weather forecasts: weighted average wind speed, solar irradiance and temperature. These forecasted values were obtained with the numerical weather prediction (NWP) mesoscale model WRF NMM [44], initialized with the forecasts provided by the global NWP model GFS [45].

d   Hourly variable forecasts of the power system: power demand forecasts and wind power generation forecasts. These forecasts were obtained by aggregating forecast information from the mentioned TSOs.

e   Chronological variables (hour, week day).

The data recorded, corresponding to years 2012 and 2013, were divided into an in-sample data set used for training and an out-sample data set used for testing DSMPF, RMPE and ISMPF models. The out-sample data set was composed of complete weeks extracted along the two years of data in order to have a good representation of the different price behaviours along the year. The in-sample and out-sample data sets were defined as follows:

i   In-sample data set: all the hours of the days in 2012 and 2013, except those included in the out-sample data set, totalizing 14,184 cases (h).

ii   Out-sample data set: all the hours of the weeks with numbers 5, 10, 15, 20, 25, 30, 35, 40, 45, 50 in 2012, and weeks number 2, 7, 12, 17, 22, 27, 32, 37, 42, 47 in 2013; a total of 3360 cases (h).

The descriptive statistics of the price variable of the data sets used for each forecasting model are shown in Table 2, including their mean value, standard deviation, maximum values (minimum values are 0), and the total of cases (count of hours). For the total cases of intraday session 1, we have only considered the 24 h of the following day instead of the 27 h covered by this session.

Table 2. Descriptive statistics of price data.

Dataset	Statistic	Daily	Intraday 1	Intraday 2	Intraday 3	Intraday 4	Intraday 5	Intraday 6
IN-SAMPLE	Mean (€/MWh)	45.02	44.73	44.75	45.69	47.30	48.08	47.90
	Standard Deviation (€/MWh)	16.44	16.19	16.54	16.82	16.60	17.08	18.11
	Maximum (€/MWh)	110.00	132.22	180.30	180.30	131.57	180.30	180.30
	Count (h)	14,184	14,184	14,184	11,820	10,047	7683	5319
OUT-SAMPLE	Mean (€/MWh)	45.88	45.80	46.93	46.91	48.48	49.10	49.03
	Standard Deviation (€/MWh)	15.37	14.69	14.03	14.59	14.37	14.94	16.54
	Maximum (€/MWh)	86.01	129.65	180.30	119.25	129.87	137.23	148.20
	Count (h)	3360	3360	3360	2800	2380	1820	1260

Figures 4 and 5, for years 2012 and 2013 respectively, represent the hourly average values of power generation for each power generation type, power demand and price in the MIBEL. The hourly average power demand was quite similar in these years, although there were slight changes in the generation-mix production. In 2013, the renewable power generation (hydro, solar and wind power productions) was higher than in year 2012. For thermal power generation (combined cycles, coal and nuclear power productions), the electricity generation from nuclear power production was almost the same in both years, but there was a reduction in the other two productions, with a more significant reduction in the combined cycles, caused by a difference between coal and natural gas prices favourable to coal prices in 2013. In 2013, mainly due to a higher renewable proportion of power generation, prices decreased an average of around 4.4 €/MWh.

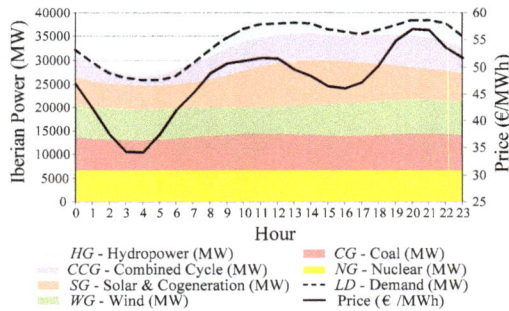

Figure 4. Hourly average power generations, power demand and price in year 2012.

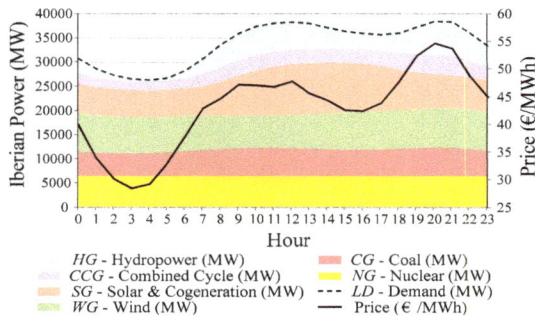

Figure 5. Hourly average power generations, power demand and price in year 2013.

3. Reference Models for Price Estimation

RMPE models are hourly price estimation models that utilize the variables shown in Table 3 as inputs. These input variables include the chronological variables "hour" and "week day" (variables V1 and V2), the prices on previous days at the same hour h (variables V3 and V4), the actual power demand (load) and forecasted power demand variables of the power system for hour h of day $D + 1$ (variables V5R and V6R), the actual and forecasted wind power generation variables for hour h of day $D + 1$ (variables V7R and V8R), the weather forecasts of wind speed, temperature and irradiance for hour h of day $D + 1$ (variables V9R to V11R) and the actual power generations corresponding to hour h of day $D + 1$ (variables V12R to V16R).

Table 3. Variables of reference models for price estimation (RMPE models).

Variable	Description	Range
V1	Hour	0–23
V2	Week day	0 (Monday)–7(special day)
V3	Hourly price D	0–110 €/MWh
V4	Hourly price $D - 6$	0–110 €/MWh
V5R	Hourly power demand $D + 1$	20,338–52,853 MW
V6R	Forecasted hourly power demand $D + 1$	18,200–51,839 MW
V7R	Hourly wind power generation $D + 1$	161–20,198 MW
V8R	Forecasted hourly wind power generation $D + 1$	427–19,688 MW
V9R	Forecasted hourly temperature $D + 1$	−0.8–35.8 °C
V10R	Forecasted hourly wind speed $D + 1$	1.48–11.15 m/s
V11R	Forecasted hourly irradiance $D + 1$	0–1031.8 W/m^2
V12R	Hourly hydropower generation $D + 1$	−3957–15,384 MW
V13R	Hourly cogeneration and solar power generation $D + 1$	3824–13,668 MW
V14R	Hourly coal power generation $D + 1$	615–11,604 MW
V15R	Hourly nuclear power generation $D + 1$	3391–7525 MW
V16R	Hourly combined cycled power generation $D + 1$	336–15,172 MW

We did not explore the relative importance of the different price explanatory variables considered in this paper since it was extensively presented in a previous publication [33]. Instead, in the present article, a set of studies of suitable combinations of input variables for the different type of models (DSMPF, RMPE and ISMPF models) are being described in order to determine the best DSMPF model, the best RMPE model and the best ISMPF model for each intraday session in the MIBEL.

RMPE models contain the variables of the REMPE model introduced by the authors in [33], but now two additional variables are considered: the forecasted hourly power demand for day $D + 1$ (variable V6R) and the forecasted hourly wind power generation for day $D + 1$ (variable V8R). Then, we formulated the question: "How relevant are these forecasted variables (V6R and V8R) compared to the variables of the actual demand $D + 1$ and the actual wind generation $D + 1$ (V5R and V7R)?". In order to answer this question, four RMPE models (model REF1 to model REF4) were built, shown in Table 4. In this table, PG means power generation. Please observe that model REF1 is the REMPE model.

RMPE models were implemented with a multilayer perceptron neural network (MLP) [46], using one hidden layer with $2n + 1$ neurons, where n is the number of input variables (explanatory variables). These models were trained and tested with the in-sample and out-sample data sets previously described in Section 2, which were utilized in all computing experiences presented in this paper. Since we used random weight initiation in these neural networks, different training of the same MLP resulted in slightly different computer results (outputs). In order to avoid this inconvenience, we used, as a final forecasting result, the ensemble averaging [47] of the outputs of 20 training processes of the same MLP, thus achieving a more stable response and a lower error.

Table 4. RMPE models and their mean absolute percentage errors (*MAPEs*).

Explanatory Variables		Description	REF1	REF2	REF3	REF4
Chronological	V1	Hour (0–23 h)	●	●	●	●
	V2	Week day (1–7)	●	●	●	●
Price	V3	Hourly price (D)	●	●	●	●
	V4	Hourly price ($D - 6$)	●	●	●	●
Demand	V5R	Hourly power demand $D + 1$	●	-	●	●
	V6R	Forecasted hourly power demand $D + 1$	-	●	-	●
Wind PG	V7R	Hourly wind power generation $D + 1$	●	●	-	●
	V8R	Forecasted hourly wind power generation $D + 1$	-	-	●	●
Weather	V9R	Forecasted hourly temperature $D + 1$	●	●	●	●
	V10R	Forecasted hourly wind speed $D + 1$	●	●	●	●
	V11R	Forecasted hourly irradiance $D + 1$	●	●	●	●
Other PG	V12R	Hourly hydropower generation $D + 1$	●	●	●	●
	V13R	Hourly cogeneration and solar power generation $D + 1$	●	●	●	●
	V14R	Hourly coal power generation $D + 1$	●	●	●	●
	V15R	Hourly nuclear power generation $D + 1$	●	●	●	●
	V16R	Hourly combined cycled power generation $D + 1$	●	●	●	●
MAPE (%)			10.23	9.89	10.43	9.88

An error analysis for RMPE models using the *MAPE* was carried out in the price estimations corresponding to the out-sample data set, where the *MAPE* is defined by Equation (1):

$$MAPE = \frac{1}{N} \sum_{T=1}^{N} \frac{|P_{\text{real\_}T} - P_{\text{estimation\_}T}|}{P_{\text{real\_M}}} 100 \tag{1}$$

where $P_{\text{real\_}T}$ is the real hourly price value, $P_{\text{estimation\_}T}$ is the estimation of the hourly price value obtained from each RMPE model, N is the number of elements of the out-sample data set and $P_{\text{real\_M}}$ is the mean real hourly price value corresponding to that data set.

Variable V6R (forecasted power demand) is used in model REF2 instead of variable V5R (actual power demand) of model REF1. Thus, the *MAPEs* of Table 4 indicate that the forecasted power demand (for day $D + 1$) explains electricity prices better than the actual power demand (of day $D + 1$).

We repeated the experience using variable V8R (forecasted wind power generation) in model REF3 instead of variable V7R (actual wind power generation) in model REF1. Then, the *MAPEs* of Table 4 indicate that the actual wind power generation (of day $D + 1$) explains electricity prices better than the forecasted wind power generation (of day $D + 1$). Furthermore, model REF4 including all variables leads to a *MAPE* value almost equal to that of model REF2.

Therefore, RMPE models can achieve the best *MAPE* value of approximately 9.9%. It represents the lowest error ("minimum error") using the considered explanatory variables, that is, the lowest limit of the possible performance of any model for price estimation or for price forecast belonging to the same class of models with a similar kind of variables to those used in RMPE models.

4. Daily Session Models for Price Forecasts

The inputs of DSMPF models are:

a   Chronological variables ("hour" and "week day");
b   Hourly prices of days D and $D - 6$;
c   Recorded hourly power demand and hourly power generations of days $D - 1$;
d   Hourly power demand forecasts and hourly wind power generation forecast for day $D + 1$; and
e   Hourly weather forecasts of wind speed, temperature and irradiance for day $D + 1$.

Then, DSMPF models take into consideration the sets of input variables shown in Table 5. Obviously, variables of DSMPF models V6, V7, V8, V9 and V10 correspond to variables V6R, V9R, V10R, V11R and V8R used in RMPE models (Table 3).

Table 5. Variables of daily session models for price forecasts (DSMPF models).

Variable	Description	Range
V1	Hour	0–23
V2	Week day	0 (Monday)–7 (special day)
V3	Hourly price D	0–110 €/MWh
V4	Hourly price $D - 6$	0–110 €/MWh
V5	Hourly power demand $D - 1$	20,338–52,853 MW
V6	Forecasted hourly power demand $D + 1$	18,200–51,839 MW
V7	Forecasted hourly temperature $D + 1$	−0.8–35.8 °C
V8	Forecasted hourly wind speed $D + 1$	1.48–11.15 m/s
V9	Forecasted hourly irradiance $D + 1$	0–1031.8 W/m^2
V10	Forecasted hourly wind power generation $D + 1$	427–19,688 MW
V11	Hourly hydropower generation $D - 1$	−3957–15,384 MW
V12	Hourly cogeneration and solar power generation $D - 1$	3824–13,668 MW
V13	Hourly coal power generation $D - 1$	615–11,604 MW
V14	Hourly nuclear power generation $D - 1$	3391–7525 MW
V15	Hourly combined cycled power generation $D - 1$	336–15,172 MW

DSMPF models were implemented with MLPs with the same structure used for the RMPE models, that is, one hidden layer with $2n + 1$ neurons, where n is the number of input explanatory variables. For the training and testing of the MLP, in-sample and out-sample data sets previously described in Section 2 were used again, as well as the abovementioned ensemble technique for the corresponding computer results.

In a similar way than that followed for RMPE models, the *MAPE* was calculated for the price forecasts corresponding to the out-sample data set for DSMPF models. In this case, the *MAPE* is defined by Equation (2):

$$MAPE = \frac{1}{N} \sum_{T=1}^{N} \frac{|P_{real\_T} - P_{forecast\_T}|}{P_{real\_M}} 100 \tag{2}$$

where P_{real_T} is the real hourly price value, $P_{forecast\_T}$ is the forecasted hourly price value of the forecasting model, N is the number of elements in the out-sample data set, and P_{real_M} is the mean real hourly price value corresponding to that data set.

In the following paragraphs, a summary of variable selection studies for DSMPF models is shown corresponding to reasonable combinations of variables (grouped by their common characteristics) in order to look for the best *MAPE*, that is, the best DSMPF model.

The *MAPE*s for DSMPF models (M1 to M18), with different input variables, are presented in Table 6. The selection of variables follows an ordered analysis, such that only some DSMPF models are presented in the table for conclusive purposes. The construction of Table 6 corresponds to a selection process with the following sequence:

i   Models M1 to M3 for price variables selection;
ii   Models M4 and M5 for selection of power demand and forecasted power demand variables;
iii Models M6 to M11 for selection of forecasted weather and wind generation variables; and
iv   Models M12 to M18 for power generation variables selection.

Model M1 is a simple baseline model with a *MAPE* value of 20.83%, slightly higher than the double of 9.9% achieved by the best RMPE model. As mentioned in the previous section, 9.9% is the lowest limit error value that RMPE models or DSMPF models could obtain.

Table 6. DSMPF models and their *MAPEs*.

Explanatory Variables		M1	M2	M3	M4	M5	M6	M7	M8	M9	M10	M11	M12	M13	M14	M15	M16	M17	M18
Chronological	V1	•	•	•	•	•	•	•	•	•	•	•	•	•	•	•	•	•	•
	V2	•	•	•	•	•	•	•	•	•	•	•	•	•	•	•	•	•	•
Price	V3	-	•	•	•	•	•	•	•	•	•	•	•	•	•	•	•	•	•
	V4	-	-	•	•	•	•	•	•	•	•	•	•	•	•	•	•	•	•
Demand	V5	-	-	-	•	•	•	•	•	•	•	•	•	•	•	•	•	•	•
	V6	-	-	-	-	•	•	•	•	•	•	•	•	•	•	•	•	•	•
Weather	V7	-	-	-	-	-	•	-	-	-	-	•	•	•	•	•	•	•	•
	V8	-	-	-	-	-	-	•	-	-	•	-	-	-	-	-	-	-	-
	V9	-	-	-	-	-	-	-	•	-	-	-	-	-	-	-	-	-	-
Power generation	V10	-	-	-	-	-	-	-	-	•	•	•	•	•	•	•	•	•	•
	V11	-	-	-	-	-	-	-	-	-	-	-	•	-	-	-	-	•	•
	V12	-	-	-	-	-	-	-	-	-	-	-	-	•	-	-	-	•	-
	V13	-	-	-	-	-	-	-	-	-	-	-	-	-	•	-	-	-	•
	V14	-	-	-	-	-	-	-	-	-	-	-	-	-	-	•	-	-	•
	V15	-	-	-	-	-	-	-	-	-	-	-	-	-	-	-	•	-	•
MAPE (%)		20.83	17.00	16.81	16.70	16.33	16.37	12.85	16.40	11.64	11.86	11.58	10.86	11.41	11.54	11.43	11.54	11.00	10.69

If variable V3 (hourly price D) is added to those used by model M1 leading to model M2, then the *MAPE* decreases to 17%. The inclusion of variable V5 (hourly power demand $D − 1$) in model M4 slightly reduces the *MAPE* to 16.70%. Additionally, if variable V6 (hourly forecasted power demand $D + 1$) is included (model M5), the *MAPE* is reduced to 16.33%.

Models M6–M9, compared to model M5, show an added price explicability by including variables V7–V10 (forecasted weather variables and forecasted wind power generation variable). The forecasted wind power generation variable (V10) in model M9 provides a better performance (*MAPE* of 11.64%). Alternatively, the forecasted wind speed variable (V8) in model M7 obtains a *MAPE* of 12.85%.

The two variables V8 and V10 have collinear information, leading to an error of 11.86% (in model M10), which is worse than error 11.64% considering only variable V10 in model M9. However, the use of variable V10 together with variable V7 (forecasted temperature $D + 1$) improves the performance to a better *MAPE* value (11.58%) in model M11.

Models M12 to M16 allow the evaluation of the improvement in the *MAPE* by adding variables V11 to V15 (power generation variables $D − 1$). Variable V11 (hydropower generation $D − 1$) is the one that achieves a lower error, 10.86% in model M12. The inclusion of variables V13 to V15 (thermal power generation $D − 1$) to the previous model M12 results in the best performance of DSMPF models (model M18), reaching a *MAPE* value of 10.69%, which is very satisfactory in comparison to the aforementioned lowest limit error value of 9.9% of RMPE models.

Figure 6 shows an example of the hourly evolution of actual price values and forecasts of model M18 for week 7 of year 2013.

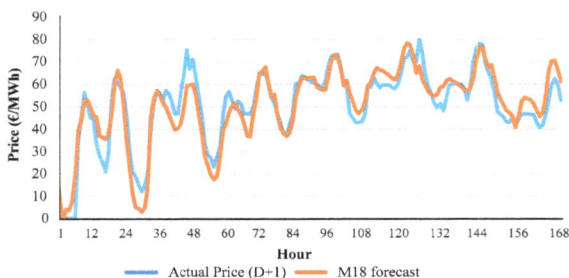

Figure 6. Actual price values and forecast values of model M18 for week 7 of year 2013.

5. Intraday Session Models for Price Forecasts

As indicated in the "Introduction" section, this paper is focused on the ISMPF models corresponding to the six intraday sessions of the MIBEL.

ISMPF models of each intraday session are short-term hourly price forecasting models that can utilize hourly prices on previous days D and $D - 6$, price values of the daily session and price values of previous intraday sessions of the MIBEL. They also include recorded explanatory variables mainly corresponding to days $D - 1$ and weather forecasts of day $D + 1$, as well as power demand forecasts and wind power generation forecasts for day $D + 1$ in order to forecast the electricity price values of the intraday sessions of the MIBEL.

Thus, six types of explanatory variables were considered in ISMPF models for a given intraday session:

- Chronological variables ("hour" and "week day");
- Hourly prices of days D and $D - 6$;
- Hourly prices of the daily session $D + 1$ and hourly prices of previous intraday sessions of the MIBEL;
- Hourly power demand and hourly power generations of days $D - 1$;
- Hourly power demand forecasts and hourly wind power generation forecasts for day $D + 1$;
- Hourly weather forecasts of wind speed, temperature and irradiance for day $D + 1$.

Then, ISMPF models consider the sets of input variables shown in Table 7. Obviously, variables V6I to V16I of ISMPF models correspond to variables V5 to V15 used in DSMPF models (Table 5).

ISMPF models were implemented with MLPs with the same structure used for DSMPF models, that is, one hidden layer with $2n + 1$ neurons, where n is the number of input explanatory variables. For the training and testing of the MLP, in-sample and out-sample data sets previously described in Section 2 were used again as well as the ensemble technique for the corresponding computer results.

In a similar way to that used for DSMPF models, the *MAPE* was calculated for the price forecasts corresponding to the out-sample data set for DSMPF models. In this case, the *MAPE* for intraday session k is defined by Equation (3):

$$MAPE_k = \frac{1}{N_k} \sum_{T=1}^{N_k} \frac{|P_{k \text{ real}\_T} - P_{k \text{ forecast}\_T}|}{P_{\text{real\_MN}k}} 100 \tag{3}$$

where $P_{k\text{real}\_T}$ is the real hourly price value, $P_{k\text{forecast}\_T}$ is the forecasted hourly price value of the forecasting model of intraday session k, N_k is the number of elements (h) in the out-sample data set and $P_{\text{real\_MN}k}$ is the mean real hourly price value corresponding to that data set. The values for N_k were 3360 h for intraday sessions 1 and 2, 2800 h for intraday session 3, 2380 h for intraday session 4, 1820 h for intraday session 5, and 1260 h for intraday session 6.

Table 7. Variables of ISMPF models.

Variable	Description	Range
V1	Hour	0–23
V2	Week day	0 (Monday)–7 (special day)
V3	Hourly price D	0–110 €/MWh
V4	Hourly price $D - 6$	0–110 €/MWh
V5I	Hourly price $D + 1$ of daily session	0–110 €/MWh
V6I	Hourly power demand $D - 1$	20,338–52,853 MW
V7I	Forecasted hourly power demand $D + 1$	18,200–51,839 MW
V8I	Forecasted hourly temperature $D + 1$	−0.8–35.8 °C
V9I	Forecasted hourly wind speed $D + 1$	1.48–11.15 m/s
V10I	Forecasted hourly irradiance $D + 1$	0–1031.8 W/m^2
V11I	Forecasted hourly wind power generation $D + 1$	427–19,688 MW
V12I	Hourly hydropower generation $D - 1$	−3957–15,384 MW
V13I	Hourly cogeneration and solar power generation $D - 1$	3824–13,668 MW
V14I	Hourly coal power generation $D - 1$	615–11,604 MW
V15I	Hourly nuclear power generation $D - 1$	3391–7525 MW
V16I	Hourly combined cycled power generation $D - 1$	336–15,172 MW
V17I	Hourly price $D + 1$ from intraday session 1	0–132.22 €/MWh
V18I	Hourly price $D + 1$ from intraday session 2	0–80.30 €/MWh
V19I	Hourly price $D + 1$ from intraday session 3	0–180.30 €/MWh
V20I	Hourly price $D + 1$ from intraday session 4	0–131.57 €/MWh
V21I	Hourly price $D + 1$ from intraday session 5	0–180.30 €/MWh

A summary of variable selection studies of ISMPF models for each intraday session will be presented in Sections 5.1–5.6, corresponding to reasonable combinations of explanatory variables (grouped by their common characteristics) that allow to achieve the best ISMPF model of such intraday session.

Following a similar process to that used to select suitable variables for DSMPF models (in previous Section 4), a significant number of combinations of variables were analysed for the six intraday sessions, but only ISMPF models that led to relevant conclusions are going to be presented.

The procedure of analysis was based on a sequential integration of a different kind of variables by the following order of importance:

i   Chronological variables V1 and V2 (hour and week day);
ii Price variables, including price $D - 1$, price $D - 6$, price $D + 1$ of daily session, and price $D + 1$ of previous intraday sessions;
iii Power demand $D - 1$ and forecasted power demand $D + 1$;
iv Forecasted weather $D + 1$ and forecasted wind power generation $D + 1$;
v   Power generations $D - 1$.

An ISMPF model that includes the price $D + 1$ of the daily session (clearing hourly price for the day ahead $D + 1$) and the price $D + 1$ of previous intraday sessions (clearing hourly price $D + 1$ from previous intraday sessions) can be used immediately after the clearing market for the sessions whose prices are included as inputs in the ISMPF model. Simpler ISMPF models that do not include these variables can be utilized at the first hours of day D in a similar way to DSMPF models presented in Section 4.

Descriptions of ISMPF models for each intraday session are presented in the next paragraphs.

5.1. Intraday Market Session 1

The baseline model S1M1 in Table 8 that uses only chronological information (variables V1 and V2) has a *MAPE* close to 20%. This order of magnitude of the error was similar for most of the intraday sessions (models S1M1, S2M1, S3M1, S4M1 and S5M1) and it was slightly lower than the error of model M1 (Table 6) for the daily session. Notice that Tables 9–13 give the *MAPEs* of baseline models S2M1, S3M1, S4M1, S5M1 and S6M1 for intraday sessions 2–6.

Table 8. ISMPF models for intraday session 1 and their *MAPEs*.

Explanatory Variables		Description	S1M1	S1M2	S1M3	S1M4	S1M5	S1M6	S1M7	S1M8	S1M9
Chronological	V1	Hour (0–23 h)	●	●	●	●	●	●	●	●	●
	V2	Week day (1–7)	●	●	●	●	●	●	●	●	●
Price	V3	Hourly price (D)	-	●	-	-	●	●	-	-	-
	V4	Hourly price ($D - 6$)	-	-	●	-	-	●	-	-	-
	V5I	Hourly price ($D + 1$) from daily session	-	-	-	●	●	●	●	●	●
Demand	V6I	Hourly power demand ($D - 1$)	-	-	-	-	-	-	●	-	-
	V7I	Forecasted hourly power demand ($D + 1$)	-	-	-	-	-	-	●	-	-
Weather	V8I	Forecasted hourly temperature ($D + 1$)	-	-	-	-	-	-	-	●	-
	V9I	Forecasted hourly wind speed ($D + 1$)	-	-	-	-	-	-	-	●	-
	V10I	Forecasted hourly irradiance ($D + 1$)	-	-	-	-	-	-	-	●	-
Power generation	V11I	Forecasted hourly wind power generation ($D + 1$)	-	-	-	-	-	-	-	●	-
	V12I	Hourly hydropower generation ($D - 1$)	-	-	-	-	-	-	-	-	●
	V13I	Hourly cogeneration and solar power generation ($D - 1$)	-	-	-	-	-	-	-	-	●
	V14I	Hourly coal power generation ($D - 1$)	-	-	-	-	-	-	-	-	●
	V15I	Hourly nuclear power generation ($D - 1$)	-	-	-	-	-	-	-	-	●
	V16I	Hourly combined cycled power generation ($D - 1$)	-	-	-	-	-	-	-	-	●
		MAPE (%)	19.95	16.56	19.64	7.49	7.84	7.51	7.59	7.49	7.48

In order to consider the price variables (variables V3, V4 and V5I), the performances of models S1M2, S1M3 and S1M4 of Table 8 were evaluated. Comparing model S1M3 (error of 19.64%) and model

S1M1 (error of 19.95%), we can observe that variable V4, price $D - 6$, led to a small improvement. The integration of variable V3, price D, in model S1M2 has a more significant improvement (error of 16.56%). If prices $D + 1$ of the daily session are available, then this variable V5I can be used in ISMPF models. By including this variable V5I, model S1M4 achieved a *MAPE* of 7.49%. Therefore, with a relatively simple model (with only three variables V1, V2 and V5I), it is possible to obtain price forecasts of the intraday session 1 with a relatively low error value. The error for the intraday session 1 (7.49%) was clearly lower than the error for the daily session (10.69%), obtained with the best DSMPF model (model M18) of Table 6. Models S1M5 and S1M6 allowed to test combinations of price explanatory variables; in both models, the performances were worse than that of model S1M4, since model S1M5 obtained an error of 7.84% and model S1M6 an error of 7.51%, whereas model S1M4 achieved an error of 7.49% (using only variable V5I).

The integration of power demand variables was tested for different combinations of variables V6I and V7I, with no improvement in performances with respect to model S1M4. Model S1M7 is one of these tested models; it uses these variables V6I and V7I combined with the variables used in model S1M4; and model S1M7 reached a worse result (error of 7.59%) than that of model S1M4 (error of 7.49%).

The consideration of forecasted weather variables (variables V8I, V9I and V10I) and also the forecasted hourly wind power generation (variable V11I), all for day $D + 1$, were also studied (in model S1M8) with no improvement of the error with respect to model S1M4. Finally, using the variables of model S1M4, the integration of variables V12I to V16I (power generations $D - 1$) was carried out in model S1M9, which obtained a satisfactory performance with an error of 7.48%, almost equal to that of the simpler model S1M4. Thus, this simpler model S1M4 was preferred.

Figure 7 shows an example of the hourly evolution of actual price values and forecast values of model S1M4 for week 7 of year 2013.

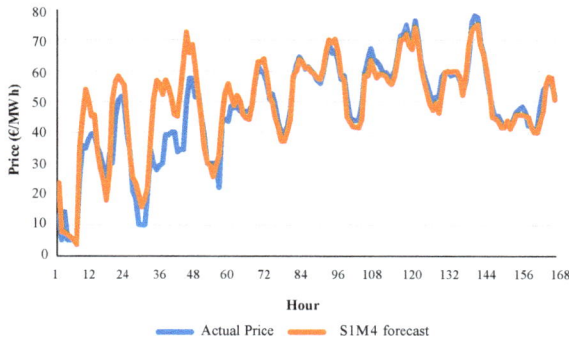

Figure 7. Actual price values and forecast values of model S1M4 of intraday session 1 for week 7 of year 2013.

5.2. Intraday Market Session 2

Let consider the possibility that variable V17I (the price $D + 1$ of intraday session 1 of the MIBEL) is available for ISMPF models for intraday market session 2. Models S2M2 to S2M5 of Table 9 tested the integration of each individual price variable (variables V3, V4, V5I and V17I) within baseline model S2M1. Thus, variable V3 (hourly price D) in model S2M2 (error of 16.71%) is clearly preferred with respect to variable V4 (hourly price $D - 6$) in model S2M3 (error of 19.61%). By integrating variables V5I (hourly price $D + 1$ of daily session) or V17I (hourly price $D + 1$ of intraday session 1) in model S2M1, model S2M4 led to an error of 9.84% and model S2M5 to 9.95%. Combining the two variables V5I and V17I with the variables of model S2M1, model S2M6 achieved the best *MAPE* of 8.85% among ISMPF models of intraday session 2. Please note that the performance of this model S2M6 (error of 8.85%) was higher than the best model S1M4 of intraday market session 1 (error of 7.49%).

Table 9. ISMPF models for intraday session 2 and their *MAPEs*.

Explanatory Variables		Description	S2M1	S2M2	S2M3	S2M4	S2M5	S2M6	S2M7	S2M8	S2M9
Chronological	V1	Hour (0–23 h)	●	●	●	●	●	●	●	●	●
	V2	Week day (1–7)	●	●	●	●	●	●	●	●	●
Price	V3	Hourly price (D)	-	●	-	-	-	-	-	-	-
	V4	Hourly price ($D - 6$)	-	-	●	-	-	-	-	-	-
	V5I	Hourly price ($D + 1$) from daily session	-	-	-	●	-	●	●	●	●
	V17I	Hourly price ($D + 1$) from intraday session 1	-	-	-	-	●	●	●	●	●
Demand	V6I	Hourly power demand ($D - 1$)	-	-	-	-	-	-	●	-	-
	V7I	Forecasted hourly power demand ($D + 1$)	-	-	-	-	-	-	●	-	-
Weather	V8I	Forecasted hourly temperature ($D + 1$)	-	-	-	-	-	-	-	●	-
	V9I	Forecasted hourly wind speed ($D + 1$)	-	-	-	-	-	-	-	●	-
	V10I	Forecasted hourly irradiance ($D + 1$)	-	-	-	-	-	-	-	●	-
	V11I	Forecasted hourly wind power gen. ($D + 1$)	-	-	-	-	-	-	-	●	-
Power generation	V12I	Hourly hydropower generation ($D - 1$)	-	-	-	-	-	-	-	-	●
	V13I	Hourly cogeneration and solar power generation ($D - 1$)	-	-	-	-	-	-	-	-	●
	V14I	Hourly coal power generation ($D - 1$)	-	-	-	-	-	-	-	-	●
	V15I	Hourly nuclear power generation ($D - 1$)	-	-	-	-	-	-	-	-	●
	V16I	Hourly combined cycled power generation ($D - 1$)	-	-	-	-	-	-	-	-	●
	MAPE (%)		19.84	16.71	19.61	9.84	9.95	8.85	8.90	9.03	9.16

The variable sets related to power demand (V6I and V7I) were tested following a procedure similar to that used for ISMPF models for intraday session 1; an error of 8.90% was obtained with model S2M7. Variables V8I to V11I (forecasted hourly weather variables $D + 1$ and forecasted hourly wind power generation $D + 1$) led to an error of 9.03% with model S2M8; and variables V12I to V16I (power generations $D - 1$) obtained an error of 9.16% with model S2M9. All these variables (V6I, V7I to V11I and V12I to V16I) did not achieve a better performance than the best model S2M6 (with a *MAPE* of 8.85%).

The best forecast error of 8.85% in model S2M6 for intraday market session 2 was higher than the best forecast error of 7.49% in model S1M4 for intraday market session 1. However, model S2M6 for intraday market session 2 still presented a significantly better performance than the best model for the daily session (model M18, with an error of 10.69% as shown in Table 5).

Figure 8 shows an example of the hourly evolution of actual price values and forecast values of model S2M6 for week 7 of year 2013.

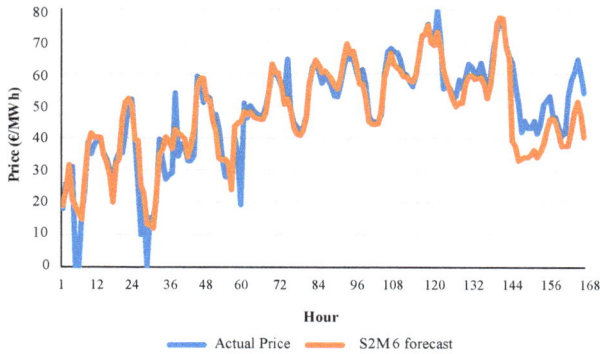

Figure 8. Actual price values and forecast values of model S2M6 of intraday session 2 for week 7 of year 2013.

5.3. Intraday Market Session 3

Let us consider that variable V17I (hourly price $D + 1$ of intraday session 1) and variable V18I (hourly price $D + 1$ of intraday session 2) can be available for ISMPF models for intraday market session 3. By applying similar procedures to those used for intraday sessions 1 and 2, the integration of individual price variables (variables V3, V4, V5I, V17I and V18I) of Table 10 within baseline model S3M1 was evaluated. Variables V5I, V17I and V18I led to better *MAPEs* of models S3M4 to S3M6 (*MAPEs* between 10.98% and 9.40%) with respect to models S3M2 and S3M3. The best combination of price variables, that is, variables V5I, V17I and V18I, was obtained in model S3M7 for intraday session 3 with a *MAPE* of 9.30%.

Table 10. ISMPF models for intraday session 3 and their *MAPEs*.

Explanatory Variables		Description	S3M1	S3M2	S3M3	S3M4	S3M5	S3M6	S3M7	S3M8	S3M9	S3M10
Chronological	V1	Hour (0–23 h)	●	●	●	●	●	●	●	●	●	●
	V2	Week day (1–7)	●	●	●	●	●	●	●	●	●	●
Price	V3	Hourly price (D)	-	●	-	-	-	-	-	-	-	-
	V4	Hourly price (D − 6)	-	-	●	-	-	-	-	-	-	-
	V5I	Hourly price (D + 1) from daily session	-	-	-	●	-	-	●	●	●	●
	V17I	Hourly price (D + 1) from intraday session 1	-	-	-	-	●	-	●	●	●	●
	V18I	Hourly price (D + 1) from intraday session 2	-	-	-	-	-	●	●	●	●	●
Demand	V6I	Hourly power demand (D − 1)	-	-	-	-	-	-	-	●	-	-
	V7I	Forecasted hourly power demand (D + 1)	-	-	-	-	-	-	-	●	-	-
Weather	V8I	Forecasted hourly temperature (D + 1)	-	-	-	-	-	-	-	-	●	-
	V9I	Forecasted hourly wind speed (D + 1)	-	-	-	-	-	-	-	-	●	-
	V10I	Forecasted hourly irradiance (D + 1)	-	-	-	-	-	-	-	-	●	-

Table 10. *Cont.*

Explanatory Variables		Description	S3M1	S3M2	S3M3	S3M4	S3M5	S3M6	S3M7	S3M8	S3M9	S3M10
Power generation	V11I	Forecasted hourly wind power gen. $(D+1)$	-	-	-	-	-	-	-	-	●	-
	V12I	Hourly hydropower generation $(D-1)$	-	-	-	-	-	-	-	-	-	●
	V13I	Hourly cogeneration and solar power generation $(D-1)$	-	-	-	-	-	-	-	-	-	●
	V14I	Hourly coal power generation $(D-1)$	-	-	-	-	-	-	-	-	-	●
	V15I	Hourly nuclear power generation $(D-1)$	-	-	-	-	-	-	-	-	-	●
	V16I	Hourly combined cycled power generation $(D-1)$	-	-	-	-	-	-	-	-	-	●
MAPE (%)			19.96	17.36	19.50	10.98	10.23	9.40	9.30	10.31	9.70	10.09

The inclusion of variables V6I and V16I with the variables of model S3M7 did not achieve improvements in the error (10.31% of model S3M8). The integration of variables V8I to V11I in model S3M9 obtained an error of 9.70%; the integration of variables V12I to V16I in model S3M10 led to an error of 10.09%.

Figure 9 shows an example of the hourly evolution of actual price values and forecast values of model S3M7 for week 7 of year 2013.

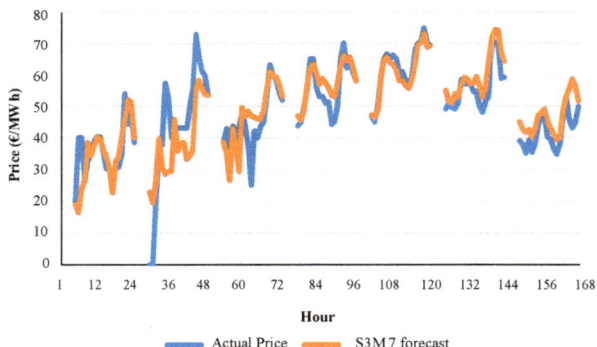

Figure 9. Actual price values and forecast values of model S3M7 of intraday session 3 for week 7 of year 2013.

5.4. Intraday Market Session 4

Let us consider that variables V17I, V18I and V19I (hourly prices $D+1$ of intraday sessions 1, 2 and 3) are available for ISMPF models for intraday market session 4. The analysis of the inclusion of individual variables V5I, V17I, V18I and V19I with chronological variables V1 and V2 of baseline model S5M1 showed that the best model is model S4M8 with a *MAPE* 9.08%, as shown in Table 11. This best model follows the pattern of variable combinations of ISMPF models for previous intraday sessions.

Table 11. ISMPF models for intraday session 4 and their *MAPEs*.

Explanatory Variables		Description	S4M1	S4M2	S4M3	S4M4	S4M5	S4M6	S4M7	S4M8	S4M9	S4M10	S4M11
Chronological	V1	Hour (0–23 h)	●	●	●	●	●	●	●	●	●	●	●
	V2	Week day (1–7)	●	●	●	●	●	●	●	●	●	●	●
Price	V3	Hourly price (D)	-	●	-	-	-	-	-	-	-	-	-
	V4	Hourly price (D − 6)	-	-	●	-	-	-	-	-	-	-	-
	V5I	Hourly price (D + 1) from daily session	-	-	-	●	-	-	●	●	●	●	●
	V17I	Hourly price (D + 1) from intraday session 1	-	-	-	-	●	-	●	●	●	●	●
	V18I	Hourly price (D + 1) from intraday session 2	-	-	-	-	-	●	●	●	●	●	●
	V19I	Hourly price (D + 1) from intraday session 3	-	-	-	-	-	-	●	●	●	●	●
Demand	V6I	Hourly power demand (D − 1)	-	-	-	-	-	-	-	-	●	-	-
	V7I	Forecasted hourly power demand (D + 1)	-	-	-	-	-	-	-	-	●	-	-
Weather	V8I	Forecasted hourly temperature (D + 1)	-	-	-	-	-	-	-	-	-	●	-
	V9I	Forecasted hourly wind speed (D + 1)	-	-	-	-	-	-	-	-	-	●	-
	V10I	Forecasted hourly irradiance (D + 1)	-	-	-	-	-	-	-	-	-	●	-
Power generation	V11I	Forecasted hourly wind power generation (D + 1)	-	-	-	-	-	-	-	-	-	●	-
	V12I	Hourly hydropower generation (D − 1)	-	-	-	-	-	-	-	-	-	-	●
	V13I	Hourly cogeneration and solar power generation (D − 1)	-	-	-	-	-	-	-	-	-	-	●
	V14I	Hourly coal power generation (D − 1)	-	-	-	-	-	-	-	-	-	-	●
	V15I	Hourly nuclear power generation (D − 1)	-	-	-	-	-	-	-	-	-	-	●
	V16I	Hourly combined cycled power generation (D − 1)	-	-	-	-	-	-	-	-	-	-	●
MAPE (%)			19.80	16.42	19.08	11.99	10.86	10.60	10.52	9.08	9.31	9.37	9.49

Again, variables V6I and V7I (power demand variables), variables V8I to V11I (forecasted variables) and variables V12I to V16I (power generations) did not improve the *MAPE* achieved by model S4M8, since they led to errors of 9.31% (model S4M9), 9.37% (model S4M10) and 9.49% (model S4M11).

Figure 10 shows an example of the hourly evolution of actual price values and forecast values of model S4M8 for week 7 of year 2013.

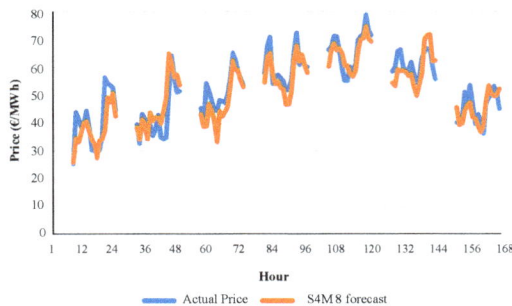

Figure 10. Actual price values and forecast values of model S4M8 of intraday session 4 for week 7 of year 2013.

5.5. Intraday Market Session 5

Let us consider that variables V17I to V20I (hourly prices $D + 1$ of intraday sessions 1 to 4) are available for ISMPF models for intraday market session 5. ISMPF models for intraday session 5 follow a similar pattern of variable combinations of ISMPF models for previous intraday sessions. The best ISMPF model is S5M9 with a *MAPE* of 9.52%, as shown in Table 12.

Table 12. ISMPF models for intraday session 5 and their *MAPEs*.

Explanatory Variables		Description	S5M1	S5M2	S5M3	S5M4	S5M5	S5M6	S5M7	S5M8	S5M9	S5M10	S5M11	S5M12
Chronological	V1	Hour (0–23 h)	•	•	•	•	•	•	•	•	•	•	•	•
	V2	Week day (1–7)	•	•	•	•	•	•	•	•	•	•	•	•
Price	V3	Hourly price (D)	-	•	-	-	-	-	-	-	-	-	-	-
	V4	Hourly price (D − 6)	-	-	•	-	-	-	-	-	-	-	-	-
	V5I	Hourly price (D + 1) from daily session	-	-	-	•	-	-	-	•	•	•	•	•
	V17I	Hourly price (D + 1) from intraday session 1	-	-	-	-	•	-	-	•	•	•	•	•
	V18I	Hourly price (D + 1) from intraday session 2	-	-	-	-	-	•	-	•	•	•	•	•
	V19I	Hourly price (D + 1) from intraday session 3	-	-	-	-	-	-	•	•	•	•	•	•
	V20I	Hourly price (D + 1) from intraday session 4	-	-	-	-	-	-	-	•	•	•	•	•
Demand	V6I	Hourly power demand (D − 1)	-	-	-	-	-	-	-	-	-	•	-	-
	V7I	Forecasted hourly power demand (D + 1)	-	-	-	-	-	-	-	-	-	•	-	-
Weather	V8I	Forecasted hourly temperature (D + 1)	-	-	-	-	-	-	-	-	-	-	•	-
	V9I	Forecasted hourly wind speed (D + 1)	-	-	-	-	-	-	-	-	-	-	•	-
	V10I	Forecasted hourly irradiance (D + 1)	-	-	-	-	-	-	-	-	-	-	•	-
	V11I	Forecasted hourly wind power generation (D + 1)	-	-	-	-	-	-	-	-	-	-	•	-
Power generation	V12I	Hourly hydropower generation (D − 1)	-	-	-	-	-	-	-	-	-	-	-	•
	V13I	Hourly cogeneration and solar power generation (D − 1)	-	-	-	-	-	-	-	-	-	-	-	•
	V14I	Hourly coal power generation (D − 1)	-	-	-	-	-	-	-	-	-	-	-	•
	V15I	Hourly nuclear power generation (D − 1)	-	-	-	-	-	-	-	-	-	-	-	•
	V16I	Hourly combined cycled power generation (D − 1)	-	-	-	-	-	-	-	-	-	-	-	•
MAPE (%)			19.62	16.16	19.37	12.17	11.47	11.07	10.40	10.18	9.52	9.53	9.69	10.16

A *MAPE* of 12.17% was obtained with model S5M4 by using information from the daily session (variable V5I) in Table 12; an error of 11.47% with model S5M5 when using information from intraday session 1 (variable V17I); 11.07% with model S5M6 by using information from intraday session 2 (variable V18I); 10.40% with model S5M7 by using information from intraday session 3 (variable V19I); and an error of 10.18% was obtained with model S5M8 when using information from intraday session 4 (variable V20I). Therefore, the evolution of the *MAPE* from model S5M4 to model S5M8 in Table 12 indicates a progressive reduction in the forecasting error with the use of more updated information (hourly price $D + 1$ of the previous intraday sessions).

The integration of variables V6I and V7I in model S5M10 resulted in a *MAPE* of 9.53%. The consideration of variables V8I to V11I in model S5M11 led to an error of 9.69%. The inclusion of variables V12I to V16I in model S5M12 achieved an error of 10.16%. All these error values obtained by including additional variables V6I, V7I, V8I to V11I, and V12I to V16I to model S5M9 were higher than the error (9.52%) of the best ISMPF model (model S5M9) for intraday session 5.

Figure 11 shows an example of the hourly evolution of actual price values and forecast values of model S5M9 for week 7 of year 2013.

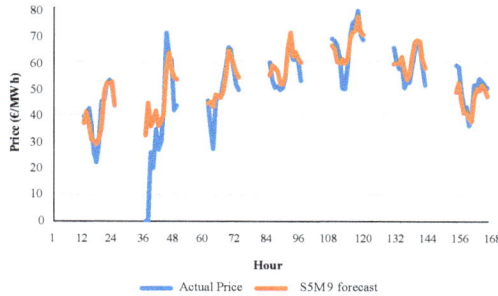

Figure 11. Actual price values and forecast values of model S5M9 of intraday session 5 for week 7 of year 2013.

5.6. Intraday Market Session 6

Let consider that variables V17I to V21I (hourly prices $D + 1$ of intraday sessions 1 to 5) are available for ISMPF models for intraday market session 6. ISMPF models for intraday session 6 follow a different pattern of variable combinations from that of ISMPF models for previous intraday sessions. As shown in Table 13, there are eight possible price variables: price D (variable V3); price $D - 6$ (variable V4); price $D + 1$ of daily session (variable V5I); and prices $D + 1$ of previous intraday sessions 1 to 5 (variables V17I to V21I). Variables V3 and V4 in models S6M2 and S6M3 improved the *MAPE* with respect to that of model S6M1. The evolution of the *MAPE* from model S6M4 to model S6M9 in Table 13, with a reduction from 15.35% of model S6M4 to 12.73% of model S6M9, indicates a progressive decrease in the *MAPE* due to the inclusion of more updated information.

Table 13. ISMPF models for intraday session 6 and their *MAPEs*.

Explanatory Variables		S6M1	S6M2	S6M3	S6M4	S6M5	S6M6	S6M7	S6M8	S6M9	S6M10	S6M11	S6M12	S6M13	S6M14	S6M15	S6M16	S6M17
Chronological	V1	●	●	●	●	●	●	●	●	●	●	●	●	●	●	●	●	●
	V2	●	●	●	●	●	●	●	●	●	●	●	●	●	●	●	●	●
Price	V3	-	●	-	-	-	-	-	-	-	-	-	-	-	-	-	-	-
	V4	-	-	●	-	-	-	-	-	-	-	-	-	-	-	-	-	-
	V5I	-	-	-	●	-	-	-	-	-	-	-	-	●	-	-	-	-
	V17I	-	-	-	-	●	-	-	-	-	-	-	●	●	-	-	-	-
	V18I	-	-	-	-	-	●	-	-	-	-	●	●	●	-	-	-	-
	V19I	-	-	-	-	-	-	●	-	-	-	●	●	●	●	●	●	●
	V20I	-	-	-	-	-	-	-	●	-	-	●	●	●	●	●	●	●
	V21I	-	-	-	-	-	-	-	-	●	●	●	●	●	●	●	●	●
Demand	V6I	-	-	-	-	-	-	-	-	-	-	-	-	-	●	-	-	-
	V7I	-	-	-	-	-	-	-	-	-	-	-	-	-	●	-	-	-
Weather	V8I	-	-	-	-	-	-	-	-	-	-	-	-	-	-	●	-	-
	V9I	-	-	-	-	-	-	-	-	-	-	-	-	-	-	●	-	-
	V10I	-	-	-	-	-	-	-	-	-	-	-	-	-	-	●	-	-
Power generation	V11I	-	-	-	-	-	-	-	-	-	-	-	-	-	-	-	●	-
	V12I	-	-	-	-	-	-	-	-	-	-	-	-	-	-	-	-	●
	V13I	-	-	-	-	-	-	-	-	-	-	-	-	-	-	-	-	●
	V14I	-	-	-	-	-	-	-	-	-	-	-	-	-	-	-	-	●
	V15I	-	-	-	-	-	-	-	-	-	-	-	-	-	-	-	-	●
	V16I	-	-	-	-	-	-	-	-	-	-	-	-	-	-	-	-	●
MAPE (%)		21.41	18.11	20.88	15.35	14.69	14.30	13.41	13.10	12.73	12.69	12.48	12.70	12.90	12.93	12.78	13.00	13.99

On the contrary to previous intraday sessions, the best ISMPF model for intraday session 6 does not use the combination of all the price $D + 1$ variables of previous intraday sessions and the price $D + 1$ of the daily session. The best *MAPE* of 12.48% was achieved with model S6M11 by including variables V19I to V21I corresponding to the previous three intraday sessions (intraday sessions 3, 4 and 5).

A suitable analysis, similar to that carried out for the other intraday sessions, allowed to determine that model S6M15 (including variables V6I and V7I) led to a higher *MAPE* (12.78%) than error 12.48% of the best model S6M11; model S6M16 (adding variables V8I to V11I) obtained an error of 13%; and model S6M17 (including variables V12I to V16I) resulted in an error of 13.99%.

Figure 12 shows an example of the hourly evolution of actual price values and forecast values of model S6M11 for week 7 of year 2013.

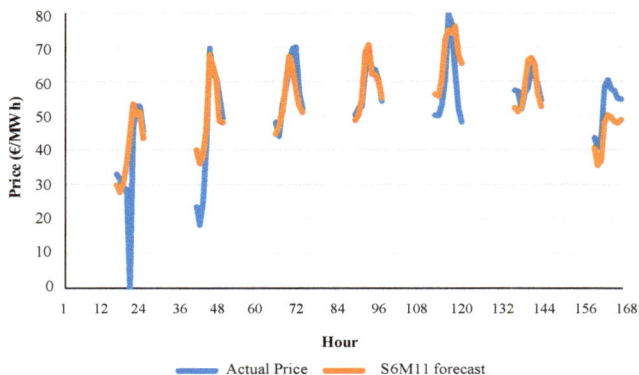

Figure 12. Actual price values and forecast values of model S6M11 of intraday session 6 for week 7 of year 2013.

5.7. Comparison of the Best ISMPF Models with the Best DSMPF Model

Table 14 shows the best ISMPF models and the best DSMPF model, including the corresponding main characteristics and the *MAPEs*. The error increased from the best model S1M4 of intraday session 1 (error of 7.49%) to the best model S6M11 for intraday session 6 (error of 12.48%). Error 10.69% of model M18 (the best DSMPF model) was higher than most of the errors of the best ISMPF models for intraday sessions (intraday sessions 1 to 5), although the main characteristics of the best ISMPF models for intraday sessions were different from those of the best DSMPF model for the daily session.

Table 14. The best ISMPF models and the best DSMPF model.

Forecasting Session for MIBEL	Forecasting Period	Best Forecasting Model	*MAPE* (%)
Daily session	0 h ($D + 1$) to 23 h ($D + 1$)	M18	10.69
Intraday session 1	21 h (D) to 23 h ($D + 1$)	S1M4	7.49
Intraday session 2	0 h ($D + 1$) to 23 h ($D + 1$)	S2M6	8.85
Intraday session 3	4 h ($D + 1$) to 23 h ($D + 1$)	S3M7	9.30
Intraday session 4	7 h ($D + 1$) to 23 h ($D + 1$)	S4M8	9.08
Intraday session 5	11 h ($D + 1$) to 23 h ($D + 1$)	S5M9	9.52
Intraday session 6	15 h ($D + 1$) to 23 h ($D + 1$)	S6M11	12.48

6. Conclusions

This paper presents novel ISMPF models for hourly price forecasting of the six intraday sessions of the MIBEL as well as a systematic analysis of the *MAPEs* corresponding to suitable combinations of their input variables in order to determine the best ISMPF model for each of the sessions, that is, the best combination of input variables for ISMPF models in each intraday session.

The methodology of the analysis is initially applied to DSMPF for the day-ahead hourly price forecasting, in the daily session of the MIBEL, and also applied to RMPE models for the estimation of the hourly prices.

DSMPF models use as input variables (price explanatory variables) recorded time series of hourly prices in previous days, regional-aggregated hourly power demands and hourly power generations of most of the types of electricity production in the previous day, forecasts of demand, wind power generation and weather (hourly wind speed, temperature and irradiation) for the day-ahead in the region, and chronological variables.

The main difference between the RMPE models and DSMPF models is that RMPE models use actual power generation values and actual power demand values of the day-ahead instead of the values of these variables on the previous day. Thus, RMPE models are not models for price forecast, but for price estimation.

Both DSMPF and RMPE models were satisfactorily applied to the real-life case study of the MIBEL that covers the mainland of Portugal and Spain. Descriptive statistics of price data have been provided. The *MAPE* of the best RMPE model was 9.9%; it represents the lowest limit of the *MAPE* of any RMPE model for price estimation or any DSMPF model for price forecast, using the same kind of input variables of RMPE models.

The *MAPE* of the best DSMPF model (model M18) was 10.7%, very close to the "minimum error" (9.9%) obtained by the best RMPE model, showing a very satisfactory performance of the best DSMPF model.

The ISMPF models consider the input variables included in DSMPF models as well as the hourly prices of the daily session and hourly prices of previous intraday sessions of the MIBEL. The methodology of the analysis used for DSMPF models is also applied to ISMPF models to find the combination of input variables that achieves the best *MAPE* for each intraday session of the MIBEL in order to determine the best ISMPF model.

The *MAPE* varied from 7.49% of the best ISMPF model (model S1M4) for the intraday session 1 to 9.52% of the best ISMPF model (model S5M9) for the intraday session 5; it raised to 12.48% of the best ISMPF model (model S6M11) for the intraday session 6. Thus, the *MAPE* of the best ISMPF models for intraday sessions 1 to 5 were clearly better than the error of 10.7% of the best DSMPF model.

The best ISMPF models for intraday sessions 1 to 5 use only the hourly prices of the daily session and hourly prices of previous intraday sessions, as well as chronological variables, and they are therefore significantly simpler than the best DSMPF model. On the other hand, the best ISMPF model of intraday session 6 exclusively utilizes the hourly prices of previous intraday sessions 3, 4 and 5, and the chronological variables, and it is also considerably simpler than the best DSMPF model.

The best Intraday Session Model for Price Forecasts of this paper, their performance in the MIBEL mainly in terms of *MAPE*, and the determination of their best input variables can be useful for agents of the electricity intraday market and the electric energy industry.

Acknowledgments: The authors would like to thank the "Ministerio de Economia y Competitividad" of the Spanish Government for supporting this research under the Project ENE2013-48517-C2-2-R, the Project ENE2013-48517-C2-1-R and the ERDF funds of the European Union; and to thank the company Smartwatt (swi.smartwatt.net) for providing data and practical experience associated with the models of this paper.

Author Contributions: All the authors contributed equally to this work.

Conflicts of Interest: The authors declare no conflict of interest.

References

1. Bello, A.; Reneses, J.; Muñoz, A. Medium-term probabilistic forecasting of extremely low prices in electricity markets: Application to the Spanish case. *Energies* **2016**, *9*, 193. [CrossRef]
2. The Iberian Energy Derivatives Exchange, OMIP. Available online: http://www.omip.pt (accessed on 25 May 2016).

3. Market Operator of the Iberian Electricity Market, OMEI. Available online: http://www.omie.es (accessed on 25 May 2016).
4. Hellmers, A.; Zugno, M.; Skajaa, A.; Morales, J.M. Operational strategies for a portfolio of wind farms and CHP plants in a two-price balancing market. *IEEE Trans. Power Syst.* **2016**, *31*, 2182–2191. [CrossRef]
5. Moreno, M.A.; Bueno, M.; Usaola, J. Evaluating risk-constrained bidding strategies in adjustment spot markets for wind power producers. *Int. J. Electr. Power Energy Syst.* **2012**, *43*, 703–711. [CrossRef]
6. Morales, J.M.; Conejo, A.J.; Pérez-Ruiz, J. Short-term trading for a wind power producer. *IEEE Trans. Power Syst.* **2010**, *25*, 554–564. [CrossRef]
7. Zugno, M.; Morales, J.M.; Pinson, P.; Madsen, H. Pool strategy of a price-maker wind power producer. *IEEE Trans. Power Syst.* **2013**, *28*, 3440–3450. [CrossRef]
8. Shafie-Khah, M.; Heydarian-Forushani, E.; Golshan, M.E.H.; Moghaddam, M.P.; Sheikh-El-Eslami, M.K.; Catalão, J.P.S. Strategic offering for a price-maker wind power producer in oligopoly markets considering demand response exchange. *IEEE Trans. Ind. Inform.* **2015**, *11*, 1542–1553. [CrossRef]
9. Ding, H.; Hu, Z.; Song, Y.; Pinson, P. Improving offering strategies for wind farms enhanced with storage capability. In Proceedings of the 2015 IEEE Eindhoven PowerTech, Eindhoven, The Netherlands, 29 June–2 July 2015.
10. Ferruzzi, G.; Cervone, G.; Delle Monache, L.; Graditi, G.; Jacobone, F. Optimal bidding in a Day-Ahead energy market for Micro Grid under uncertainty in renewable energy production. *Energy* **2016**, *106*, 194–202. [CrossRef]
11. Conejo, A.J.; Carrión, M.; Morales, J.M. *Decision Making under Uncertainty in Electricity Markets*; Springer: New York, NY, USA, 2010.
12. Chan, S.C.; Tsui, K.M.; Wu, H.C.; Hou, Y.; Wu, Y.-C.; Wu, F.F. Load/price forecasting and managing demand response for smart grids: Methodologies and challenges. *IEEE Signal Process. Mag.* **2012**, *29*, 68–85. [CrossRef]
13. Ghasemi, A.; Shayeghi, H.; Moradzadeh, M.; Nooshyar, M. A novel hybrid algorithm for electricity price and load forecasting in smart grids with demand-side management. *Appl. Energy* **2016**, *177*, 40–59. [CrossRef]
14. Chen, Z.; Wu, L.; Fu, Y. Real-time price-based demand response management for residential appliances via stochastic optimization and robust optimization. *IEEE Trans. Smart Grid* **2012**, *3*, 1822–1831. [CrossRef]
15. Research Network for Distributed Energy Resources and for Distributed Demand Resources in the Horizon of Year 2050 (Project ENE2015-70032-REDT). Available online: http://www.redyd2050-der.eu (accessed on 25 May 2016).
16. European Technology Platform on Smartgirds. Consolidated View of the ETP SG on Research, Development & Demonstration Needs in the Horizon 2020 Work Programme 2016–2017. Available online: http://www.smartgrids.eu/ETP%20Smartgrids%20View%20on%20H2020%20WP16-17.pdf (accessed on 25 May 2016).
17. Smart Energy Demand Coalition, Mapping Demand Response in Europe Today 2015. Available online: http://www.smartenergydemand.eu (accessed on 25 May 2016).
18. Contreras, J.; Espínola, R.; Nogales, F.J.; Conejo, A.J. ARIMA models to predict next-day electricity prices. *IEEE Trans. Power Syst.* **2003**, *18*, 1014–1020. [CrossRef]
19. Cruz, A.; Muñoz, A.; Zamora, J.L.; Espinola, R. The effect of wind generation and weekday on Spanish electricity spot price forecasting. *Electr. Power Syst. Res.* **2011**, *81*, 1924–1935. [CrossRef]
20. Dong, Y.; Wang, J.; Jiang, H.; Wu, J. Short-term electricity price forecast based on the improved hybrid model. *Energy Convers. Manag.* **2011**, *52*, 2987–2995. [CrossRef]
21. Tan, Z.; Zhang, J.; Wang, J.; Xu, J. Day-ahead electricity price forecasting using wavelet transform combined with ARIMA and GARCH models. *Appl. Energy* **2010**, *87*, 3606–3610. [CrossRef]
22. Shrivastava, N.A.; Khosravi, A.; Panigrahi, B.K. Prediction interval estimation of electricity prices using PSO-tuned support vector machines. *IEEE Trans. Ind. Inform.* **2015**, *11*, 322–331. [CrossRef]
23. Catalão, J.P.S.; Mariano, S.J.P.S.; Mendes, V.M.F.; Ferreira, L.A.F.M. Short-term electricity prices forecasting in a competitive market: A neural network approach. *Electr. Power Syst. Res.* **2007**, *77*, 1297–1304. [CrossRef]
24. Singhal, D.; Swarup, K.S. Electricity price forecasting using artificial neural networks. *Int. J. Electr. Power Energy Syst.* **2011**, *33*, 550–555. [CrossRef]
25. Panapakidis, I.P.; Dagoumas, A.S. Day-ahead electricity price forecasting via the application of artificial neural network based models. *Appl. Energy* **2016**, *172*, 132–151. [CrossRef]
26. Arciniegas, A.I.; Arciniegas Rueda, I.E. Forecasting short-term power prices in the Ontario Electricity Market (OEM) with a fuzzy logic based inference system. *Util. Policy* **2008**, *16*, 39–48. [CrossRef]

27. Motamedi, A.; Zareipour, H.; Rosehart, W.D. Electricity price and demand forecasting in smart grids. *IEEE Trans. Smart Grid* **2012**, *3*, 664–674. [CrossRef]
28. Pousinho, H.M.I.; Mendes, V.M.F.; Catalão, J.P.S. Short-term electricity prices forecasting in a competitive market by a hybrid PSO-ANFIS approach. *Int. J. Electr. Power Energy Syst.* **2012**, *39*, 29–35. [CrossRef]
29. Sharma, V.; Srinivasan, D. A hybrid intelligent model based on recurrent neural networks and excitable dynamics for price prediction in deregulated electricity market. *Eng. Appl. Artif. Intell.* **2013**, *26*, 1562–1574. [CrossRef]
30. Mandal, P.; Haque, A.U.; Meng, J.; Srivastava, A.K.; Martinez, R. A novel hybrid approach using wavelet, firefly algorithm, and fuzzy ARTMAP for day-ahead electricity price forecasting. *IEEE Trans. Power Syst.* **2013**, *28*, 1041–1051. [CrossRef]
31. Osório, G.J.; Matias, J.C.O.; Catalão, J.P.S. Electricity prices forecasting by a hybrid evolutionary-adaptive methodology. *Energy Convers. Manag.* **2014**, *80*, 363–373. [CrossRef]
32. Jin, C.H.; Pok, G.; Lee, Y.; Park, H.-W.; Kim, K.D.; Yun, U.; Ryu, K.H. A SOM clustering pattern sequence-based next symbol prediction method for day-ahead direct electricity load and price forecasting. *Energy Convers. Manag.* **2015**, *90*, 84–92. [CrossRef]
33. Monteiro, C.; Fernandez-Jimenez, L.A.; Ramirez-Rosado, I.J. Explanatory information analysis for day-ahead price forecasting in the Iberian electricity market. *Energies* **2015**, *8*, 10464–10486. [CrossRef]
34. Weron, R. Electricity price forecasting: A review of the state-of-the-art with a look into the future. *Int. J. Forecast.* **2014**, *30*, 1030–1081. [CrossRef]
35. Bueno, M.; Moreno, M.A.; Usaola, J.; Nogales, F.J. Strategic Wind Energy Bidding in Adjustment Markets. In Proceedings of the 45th International Universities Power Engineering Conference (UPEC), Cardiff, UK, 31 August–3 September 2010.
36. Chaves-Avila, J.P.; Hakvoort, R.A.; Ramos, A. Short-term strategies for Dutch wind power producers to reduce imbalance costs. *Energy Policy* **2013**, *52*, 573–582. [CrossRef]
37. Aïd, R.; Grue, P.; Pham, H. An optimal trading problem in intraday electricity markets. *Math. Financ. Econ.* **2016**, *10*, 49–85. [CrossRef]
38. Skajaa, A.; Edlund, K.; Morales, J.M. Intraday trading of wind energy. *IEEE Trans. Power Syst.* **2015**, *30*, 3181–3189. [CrossRef]
39. Weron, R. *Modeling and Forecasting Electricity Loads and Prices: A Statistical Approach*; John Wiley & Sons Ltd.: Oxford, UK, 2006.
40. Furió, D. A survey on the Spanish Electricity Intraday Market. *Estud. Econ. Apl.* **2011**, *29*, 1–20.
41. Ciarreta, A.; Zarraga, A. Volatility Transmissions in the Spanish Intra-Day Electricity Market. In Proceedings of the 12th International Conference on the European Energy Market (EEM), Lisbon, Portugal, 19–22 May 2015.
42. Redes Energéticas Nacionais (REN). Portuguese Transmission System Operator. Available online: http://www.centrodeinformacao.ren.pt (accessed on 25 May 2016).
43. Red Eléctrica de España (REE). Spanish Transmission System Operator. Available online: http://www.ree.es/es/actividades/demanda-y-produccion-en-tiempo-real (accessed on 25 May 2016).
44. Janjic, Z.; Black, T.; Pyle, M.; Ferrier, B.; Chuang, H.Y.; Jovic, D.; Mckee, N.; Rozumalski, R.; Michalakes, J.; Gill, D.; et al. NMM Version 3 Modeling System User's Guide. Available online: http://www.dtcenter.org/wrf-nmm/users/docs/user_guide/V3/users_guide_nmm_chap1-7.pdf (accessed on 25 May 2016).
45. Global Forecast System (GFS). Environmental Modeling Center, National Weather Service. Available online: http://www.emc.ncep.noaa.gov/index.php?branch=GFS (accessed on 25 May 2016).
46. Almeida, L.B. Multilayer Perceptrons. In *Handbook of Neural Computation*; Fiesler, E., Beale, R., Eds.; Oxford University Press: Oxford, UK, 1997.
47. Hashem, S. Optimal linear combinations of neural networks. *Neural Netw.* **1997**, *10*, 599–614. [CrossRef]

![energies logo] *energies*

MDPI

Article

Short-Term Load Forecasting Using Adaptive Annealing Learning Algorithm Based Reinforcement Neural Network

Cheng-Ming Lee [1] and Chia-Nan Ko [2,*]

[1] Department of Digital Living Innovation, Nan Kai University of Technology, Tsaotun, Nantou 542, Taiwan; t104@nkut.edu.tw

[2] Department of Automation Engineering, Nan Kai University of Technology, Tsaotun, Nantou 542, Taiwan

* Correspondence: t105@nkut.edu.tw

Academic Editor: Javier Contreras
Received: 28 September 2016; Accepted: 18 November 2016; Published: 25 November 2016

Abstract: A reinforcement learning algorithm is proposed to improve the accuracy of short-term load forecasting (STLF) in this article. The proposed model integrates radial basis function neural network (RBFNN), support vector regression (SVR), and adaptive annealing learning algorithm (AALA). In the proposed methodology, firstly, the initial structure of RBFNN is determined by using an SVR. Then, an AALA with time-varying learning rates is used to optimize the initial parameters of SVR-RBFNN (AALA-SVR-RBFNN). In order to overcome the stagnation for searching optimal RBFNN, a particle swarm optimization (PSO) is applied to simultaneously find promising learning rates in AALA. Finally, the short-term load demands are predicted by using the optimal RBFNN. The performance of the proposed methodology is verified on the actual load dataset from the Taiwan Power Company (TPC). Simulation results reveal that the proposed AALA-SVR-RBFNN can achieve a better load forecasting precision compared to various RBFNNs.

Keywords: short-term load forecasting; radial basis function neural network; support vector regression; particle swarm optimization; adaptive annealing learning algorithm

1. Introduction

Load forecasting is a crucial issue in power planning, operation, and control [1–4]. A short-term load forecasting (STLF) can be used for power maintenance scheduling, security assessment, and economic dispatch. Thus, in order to strengthen the performance of the power system, improving the load forecasting accuracy is very important [5]. An accurate forecast can reduce costs and maintain security of a power system.

In recent years, various mathematical and statistical methods have been applied to improve the accuracy of STLF. These models are roughly classified as traditional approaches and artificial intelligence (AI) based methods. Traditional approaches include exponential smoothing [6], linear regression methods [7], Box-Jenkins ARMA approaches [8], and Kalman filters [9]. In general, the traditional methods cannot correctly indicate the complex nonlinear behavior of load series. Gouthamkumar et al. [10] proposed a non-dominated sorting disruption-based gravitational search algorithm to solve fixed-head and variable-head short-term economical hydrothermal scheduling problems. With the development in AI techniques, fuzzy logic [11], PSO [12], SVM [13], and singular spectrum analysis and nonlinear multi-layer perceptron network [14] have been successfully used for STLF. AI methods have an excellent approximation ability on nonlinear functions. Therefore, it can deal with nonlinear and complex functions in the system. However, a single method cannot predict STLF efficiently.

Hybrid methods are developed to utilize the unique advantages of each approach. Adaptive ANNs for short-term load forecasting are proposed in [15], in which a PSO algorithm is employed to adjust the network's weights in the training phase of the ANNs. A modified version of the ANN already proposed for the aggregated load of the interconnected system is employed to improve the forecasting accuracy of the ANN [16]. A strategy using support vector regression machines for short-term load forecasting is proposed in [17]. A STLF algorithm based on wavelet transform, extreme learning machine (ELM) and modified artificial bee colony (MABC) algorithm is presented in [18].

As RBFNN has a single hidden layer and fast convergence speed, RBFNN has been successfully used for STLF [19,20]. When using RBFNN, one must determine the hidden layer nodes, the initial kernel parameters, and the initial network weights. A systematic approach must be established to determine the initial parameters of RBFNN. Typically, these parameters are obtained according to the designer experience, or just a random choice. However, such improper initialization usually results in slow convergence speed and poor performance of the RBFNN. An SVR method with Gaussian kernel function is adopted to determine the initial structure of the RBFNN for STLF [21].

In the training procedure, learning rates serve as an important role in the procedure of training RBFNN. The learning rate would depend on the characteristic state of inputs and outputs, in which the learning rate would be increased or decreased to match training data. Through trial and error, the learning rate is chosen to be a time-invariant constant [22,23]. However, there also exist several unstable or slow convergence problems. Many studies have been dedicated to improving the stability and convergence speed of the learning rates [24]. However, the deterministic methods for exploring appropriate learning rates are often tedious.

Recently, efficient learning algorithms for RBFNN have been developed. Besides, researchers have proposed sequential learning algorithms for resource allocation networks to enhance the convergence of the training error and computational efficiency [25]. A reinforcement learning method based on adaptive simulated annealing has been adopted to improve a decision making test problem [26]. In the literature, the learning algorithms for reduction of the training data sequence with significant information generates less computation time for a minimal network and achieves better performance. Motivated by the these learning methodologies, an adaptive learning algorithm is applied to the annealing learning procedure to promote the performance of RBFNNs. Adaptive annealing learning algorithm-based robust wavelet neural networks have been used to approximate function with outliers [27]. In [28], the author proposed time-varying learning algorithm based neural networks to identify nonlinear systems. Ko [29] proposed an adaptive annealing learning algorithm (AALA) to push forward the performance of RBFNN.

In this research, AALA is adopted to train the initial structure of RBFNN using an SVR method. In AALA, PSO approach is applied to simultaneously determine a set of suitable learning rates to improve the training RBFNN performance of STLF.

This paper is organized as follows. Section 2 introduces the architecture of RBFNN. In Section 3, the proposed algorithm, AALA-SVR-RBFNN with an adaptive annealing learning algorithm is introduced. In Section 4, simulation results for three different cases are presented and discussed. Section 5 provides conclusions for the proposed AALA-SVR-RBFNN.

2. Architecture of RBFNN

Generally, an RBFNN has a feed forward architecture with three layers: input layer, hidden layer, and output layer. The basic structure of an RBFNN is shown in Figure 1. The output of the RBFNN can be expressed as follows

$$\hat{y}_j(t+1) = \sum_{i=1}^{L} G_i \omega_{ij} = \sum_{i=1}^{L} \omega_{ij} \exp\left(-\frac{\|x - x_i^c\|^2}{2w_i^2}\right) \text{ for } j = 1, 2, \cdots, p \tag{1}$$

where $x(t) = [x_1(t) \cdots x_m(t)]^T$ is the input vector, $\hat{y}(t) = [\hat{y}_1(t) \cdots \hat{y}_p(t)]^T$ is the output vector of RBFNN, ω_{ij} is the vector of the synaptic weights in the output layer, G_i is the vector of the Gaussian function, denote the RBFNN activation function of the hidden layer, x_i^c and w_i are the vector of the centers and widths in G_i, respectively, and L is the number of neurons in the hidden layer.

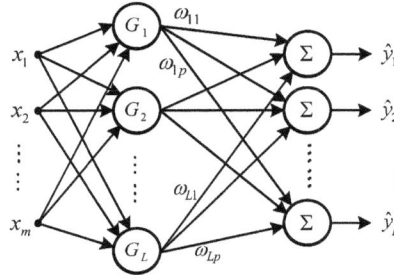

Figure 1. The structure of RBFNN.

In the training procedure, the initial values of parameters in (1) must be selected first. Then a training method is used to adjust these values iteratively to obtain their optimal combination. However, there is no way to systematically select the initial value of parameters. In the following section, an SVR is applied to perform this work.

3. AALA-SVR-RBFNN for STLF

3.1. SVR-Based Initial Parameters Estimation of RBFNN

An SVR-based algorithm is able to approximate an unknown function from a set of data, $(x^{(k)}, y^{(k)}), k = 1, 2, \cdots, N$.

In SVR, the Gaussian function is adopted as the kernel function [30]. Therefore, the approximating function can be rewritten as

$$f(x, \lambda) = \sum_{l=1}^{N_{SV}} \alpha_l \exp\left(-\frac{\| x - x_l \|^2}{2\sigma_l^2}\right) + b \tag{2}$$

where N_{SV} is the number of *support vectors* (SVs) and x_l are SVs. Comparing (2) with (1), N_{SV}, l, α_l, σ_l, and x_l in (2) can be considered to be the L, i, ω_{ij}, w_i, and x_i^c in (1), respectively. From the above derivation, the parameters of RBFNN in (1) can be obtained through the above $\varepsilon - $SVR method.

3.2. AALA-Based SVR-RBFNN

When computing the initial parameters of the SVR-RBFNN, one must establish a learning algorithm to get the optimal parameters of SVR-RBFNN. Based on time-varying learning rates, an AALA is used to train the SVR-RBFNN for conquering the drawbacks of low convergence and local minimum faced by the back-propagation learning method [31]. A cost function for the AALA is defined as

$$\Gamma_j(h) = \frac{1}{N} \sum_{k=1}^{N} \rho\left[e_j^{(k)}(h); \xi(h)\right] \text{ for } j = 1, 2, \cdots, p \tag{3}$$

where

$$e_j^{(k)}(h) = y_j^{(k)} - \hat{f}_j(x^{(k)}) = y_j^{(k)} - \sum_{i=1}^{L} \omega_{ij} \exp\left(-\frac{\| x^{(k)} - x_i^c \|^2}{2w_i^2}\right) \tag{4}$$

h denotes the epoch number, $e_j^{(k)}(h)$ denotes the error between the jth desired output and the jth output of RBFNN at epoch h for the kth input-output training data, $\xi(h)$ denotes a deterministic annealing schedule acting like the cut-off point, and $\rho(\cdot)$ denotes a logistic loss function given by

$$\rho\left[e_j^{(k)};\xi\right] = \frac{\xi}{2}\ln\left[1 + \frac{\left(e_j^k\right)^2}{\xi}\right] \text{ for } j = 1,2,\cdots,p \tag{5}$$

The root mean square error (*RMSE*) is adopted to assess the performance of training RBFNN, given by

$$RMSE = \sqrt{\frac{1}{N}\sum_{k=1}^{N}\left(e_j^{(k)}\right)^2} \text{ for } j = 1,2,\cdots,p \tag{6}$$

According to the gradient-descent learning algorithms, the synaptic weights of ω_{ij}, the centers of x_i^c, and the widths of w_i in Gaussian function are adjusted as

$$\Delta\omega_{ij} = -\gamma_\omega\frac{\partial\Gamma_j}{\partial\omega_{ij}} = -\frac{\gamma_\omega}{N}\sum_{k=1}^{N}\phi_j\left(e_j^{(k)};\xi\right)\frac{\partial e_j^{(k)}}{\partial\omega_{ij}} \tag{7}$$

$$\Delta x_i^c = -\gamma_c\frac{\partial\Gamma_j}{\partial x_i^c} = -\frac{\gamma_c}{N}\sum_{j=1}^{p}\sum_{k=1}^{N}\phi_j\left(e_j^{(k)};\xi\right)\frac{\partial e_j^{(k)}}{\partial x_i^c} \tag{8}$$

$$\Delta w_i = -\gamma_w\frac{\partial\Gamma_j}{\partial w_i} = -\frac{\gamma_w}{N}\sum_{j=1}^{p}\sum_{k=1}^{N}\phi_j\left(e_j^{(k)};\xi\right)\frac{\partial e_j^{(k)}}{\partial w_i} \tag{9}$$

$$\phi_j\left(e_j^{(k)};\xi\right) = \frac{\partial\rho\left(e_j^{(k)};\xi\right)}{\partial e_j^{(k)}} = \frac{e_j^{(k)}}{1 + \left(e_j^{(k)}\right)^2/\xi(h)} \tag{10}$$

where γ_ω, γ_c, and γ_w are the learning rates for the synaptic weights ω_{ij}, the centers x_i^c, and the widths w_i, respectively; and $\phi(\cdot)$ is usually called the influence function. In ARLA, the annealing schedule $\xi(h)$ has the capability of progressive convergence [32,33].

In the annealing schedule, complex modifications to the sampling method have been proposed, which can use a higher learning rate to reduce the simulation cost [34–37]. Based on this concept, the non-uniform sampling rates of AALA are used to train BRFNN in this work. According to the relative deviation of the late epochs, the annealing schedule $\xi(h)$ can be adjusted. The annealing schedule is updated as

$$\xi(h) = \Psi\Delta\cdot\frac{h}{h_{\max}-1} \text{ for epoch } h \tag{11}$$

$$\Delta = -2\log\left(\frac{S}{RMSE}\right) \tag{12}$$

$$S = \sqrt{\frac{1}{m}\sum_{i=1}^{m}\left(RMSE_i - \overline{RMSE}\right)^2} \tag{13}$$

where Ψ is a constant, S is the standard deviations, and $\overline{RMSE}$ is the average of *RMSE* (6) for m late epochs.

When the learning rates keep constant, choosing an appropriate learning rates γ_ω, γ_c, and γ_w is tedious; furthermore, several problems of getting stuck in a near-optimal solution or slow convergence still exist. Therefore, an AALA is used to overcome the stagnation in the search for a global optimal solution. At the beginning of the learning procedure, a large learning rate is chosen in the search space.

Once the algorithm converges progressively to the optimum, the evolution procedure is gradually tuned by a smaller learning rate in later epochs. Then, a nonlinear time-varying evolution concept is used in each iteration, in which the learning rates γ_ω, γ_c, and γ_w have a high value γ_{max}, nonlinearly decreases to γ_{min} at the maximal number of epochs, respectively. The mathematical formula can be expressed as

$$\gamma_\omega = \gamma_{min} + (epoch\,(h))^{p\omega}\,\Delta\gamma \tag{14}$$

$$\gamma_c = \gamma_{min} + (epoch\,(h))^{pc}\,\Delta\gamma \tag{15}$$

$$\gamma_w = \gamma_{min} + (epoch\,(h))^{pw}\,\Delta\gamma \tag{16}$$

$$\Delta\gamma = (\gamma_{max} - \gamma_{min}) \tag{17}$$

$$epoch(h) = \left(1 - \frac{h}{epoch_{max}}\right) \tag{18}$$

where $epoch_{max}$ is the maximal number of epochs and h is the present number of epochs. During the updated process, the performance of RBFNN can be improved using suitable functions for the learning rates of γ_ω, γ_c, and γ_w. Furthermore, simultaneously determining the optimal combination of $p\omega$, pc, and pw in (14) to (16) is a time-consuming task. A PSO algorithm with linearly time-varying acceleration coefficients will be used to obtain the optimal combination of $(p\omega, pc, pw)$. A PSO algorithm with linearly time-varying acceleration coefficients for searching the optimal combination of $(p\omega, pc, pw)$ is introduced in the following section.

3.3. Particle Swarm Optimization

PSO is a population-based stochastic searching technique developed by Kennedy and Eberhart [38]. The searching process behind the algorithm was inspired by the social behavior of animals, such as bird flocking or fish schooling. It is similar to the continuous genetic algorithms, in which it begins with a random population matrix and searches for the optimum by updating generations. However, the PSO has no evolution operations such as crossover and mutation. The potential of this technique makes it a powerful optimization tool which has been successfully applied to many fields. Nowadays, PSO has been developed to be a real competitor with other well-established techniques for population-based evolutionary computation [39,40].

In this paper, the PSO method is adopted to find an optimal combination $(p\omega, pc, pw)$ of learning rates in (14) to (16). When applying the PSO method, possible solutions must be encoded into particle positions and a fitness function must be chosen. In the optimizing procedure, the goal is to minimize the error between desired outputs and trained outputs, and then root mean square error (*RMSE*) will be defined as the fitness function.

In the PSO, a particle position is represented as

$$\mathbf{P} = [p_1, p_2, p_3] = [p\omega,\ pc,\ pw] \tag{19}$$

At each iteration (generation), the particles update their velocities and positions based on the local best and global best solutions as follows [40]:

$$\mathbf{V}(k+1) = \lambda(k+1)\cdot\mathbf{V}(k) + c_1(k+1)\cdot\mathbf{r}_1\cdot(\mathbf{P}^{lbest}(k) - \mathbf{P}(k)) + c_2(k+1)\cdot\mathbf{r}_2\cdot(\mathbf{P}^{gbest}(k) - \mathbf{P}(k)) \tag{20}$$

$$\mathbf{P}(k+1) = \mathbf{P}(k) + \mathbf{V}(k+1) \tag{21}$$

where $\mathbf{V}(k)$ and $\mathbf{V}(k+1)$ denote the particle velocities at iterations k and $(k+1)$, respectively, $\mathbf{P}(k)$ and $\mathbf{P}(k+1)$ denote the particle positions at iteration k and $(k+1)$, respectively, $\lambda(k+1)$ denotes the inertia weight at iteration $(k+1)$, r_1 and r_2 are random numbers between 0 and 1, $c_1(k+1)$ is the cognitive parameter, $c_2(k+1)$ is the social parameter at iteration $(k+1)$, $\mathbf{P}^{lbest}(k)$ is the local best solution at iteration k, and $\mathbf{P}^{gbest}$ is the global best solution of the group.

Considering the computational efficiency, a linearly adaptable inertia weight [38] and linearly time-varying acceleration coefficients [40] over the evolutionary procedure of PSO method are adopted in this paper. The inertia weight λ starts with a high value λ_{max} and linearly decreases to λ_{min} at the maximal number of iterations. The cognitive parameter c_1 starts with a high value c_{1max} and linearly decreases to c_{1min}. Whereas the social parameter c_2 starts with a low value c_{2min} and linearly increases to c_{2max}. Therefore, the inertia weight $\lambda(k+1)$ and the acceleration coefficients $c_1(k+1)$ and $c_2(k+1)$ can be expressed as follows:

$$\lambda(k+1) = \lambda_{max} - \frac{\lambda_{max} - \lambda_{min}}{iter_{max}} \cdot iter \tag{22}$$

$$c_1(k+1) = c_{1max} - \frac{c_{1max} - c_{1min}}{iter_{max}} \cdot iter \tag{23}$$

$$c_2(k+1) = c_{2max} - \frac{c_{2max} - c_{2min}}{iter_{max}} \cdot iter \tag{24}$$

where $iter_{max}$ is the maximal number of iterations (generations) and $iter$ is the current number of iterations.

3.4. Procedure of Hybrid Learning Algorithm

The proposed AALA-SVR-RBFNN using PSO can be summarized as follows:

Algorithm 1. AALA-SVR-RBFNN

Step 1:   Given a set of input-output data, $(x^{(k)}, y^{(k)})$, $k = 1, 2, \cdots, N$ for STLF.
Step 2:   Formulate and solve an SVR problem as described in Section 3.1 to determine the initial structure of the RBFNN in (1) based on the given data obtained in Step 1..
Step 3:   Adopt PSO method to generate different optimal sets of (pw, pc, pw).
Step 4: $K = 0$
Step 5: $K = K + 1$
Step 6:   Produce initial populations of position and velocity particles randomly within the feasible range.
Step 7:   Perform the AALA.

  Step 7-1:   Calculate the corresponding errors by (4) for all training data.
  Step 7-2:   Determine the values of the AALA schedule $\xi(h)$ in (11) for each epoch.
  Step 7-3:   Update the synaptic weights ω_{ij}, the centers x_i^c, and the widths w_i of Gaussian functions iteratively according to (7) through (10) and (14) through (16).
  Step 7-4:   Repeat the procedure from Step 7-1 to Step 7-3 until the current number h of epochs reaches $epoch_{max}$.

Step 8:   Evaluate fitness value for each population using the fitness function (6).
Step 9:   Select the local best for each particle by ranking the fitness values. If the best value of all current local best solutions is better than the previous global best solution, then update the value of the global best solution.
Step 10:   Update the velocity and position of each particle by using linearly the updated inertia weight, the local best particle, and the global best particle.
Step 11:   Repeat the procedure in Step 7 through Step 10 until the current number of iterations reaches the maximal iteration number in PSO.
Step 12:   If $K < m$ (m is the number of times for performing PSO method) then go to Step 6. Otherwise, determine the average optimal value of (pw, pc, pw) in (14) to (16).
Step 13:   Use the learning rates with the average optimal value of (pw, pc, pw) in (14) to (16) to train the proposed RBFNN.

In the Algorithm 1, the solution obtained in Step 13 is considered the average optimal value of $(p\omega, pc, pw)$ decided by PSO through M times independently; meanwhile the optimal structure of the proposed AALA-SVR-RBFNN is determined.

The flowchart of AALA-SVR-RBFNN using PSO is illustrated in Figure 2. The solution of the average optimal value of $(p\omega, pc, pw)$ is determined using PSO through m times independently. Then, the optimal structure of the proposed AALA-SVR-RBFNN is obtained.

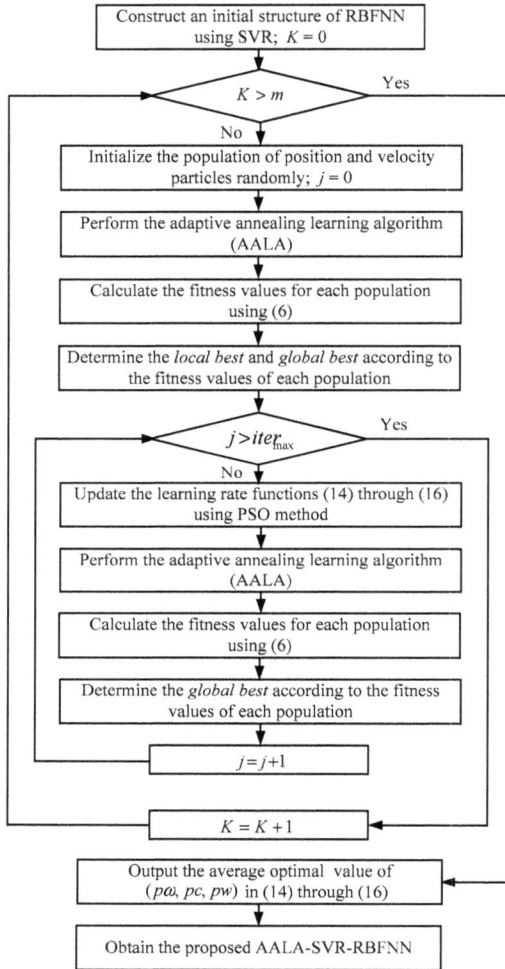

Figure 2. The flowchart of AALA-SVR-RBFNN.

4. Case Studies

The 24-h-ahead forecasting performance of the proposed AALA-SVR-RBFNN is verified on a real-word load dataset from Taiwan Power Company (TPC) in 2007. Three different patterns of load data, such as the working days (Monday through Friday), the weekends (Saturday), and holidays (Sunday and national holiday), are utilized to evaluate the effectiveness of the proposed algorithm. Table 1 lists the periods of the training and testing load data on TPC. The forecasting performance is also compared to those of DEKF-RBFNN [19], GRD-RBFNN [19], SVR-DEKF-RBFNN [19], and ARLA-SVR-RBFNN. All our simulations are carried out in Matlab 2013a using a personal computer with Intel i3-6100 and 4G RAM, Windows 7 as operating system.

The initial parameters in the simulations must be specified first. For the PSO parameters, the population size is set to be 40 and the maximal iteration number is set to be 200. Meanwhile, the number of generating different optimal sets of $(p\omega, pc, pw)$ is set to 10 ($m = 10$). The values of $(p\omega, pc, pw)$ in learning rate functions (14) to (16) are all set to real numbers in the interval $[0.1, 5]$. Furthermore, the value of γ_{max} is chosen to be 2.0 and the value of γ_{min} is chosen to be 0.05.

Table 1. The periods of the training and testing load data on TPC in 2007.

Case	Data Type	Training Data	Testing Data
1	Weekdays	2 February–15 March	16 March
2	Weekends	12 May–27 October	3 November
3	Holidays	1 July–9 December	16 December

In order to evaluate the forecasting performance of the models, two forecast error measures, such as mean absolute percentage error (*MAPE*), and standard deviation of absolute percentage error (*SDAPE*), which are utilized for model evaluation, and their definitions are shown as follows:

$$MAPE = \frac{1}{N}\sum_{k=1}^{N} \frac{\left|A^{(k)} - F^{(k)}\right|}{A^{(k)}} \times 100 \tag{25}$$

$$SDAPE = \sqrt{\frac{1}{N}\sum_{k=1}^{N}\left(\frac{\left|A^{(k)} - F^{(k)}\right|}{A^{(k)}} \times 100 - MAPE\right)^2} \tag{26}$$

where N is the number of forecast periods, $A^{(k)}$ is the actual value, and $F^{(k)}$ is the forecast value. Moreover, the *RMSE* in (6) is employed to verify the performance of training RBFNN.

Case 1: Load prediction of weekdays

The training hourly actual load data are shown in Table 1 and Figure 3. After 1000 training epochs, the initial parameters of RBFNN are determined by using SVR. The value of L in (1) is found to be 5 for parameters $C = 1$ and $\varepsilon = 0.05$ in SVR. Meanwhile, the average optimal learning rate set of $(p\omega, pc, pw)$ is determined by PSO, which is found to be (2.1711, 1.4654, 3.6347).

Table 2 lists the comparison results of *RMSE* in (6) between ARLA and AALA. In Table 2, it can be observed that AALA is superior to ARLA. After training, the proposed approach is evaluated on 24-h-ahead load forecasting on 16 March 2007. From Figure 4, it can be observed that the predicted values of the proposed AALA-SVR-RBFNN are close to the actual values.

The comparisons of *MAPE* and *SDAPE* using the five prediction models are shown in Table 3. From the comparisons, the value of *MAPE* of the proposed method is the smallest among the prediction approaches and has improvements of 48.10%, 51.19%, 31.67%, and 4.65% over DEKF-RBFNN [19], GRD-RBFNN [19], SVR-DEKF-RBFNN [19], and ARLA-SVR-RBFNN, respectively. Moreover, the *SDAPE* value of the proposed AALA-SVR-RBFNN is 0.34%, less than those obtained by the four methods.

Table 2. Results of *RMSE* in (6) of the ARLA ($0.002 \leq \gamma \leq 2$) and AALA after 1000 training epochs for three cases.

Case	ε	AALA	ARLA (γ)									
			2	1.5	1	0.5	0.1	0.05	0.02	0.01	0.005	0.002
1	0.05	**0.0072774**	0.010025	0.009653	0.009227	0.008514	0.008452	0.008386	**0.008379**	0.008426	0.008554	0.008808
2	0.06	**0.0109520**	0.015183	0.014111	0.012962	0.011622	**0.011525**	0.011596	0.011681	0.011708	0.011861	0.012267
3	0.05	**0.0097116**	0.012594	0.011968	0.01132	0.011636	0.010698	0.010365	0.010186	**0.010104**	0.010127	0.01018

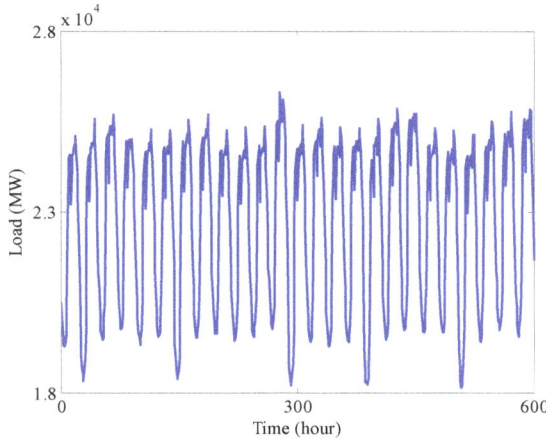

Figure 3. The training hourly actual load data in Case 1.

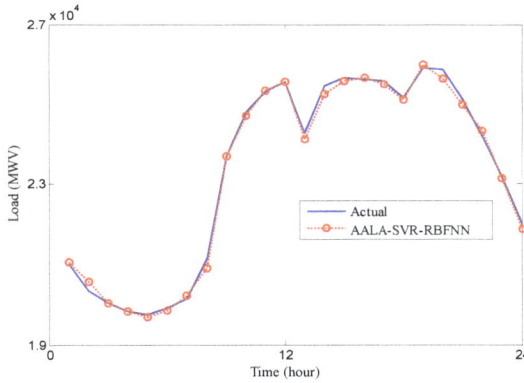

Figure 4. The forecasting results of the proposed AALA-SVR-RBFNN in Case 1.

Table 3. *MAPE* (%) and *SDAPE* (%) results for prediction methods in Case 1.

Method	*MAPE*	*SDAPE*
DEKF-RBFNN [19]	0.79	0.6
GRD-RBFNN [19]	0.84	0.6
SVR-DEKF-RBFNN [19]	0.6	0.37
ARLA-SVR-RBFNN	0.43	0.34
AALA-SVR-RBFNN	0.41	0.34

Case 2: Load prediction of weekends

The training hourly load data are shown in Table 1 and Figure 5. After 1000 training epochs, the initial parameters of RBFNN are obtained by using SVR. The value of L in (1) is found to be 7 for parameters $C = 1$ and $\varepsilon = 0.06$ in SVR. Meantime, the average optimal learning rate set of $(p\omega, pc, pw)$ is determined by PSO, which is found to be (2.6007, 0.7976, 4.0978).

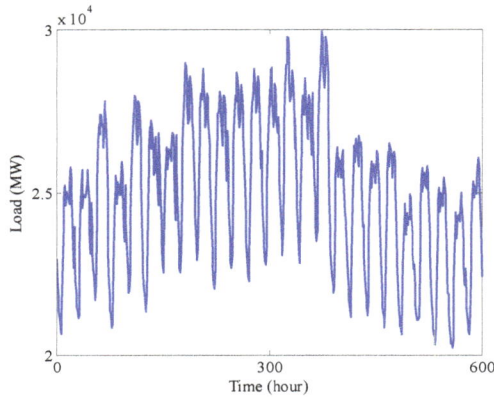

Figure 5. The training hourly actual load data in Case 2.

Table 2 shows the *RSME* in (6) ARLA and AALA. As seen from Table 2, the AALA can obtain an encouraging result than ARLA. After training, the proposed approach is evaluated on 24-h-ahead load forecasting. Figure 6 shows that the predicted values of the proposed method are very close to the actual values.

The comparison results of *MAPE* and *SDAPE* using the five prediction models are shown in Table 4. From the comparisons, the proposed AALA-SVR-RBFNN has the minimum value of *MAPE*. It proves the *MAPE* of AALA-SVR-RBFNN over DEKF-RBFNN [19], GRD-RBFNN [19], SVR-DEKF-RBFNN [19], and ARLA-SVR-RBFNN by 58.76%, 62.63%, 44.33%, and 24.53%. Moreover, the value of *SDAPE* of the proposed algorithm is smaller than those of other four methods.

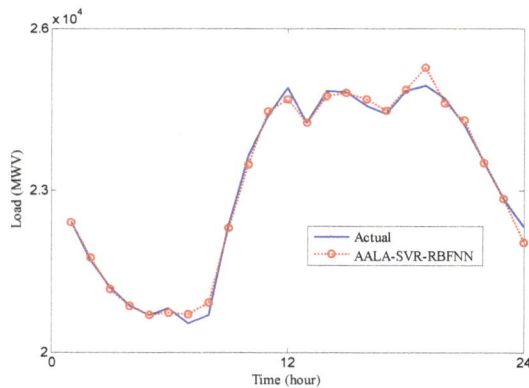

Figure 6. The forecasting results of the proposed AALA-SVR-RBFNN) in Case 2.

Table 4. *MAPE* (%) and *SDAPE* (%) results for prediction methods in Case 2.

Method	*MAPE*	*SDAPE*
DEKF-RBFNN [19]	0.97	0.6
GRD-RBFNN [19]	1.07	0.54
SVR-DEKF-RBFNN [19]	0.72	0.44
ARLA-SVR-RBFNN	0.53	0.5
AALA-SVR-RBFNN	0.4	0.4

Case 3: Load prediction of Holidays

The training hourly load data are shown in Table 1 and Figure 7. After 1000 training epochs, the initial parameters of RBFNN are determined by using SVR. The value of L in (1) is found to be 11 for $C = 1$ and $\varepsilon = 0.05$ in SVR. Meanwhile, the average optimal learning rate set of (pw, pc, pw) is determined by PSO, which is found to be (1.9640, 0.8327, 4.9085).

Table 2 illustrates the comparison results of *RSME* in (6) between ARLA and AALA. As seen from Table 2, the AALA can produce results superior to ARLA. After training, the proposed approach is tested on 24-h-ahead load forecasting. Figure 8 shows that the predicted values of the proposed AALA-SVR-RBFNN are quite close to the actual values.

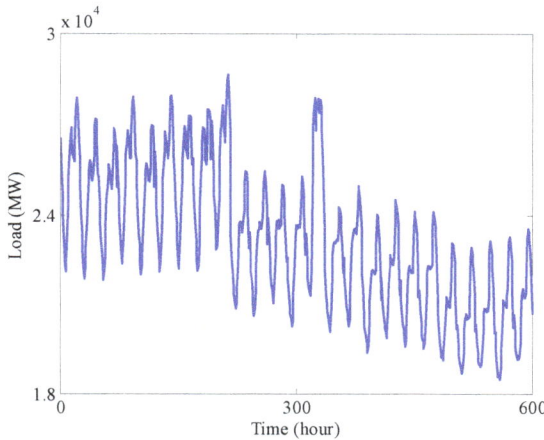

Figure 7. The training hourly actual load data in Case 3.

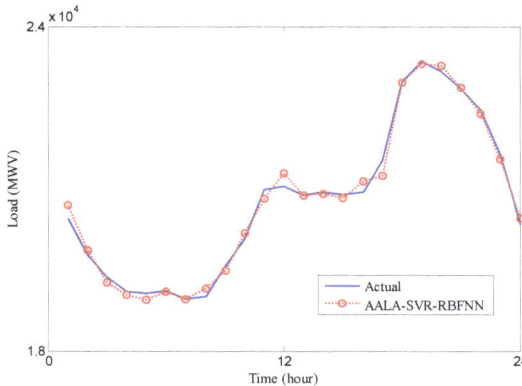

Figure 8. The forecasting results of the proposed AALA-SVR-RBFNN in Case 3.

The comparisons of *MAPE* and *SDAPE* using the five prediction models are shown in Table 5. Comparing DEKF-RBFNN [19], GRD-RBFNN [19], SVR-DEKF-RBFNN [19], and ARLA-SVR-RBFNN with the proposed AALA-SVR-RBFNN, the error of *MAPE* is reduced by 44.94%, 68.79%, 12.50%, and 20.97%, respectively. Moreover, the *SDAPE* value of the proposed algorithm is 0.38%, smaller than those obtained by using the four approaches. These results verify the superiority of the proposed AALA-SVR-RBFNN over other prediction methods.

Table 5. *MAPE* (%) and *SDAPE* (%) results for prediction methods in Case 3.

Method	*MAPE*	*SDAPE*
DEKF-RBFNN [19]	0.89	0.52
GRD-RBFNN [19]	1.57	1.04
SVR-DEKF-RBFNN [19]	0.56	0.52
ARLA-SVR-RBFNN	0.62	0.51
AALA-SVR-RBFNN	0.49	0.38

5. Conclusions

In this paper, a reinforcement neural network of AALA-SVR-RBFNN is developed for predicting STLF accurately. An SVR method is first used to find the initial parameters of RBFNN. After initializations, the parameters of RBFNN are adjusted by using AALA to obtain an optimal combination. When performing the AALA, the optimal nonlinear learning rates are simultaneously determined by using PSO. Meantime, the stagnation of training RBFNN can be overcome during the adaptive annealing learning procedure. Once the optimal RBFNN is established, the 24-h-ahead load forecasting is performed. Three different patterns of load are considered. Simulation results indicate that the proposed AALA-SVR-RBFNN can yield excellent forecasting results over DEKF-RBFNN, GRD-RBFNN, SVR-DEKF-RBFNN, and ARLA-SVR-RBFNN.

Acknowledgments: This work was supported in part by the National Science Council, Taiwan, R.O.C., under grants MOST 105-2221-E-252-002.

Author Contributions: Cheng-Ming Lee and Chia-Nan Ko conceived and designed the experiments; Cheng-Ming Lee performed the experiments; Cheng-Ming Lee and Chia-Nan Ko analyzed the data; and Chia-Nan Ko wrote the paper.

Conflicts of Interest: The authors declare no conflict of interest.

References

1. Bozic, M.; Stojanovic, M.; Stajic, Z.; Tasic, D. A New Two-Stage Approach to Short Term Electrical Load Forecasting. *Energies* **2013**, *6*, 2130–2148. [CrossRef]
2. Hernandez, L.; Baladrón, C.; Aguiar, J.M.; Carro, B.; Sanchez-Esguevillas, A.J.; Lloret, J. Short-Term Load Forecasting for Microgrids Based on Artificial Neural Networks. *Energies* **2013**, *6*, 1385–1408. [CrossRef]
3. Amjady, N.; Keynia, F. A New Neural Network Approach to Short Term Load Forecasting of Electrical Power Systems. *Energies* **2011**, *4*, 488–503. [CrossRef]
4. Wang, J.; Jiang, H.; Wu, Y.; Dong, Y. Forecasting solar radiation using an optimized hybrid model by Cuckoo Search algorithm. *Energy* **2015**, *81*, 627–644. [CrossRef]
5. Shahidehpour, M.; Yamin, H.; Li, Z. *Market Operations in Electric Power Systems: Forecasting, Scheduling, and Risk Management*; John Wiley & Sons: Hoboken, NJ, USA, 2002.
6. Christiaanse, W.R. Short term load forecasting using general exponential smoothing. *IEEE Trans. Power Appar. Syst.* **1971**, *90*, 900–911. [CrossRef]
7. Dudek, G. Pattern-based local linear regression models for short-term load forecasting. *Electr. Power Syst. Res.* **2016**, *130*, 139–147. [CrossRef]
8. Darbellay, G.A.; Slama, M. Forecasting the short-term demand for electricity. *Int. J. Forecast.* **2000**, *16*, 71–83. [CrossRef]
9. Zheng, T.; Girgis, A.A.; Makram, E.B. A hybrid wavelet-Kalman filter method for load forecasting. *Electr. Power Syst. Res.* **2000**, *54*, 11–17. [CrossRef]
10. Gouthamkumar, N.; Sharma, V.; Naresh, R. Non-dominated sorting disruption-based gravitational search algorithm with mutation scheme for multi-objective short-term hydrothermal scheduling. *Electr. Power Compon. Syst.* **2016**, *44*, 990–1004.
11. Mori, H.; Kobayashi, H. Optimal fuzzy inference for short-term load forecasting. *IEEE Trans. Power Syst.* **1996**, *11*, 390–396. [CrossRef]

12. El-Telbany, M.; El-Karmi, F. Short-term forecasting of Jordanian electricity demand using particle swarm optimization. *Expert Syst. Appl.* **2008**, *78*, 425–433. [CrossRef]
13. Hong, W.C. Electric load forecasting by support vector model. *Appl. Math. Model.* **2009**, *33*, 2444–2454. [CrossRef]
14. Niu, M.F.; Sun, S.L.; Wu, J.; Yu, L.; Wang, J. An innovative integrated model using the singular spectrum analysis and nonlinear multi-layer perceptron network optimized by hybrid intelligent algorithm for short-term load forecasting. *Appl. Math. Model.* **2016**, *40*, 4079–4093. [CrossRef]
15. Bashir, Z.A.; El-Hawary, M.E. Applying Wavelets to Short-Term Load Forecasting Using PSO-Based Neural Networks. *IEEE Trans. Power Syst.* **2009**, *24*, 20–27. [CrossRef]
16. Panapakidis, I.P. Application of hybrid computational intelligence models in short-term bus load forecasting. *Expert Syst. Appl.* **2016**, *54*, 105–120. [CrossRef]
17. Ceperic, E.; Ceperic, V.; Baric, A. A strategy for short-term load forecasting by support vector regression machines. *IEEE Trans. Power Syst.* **2013**, *28*, 4356–4364. [CrossRef]
18. Li, S.; Wang, P.; Goel, L. Short-term load forecasting by wavelet transform and evolutionary extreme learning machine. *Electr. Power Syst. Res.* **2015**, *122*, 96–103. [CrossRef]
19. Ko, C.N.; Lee, C.M. Short-term load forecasting using SVR (support vector regression)-based radial basis function neural network with dual extended Kalman filter. *Energy* **2013**, *49*, 413–422. [CrossRef]
20. Lin, Z.; Zhang, D.; Gao, L.; Kong, Z. Using an adaptive self-tuning approach to forecast power loads. *Neurocomputing* **2008**, *71*, 559–563. [CrossRef]
21. Wang, J.; Li, L.; Niu, D.; Tan, Z. An annual load forecasting model based on support vector regression with differential evolution algorithm. *Appl. Energy* **2012**, *94*, 65–70. [CrossRef]
22. Attoh-Okine, N.O. Analysis of learning rate and momentum term in backpropagation neural network algorithm trained to predict pavement performance. *Adv. Eng. Softw.* **1999**, *30*, 291–302. [CrossRef]
23. Nied, A.; Seleme, S.I., Jr.; Parma, G.G.; Menezes, B.R. On-line neural training algorithm with sliding mode control and adaptive learning rate. *Neurocomputing* **2007**, *70*, 2687–2691. [CrossRef]
24. Li, Y.; Fu, Y.; Li, H.; Zhang, S.-W. The improved training algorithm of back propagation neural network with self-adaptive learning rate. In Proceedings of the International Conference on Computational Intelligence and Natural Computing, Wuhan, China, 6–7 June 2009; pp. 73–76.
25. Suresh, S.; Savitha, R.; Sundararajan, N. Sequential learning algorithm for complex-valued self-regulating resource allocation network-CSRAN. *IEEE Trans. Neural Netw.* **2011**, *22*, 1061–1072. [CrossRef] [PubMed]
26. Atiya, A.F.; Parlos, A.G.; Ingber, L. A reinforcement learning method based on adaptive simulated annealing. In Proceedings of the 2003 IEEE 46th Midwest Symposium on Circuits and Systems, Cairo, Egypt, 27–30 December 2004; Volume 1, pp. 121–124.
27. Kuo, S.S.; Ko, C.N. Adaptive annealing learning algorithm-based robust wavelet neural networks for function approximation with outliers. *Artif. Life Robot.* **2014**, *19*, 186–192. [CrossRef]
28. Ko, C.N. Identification of nonlinear systems using RBF neural networks with time-varying learning algorithm. *IET Signal Process.* **2012**, *6*, 91–98. [CrossRef]
29. Ko, C.N. Reinforcement radial basis function neural networks with an adaptive annealing learning algorithm. *Appl. Math. Model.* **2013**, *221*, 503–513. [CrossRef]
30. Vapnik, V. *The Nature of Statistic Learning Theory*; Springer: New York, NY, USA, 1995.
31. Chuang, C.C.; Jeng, J.T.; Lin, P.T. Annealing robust radial basis function networks for function approximation with outliers. *Neurocomputing* **2004**, *56*, 123–139. [CrossRef]
32. Fu, Y.Y.; Wu, C.J.; Jeng, J.T.; Ko, C.N. Identification of MIMO systems using radial basis function networks with hybrid learning algorithm. *Appl. Math. Comput.* **2009**, *213*, 184–196. [CrossRef]
33. Chuang, C.C.; Su, S.F.; Hsiao, C.C. The annealing robust backpropagation (BP) learning algorithm. *IEEE Trans. Neural Netw.* **2000**, *11*, 1067–1077. [CrossRef] [PubMed]
34. Alrefaei, M.H.; Diabat, A.H. A simulated annealing technique for multi-objective simulation optimization. *Appl. Math. Comput.* **2009**, *215*, 3029–3035. [CrossRef]
35. Ingber, L. Very fast simulated re-annealing. *Math. Comput. Model.* **1989**, *12*, 967–983. [CrossRef]
36. Ingber, L. *Adaptive Simulated Annealing (ASA)*; Lester Ingber Research: McLean, VA, USA, 1993.
37. Shieh, H.L.; Kuo, C.C.; Chiang, C.M. Modified particle swarm optimization algorithm with simulated annealing behavior and its numerical verification. *Appl. Math. Comput.* **2011**, *218*, 4365–4383. [CrossRef]

38. Kennedy, J.; Eberhart, R. Particle swarm optimization. In Proceedings of the IEEE International Conference on Neural Networks, Perth, Australia, 27 November–1 December 1995; pp. 1942–1948.

39. Elbeltagi, E.; Hegazy, T.; Grierson, D. Comparison among five evolutionary-based optimization algorithms. *Adv. Eng. Inform.* **2005**, *19*, 43–53. [CrossRef]

40. Ratnaweera, A.; Halgamuge, S.K.; Watson, C. Self-organizing hierarchical particle swarm optimizer with time-varying acceleration coefficients. *IEEE Trans. Evol. Comput.* **2004**, *8*, 240–255. [CrossRef]

Chapter 4:
Market Equilibrium and Fundamental Models

![energies logo] MDPI

Article

Market Equilibrium and Impact of Market Mechanism Parameters on the Electricity Price in Yunnan's Electricity Market

Chuntian Cheng [1,*], Fu Chen [1], Gang Li [1] and Qiyu Tu [2]

[1] Institute of Hydropower System & Hydroinformatics, Dalian University of Technology, Dalian 116024, China; chenfu@mail.dlut.edu.cn (F.C.); glee@dlut.edu.cn (G.L.)
[2] Yunnan Power Exchange, Kunming 650000, China; tuqiyu@yn.csg.cn
[*] Correspondence: ctcheng@dlut.edu.cn; Tel.: +86-411-8470-8468

Academic Editor: Javier Contreras
Received: 28 March 2016; Accepted: 10 June 2016; Published: 17 June 2016

Abstract: In this paper, a two-dimensional Cournot model is proposed to study generation companies' (GENCO's) strategic quantity-setting behaviors in the newly established Yunnan's electricity market. A hybrid pricing mechanism is introduced to Yunnan's electricity market with the aim to stimulate electricity demand. Market equilibrium is obtained by iteratively solving each GENCO's profit maximization problem and finding their optimal bidding outputs. As the market mechanism is a key element of the electricity market, impacts of different market mechanism parameters on electricity price and power generation in market equilibrium state should be fully assessed. Therefore, based on the proposed model, we precisely explore the impacts on market equilibrium of varying parameters such as the number of GENCOs, the quantity of ex-ante obligatory-use electricity contracts (EOECs) and the elasticity of demand. Numerical analysis results of Yunnan's electricity market show that these parameters have notable but different effects on electricity price. A larger number of GENCOs or less EOEC contracted with GENCOs will have positive effects on reducing the price. With the increase of demand elasticity, the price falls first and then rises. Comparison of different mechanisms and relationship between different parameters are also analyzed. These results should be of practical interest to market participants or market designers in Yunnan's or other similar markets.

Keywords: market equilibrium; electricity price; Cournot model; market mechanism parameter

1. Introduction

Electricity market reforms have ocurred for decades in many regions and countries, and lots of the envisaged goals have been achieved so far. In 2015, China took steps to further reform its electric power industry [1] by introducing electricity markets in several electricity market pilot provinces, and Yunnan is one of them. The combination of insufficient demand and surplus clean energy capacity is the great driver for Yunnan to further reform its power sector [2,3]. In this context, Yunnan's electricity market was opened on 1 January 2015.

The two main purposes of Yunnan's electricity market at its early stage are: firstly to guarantee the stability and security operation of the power grid; secondly to stimulate demand in the province and make more clean energy be consumed at a lower electricity price [4]. To guarantee the stability and secure operation of the power grid [5,6], power generation contracts in Yunnan are classified into two parts: ex-ante obligatory-use electricity contracts (EOECs) which are bilateral contracts between generation companies (GENCOs) and the Yunnan Power Grid (YNPG), and generation contracts in the market. YNPG can regulate EOECs to cope with the imbalance and other problems of the power grid. Besides EOECs, generation contracts are made in the monthly electricity market. In order to promote clean energy consumption, a hybrid pricing mechanism is adopted by the Yunnan Power Exchange (YNPX). GENCOs can offer a two-part bid in a monthly market to compete for generation contracts (see Section 2.2 for more details). The adopted hybrid pricing mechanism is expected to be an effective mechanism to reduce the electricity price [7]. However, the newly established Yunnan electricity market is more akin to an oligopoly than perfect market competition. In every market environment, each GENCO's goal is to maximize its own profit rather than achieve the market designer's goal of stimulating the demand or reducing the price, so it is obvious that a GENCO could withdraw its generation output from the market so as to force the market clearing price to rise. This will be a profitable strategy if the increased revenue on the remaining power generation exceeds the lost profit on the foregone power generation. Thus, the strategic bidding behaviors of GENCOs have a great influence on the electricity price and generation output in the market. Meanwhile, the impact of market mechanism parameters such as the number of GENCOs, the quantity of EOECs and the elasticity of demand on the electricity price and generation in the market equilibrium state should be properly quantified and fully assessed.

Massive amounts of work have been done in the field of developing optimal bidding strategies for GENCOs and calculating market equilibrium states. Different types of models in different markets with different mechanisms are available in the literature. David [8] formally addressed the strategic bidding issue for competitive power suppliers first; a conceptual optimal bidding model and a dynamic programming method for England-Wales type electricity markets were developed [8]. In a pool-based market, Conejo *et al.* modeled the bidding strategy problem as a stochastic mixed-integer linear programming model [9] and Song *et al.* used a Markov decision process model to optimize the supplier's decision over a planning horizon [10]. Krause *et al.* performed a Nash equilibrium analysis by defining a pool market as a repeatedly played matrix game [11], and Kang *et al.* proposed a bidding model by using a two-player static game theory [12]. Zhang *et al.* developed optimal bidding strategies for wind power producers by three different strategies [13]. Contreras *et al.* developed a cobweb bidding model for competitive electricity markets [14]. Song *et al.* introduced a conjectural variation-based bidding strategy [15], and Song *et al.* and Day *et al.* presented conjectured supply function (CSF) models to obtain optimal bidding strategies [16,17]. Agent-based modeling methods are also used in the field of studying optimal bidding strategy [18,19]. In a bilateral market, Song *et al.* used a Nash equilibrium bidding strategy [20] and Hobbs introduced linear complementarity models into Nash-Cournot competition to generate optimal bidding strategies [21]. In a hybrid market, Bathurst *et al.* presented a strategy for bidding a few hours before the operation time for the wind producers [22]. Fujii *et al.* applied a multi-agent model to numerically analyze the price formation process of an open electricity market [23]. More methods for modeling GENCO bidding strategies are reviewed by references [24–27].

In this paper, a two-dimensional Cournot model considering generation capacity constraint is introduced to determine the optimal bidding strategies for GENCOs who choose to offer a two-part bid in the market. The Cournot model is used to describe quantity-setting behavior when GENCOs make strategic decisions. Market equilibrium is obtained by iteratively solving each GENCO's profit maximization problem and finding their optimal bidding outputs. Next, to assess the impacts of such market mechanism parameters as the number of GENCOs, the quantity of EOEC and the elasticity of demand on the market equilibrium and the electricity price, the unconstraint Cournot model is adopted. The rationale for choosing this model is twofold. First, although generation capacities of GENCOs are different, generation capacity caps of GENCOs are set by YNPG according to the inflow predictions of different reservoir, the operation of cascade hydropower stations and the power demand prediction. After EOECs are contracted with YNPG, the bidding spaces for GENCOs are almost the same in a surplus generation environment, so it is reasonable to take the generation capacity constraint out of consideration when assessing the impact of different parameters. Second, the impact of different market mechanism parameters can be assessed by calculating the two-dimensional Nash equilibrium of the unconstraint model to make the numerical analysis results accessible and easy to understand, so the different impacts of various parameters on the market outcomes can be compared. Ultimately, the objective is to find out how the varying parameters influence the market equilibrium and the electricity price and compare market outcomes in markets with different mechanisms.

The rest of this paper is organized as follows: the market assumptions and hybrid pricing mechanism adopted by Yunnan's electricity market are introduced in Section 2, followed by the two-dimensional game model, the market equilibrium state and the solution methodology. In Section 3, impacts of different market mechanism parameters on electricity price are assessed. At the end, concluding remarks are presented in Section 4.

2. Bidding Strategy and Market Equilibrium

2.1. Market Assumptions

Based on the actual situation of Yunnan's electricity market, in this paper, the inverse demand curve of energy is described as a monotone decreasing linear function. The Cournot model is used to describe quantity-setting behavior when GENCOs make strategic decisions. As the installed generation capacity in Yunnan consists of hydro power (74%), thermal power (17%) and other renewable resources (9%), it is assumed that GENCOs in the market are hydro and thermal power sellers based on the actual situation of Yunnan's electricity market. The generation cost of each GENCO can be described as a quadratic function. Moreover, the newly established Yunnan's electricity market is in an environment of severe surplus capacity and generation, so the generation capacity constraints of each GENCO are not taken into consideration when assessing the impact of market mechanism parameters on electricity price in this paper as mentioned in the previous section.

Yunnan's electricity market is a medium-term (monthly) market, and there is currently no balancing real-time market, ancillary service market or reserve market, so the stability and secure operation of the power grid are guaranteed by EOECs, which are dispatched by YNPG based on non-market-oriented deviation control principle. Time coupling issues among different months of hydropower are fully considered by YNPG when contracting EOEC and setting generation capacity caps for GENCOs in each month, so individual GENCO's bidding problem can be formulated for each month to maximize his profit. Because of the great significances of EOECs, EOECs of GENCOs were contracted and published before they offer their bids in the market. Power demand in Yunnan is first met with EOECs of all GENCOs, and the remaining demand (regarded as market demand) is then met with generation contracts in the market. Generally in the market, demand is cleared depending on the hybrid pricing mechanism and GENCOs' bidding behaviors in each submarket (see Section 2.2 for more details). Solving the problem with twelve-month data individually will give an annual bidding strategy.

2.2. Hybrid Pricing Mechanism

To make the hybrid pricing mechanism more understandable, the market is regarded as two different submarkets with different pricing mechanisms. In the first submarket, a pool-based clearing and pricing mechanism is used, so we call it the POOL submarket. In the POOL submarket, all GENCOs wishing to sell generation in it must offer their sealed bids to YNPX. Energy is cleared depending on each GENCO's strategic bidding behavior. The pricing mechanism in POOL submarket is uniform pricing, the uniform market clearing price is determined by the intersection of total supply (power generation of EOECs and bidding outputs of GENCOs) and inverse demand curve. This submarket is almost the same as a well-known pool-based market. The pricing mechanism is illustrated in Figure 1. Energy cleared Q_{POOL} and MCP in this submarket p_{POOL} are as follows:

$$Q_{POOL} = \sum_{i=1}^{n} q_{i,POOL} \tag{1}$$

$$p_{POOL} = f(q)\big|_{q=\sum_{i=1}^{n} q_{i,EOEC}+Q_{POOL}} \tag{2}$$

where Q_{POOL} represents energy cleared in POOL submarket; p_{POOL} represents MCP in POOL submarket; $f(q)$ represents function of inverse demand curve; $q_{i,EOEC}$ represents EOEC of GENCO i; $q_{i,pool}$ represents bidding outputs of GENCO i.

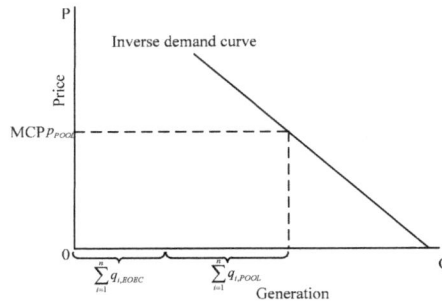

Figure 1. Pricing mechanism of POOL submarket.

Then, in the second submarket, an average purchase price (APP) clearing and pricing mechanism is used, so we call it the APP submarket. In this submarket, all GENCOs wishing to sell generation in it also need to offer their sealed bids to YNPX. Uncleared demand in the pool submarket is cleared depending on each GENCO's strategic bidding behavior in APP submarket. However, the pricing mechanism in the APP submarket is different from the one in the POOL submarket. In this submarket, the pricing mechanism is uniform pricing, but the uniform market clearing price is not determined by the intersection of total supply and inverse demand curve. For demand side, the APP submarket is regarded as a market to stimulate the demand by reducing the price. Thus, in terms of supply side, it means that if a GENCO wants to sell more, it must sell at a cheaper price. Therefore, the uniform market clearing price is determined as the average purchasing price of all buyers in this submarket, which is lower than the price in POOL submarket as illustrated in Figure 2.

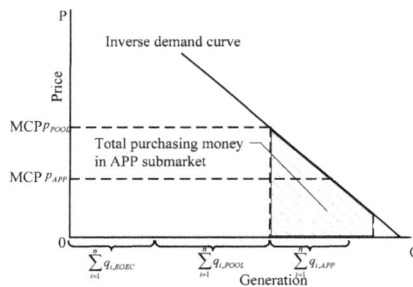

Figure 2. Pricing mechanism of the APP submarket.

Energy cleared Q_{APP} and MCP in APP submarket p_{APP} are as follows:

$$Q_{APP} = \sum_{i=1}^{n} q_{i,APP} \tag{3}$$

$$p_{APP} = \frac{\int_{\sum_{i=1}^{n} q_{i,EOEC}+Q_{POOL}}^{\sum_{i=1}^{n} q_{i,EOEC}+Q_{POOL}+Q_{APP}} f(q)dq}{Q_{APP}} \tag{4}$$

where Q_{APP} represents energy cleared in the APP submarket; p_{APP} represents MCP in APP submarket; $q_{i,APP}$ represents bidding output of GENCO i in the APP submarket.

2.3. Two-Dimensional Cournot Model

The problem faced by each GENCO is the maximization of its profit where comprises the difference between revenue and generation cost. The cost of generating is calculated as:

$$C_i(q_i) = \frac{1}{2}a_i q_i^2 + b_i q_i + c_i \tag{5}$$

Each GENCO's EOEC is contracted with YNPX before the bidding process, and the selling price for each GENCO is its on-grid price determined by National Development and Reform Committee (NDRC). So, GENCO i's revenue from EOEC can be regarded as fixed value and calculated as:

$$r_{i,EOEC} = q_{i,EOEC} \times K_i \tag{6}$$

where $r_{i,EOEC}$ represents GENCO i's revenue from EOEC; $q_{i,EOEC}$ represents the EOEC of GENCO i contracted and published by YNPX before the bidding process; K_i represents the on-grid price decided by NDRC for GENCO i.

Since each GENCO chooses to offer a two-part bid, the revenue components of each GENCO consist of selling energy in the POOL and APP submarkets. Suppose that the linear inverse demand curve is described as:

$$p = \beta - \alpha Q \tag{7}$$

where $\alpha > 0$ and $\beta > 0$ represent the slope and the intercept of the inverse demand curve respectively.

The revenues from POOL and APP submarket can be then calculated as:

$$r_{i,POOL} = q_{i,POOL} \times p_{POOL} = q_{i,POOL} \times f(q)\Big|_{q=\sum\limits_{i=1}^{n} q_{i,EOEC} + \sum\limits_{i=1}^{n} q_{i,POOL}} \tag{8}$$

$$r_{i,APP} = q_{i,APP} \times p_{APP} = q_{i,APP} \times \frac{\int_{\sum\limits_{i=1}^{n} q_{i,POOL} + \sum\limits_{i=1}^{n} q_{i,POOL}}^{\sum\limits_{i=1}^{n} q_{i,EOEC} + \sum\limits_{i=1}^{n} q_{i,POOL} + \sum\limits_{i=1}^{n} q_{i,APP}} f(q)dq}{\sum\limits_{i=1}^{n} q_{i,APP}} \tag{9}$$

where $r_{i,POOL}$ represents the GENCO i's revenue from POOL submarket; $r_{i,APP}$ represents the GENCO i's revenue from APP submarket.

The GENCO's payoff (profit) function π_i is determined by revenue minus cost, namely:

$$\pi_i = r_{i,EOEC} + r_{i,POOL} + r_{i,APP} - C_i(q) = q_{i,EOEC} \times K_i + q_{i,POOL} \times f(q)\Big|_{q=\sum\limits_{i=1}^{n} q_{i,EOEC} + \sum\limits_{i=1}^{n} q_{i,POOL}}$$

$$+ q_{i,APP} \times \frac{\int_{\sum\limits_{i=1}^{n} q_{i,EOEC} + \sum\limits_{i=1}^{n} q_{i,POOL}}^{\sum\limits_{i=1}^{n} q_{i,EOEC} + \sum\limits_{i=1}^{n} q_{i,POOL} + \sum\limits_{i=1}^{n} q_{i,APP}} f(q)dq}{\sum\limits_{i=1}^{n} q_{i,APP}} - C_i(q_{i,EOEC} + q_{i,POOL} + q_{i,APP}) \tag{10}$$

Thus, the profit maximization problem for GENCOs can be modeled as a mathematical problem with the objective as:

$$\max : \pi_i(\forall i \in N) \tag{11}$$

$$Subject\ to : q_{i,EOEC} + q_{i,POOL} + q_{i,APP} \leqslant q_{i,\max} \ (\forall i \in N) \tag{12}$$

where $q_{i,\max}$ represents the generation capacity cap of GENCO i, and N represents the number of GENCOs.

2.4. Market Equilibrium and Solution Methodology

As we can see from Equations (10) and (11), each GENCO's profit in Yunnan's electricity market depends not only on its own strategic bidding behavior, but also on those of its competitors. Moreover, GENCOs' bidding outputs in the POOL submarket affect their and others' profits not only in the POOL submarket, but also in the APP submarket. As each GENCO offers a two-part bid $(q_{i,POOL}, q_{i,APP})$ and each GENCO's payoff function is commonly known to all players, the optimal bidding strategy problem can be modeled as a two-dimensional non-cooperative game with complete information [28]. To compute the market equilibrium, we can state the problem as an n-player game: There are n GENCOs (players) in the game, each simultaneously playing with their own two-part bidding strategies. Each GENCO knows the profit (payoff) functions of its competitors and tries to maximize its own profit by taking its competitors' strategic bidding behaviors into consideration. Optimal two-part bidding strategies of GENCOs and the market equilibrium can be reached by an iterative algorithm [29]. All players play the game by submitting their optimal two-part bidding strategies iteratively until no players can improve his profit by unilaterally changing his own bidding strategy. Thus, every GENCO will finally choose its two-part strategy exactly as the two-dimensional equilibrium strategy combination. A flowchart of the iterative algorithm is shown in Figure 3, and the specific steps are described below:

(1) Initialize the GENCOs' bidding outputs in each submarket $\left(q_{i,POOL}^0, q_{i,APP}^0\right)$ ($\forall i \in N$), market structure parameters, generation capacity caps and generation cost coefficients of each GENCO. Note that each GENCO's information is available to others.

(2) For the first GENCO in the market, regard other GENCOs' bidding output as fixed values and solve the optimization problem in Equation (11) to obtain the optimal bidding output in this round, and pass this information to the second GENCO. The second GENCO regards other GENCOs' bidding output as fixed values and solves the optimization problem in Equation (11) to obtain the optimal bidding output. Go on with the process until all GENCOs obtain new bidding outputs which provide iteration.

(3) Compare each GENCO's bidding output in this round $\left(q_{i,POOL}^k, q_{i,APP}^k\right)$ with those in the previous round $\left(q_{i,POOL}^{k-1}, q_{i,APP}^{k-1}\right)$. If no GENCO's bidding output is updated, it means that no one could unilaterally improve his profit, the algorithm is convergent and the market equilibrium is obtained. Otherwise, the iteration process goes to Equation (2) for a new round.

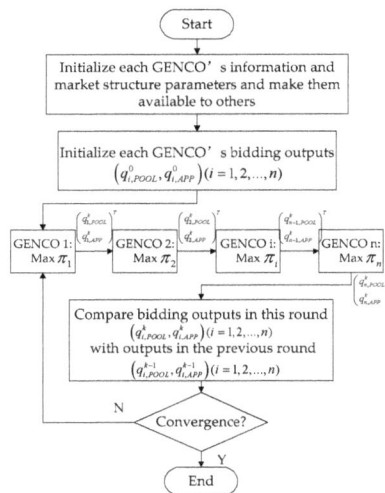

Figure 3. Flowchart of the iterative algorithm.

As discussed in reference [29], not all initial bidding outputs will lead to the convergence by using this method to solve the problem. In this paper, GENCOs' sets of rationalizable strategies are any outputs combination between outputs of EOECs and outputs of generation capacity caps. From simulation we find that any initial output combination in sets of rationalizable strategies will lead to the convergence and the same equilibrium state of the market. Nevertheless, the proper initialization for solving the model and order for solving the GENCOs should be further studied.

2.5. Numerical Results

An example with four GENCOs is shown to illustrate the equilibrium results. We first run the model for a single month to show its effectiveness, and then calculate the equilibrium results for an entire year. The generation cost coefficients, on-grid prices, EOECs and generation capacity caps in a single month are listed in Table 1. The slope and the intercept of the inverse demand curve are 0.865 and 0.017 respectively. The inverse demand curve $p = 0.865 - 0.017Q$ is thus used to demonstrate the proposed method.

Table 1. Information of power generators.

GENCO	Type	a_i	b_i	c_i	K_i (US\$/MWh)	EOEC (10^8 kWh)	Capacity Caps (10^8 kWh)
A	Hydro	0	0.1	0	48.92	6.5	12
B	Hydro	0	0.08	0	43.38	7.0	13.5
C	Hydro	0	0.11	0	36.15	6.3	13
D	Thermal	0.0012	0.18	0.08	49.85	5.1	8

Figures 4–7 show the evolution of MCPs, profits, bidding outputs in POOL submarket and bidding outputs in APP submarket of each GENCO respectively. We can see that the equilibrium state is achieved after 9 counts of iteration, because the bidding outputs in both submarkets of GENCOs no longer change. Table 2 shows the equilibrium outputs, market clearing prices and profits of GENCOs.

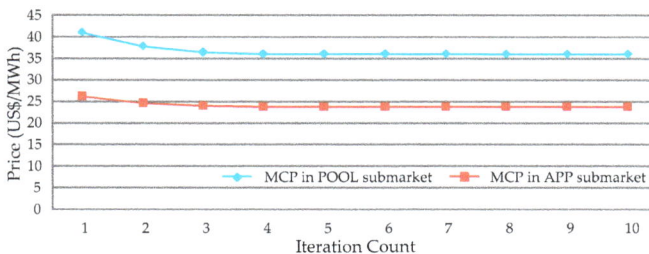

Figure 4. Evolution of MCPs in POOL and APP submarkets.

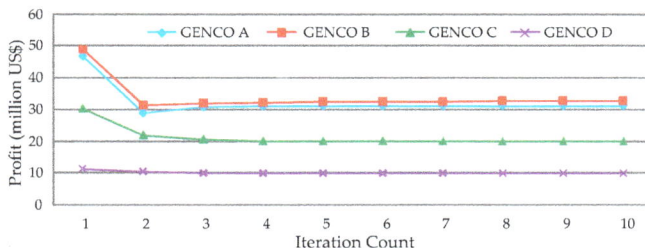

Figure 5. Evolution of the profit of GENCOs.

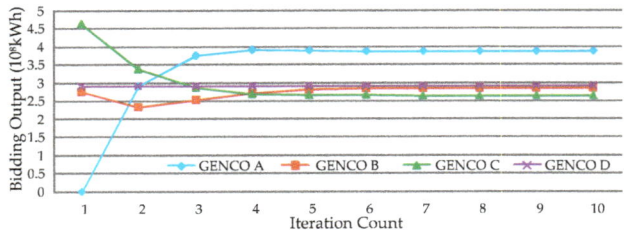

Figure 6. Evolution of bidding outputs in POOL submarket of GENCOs.

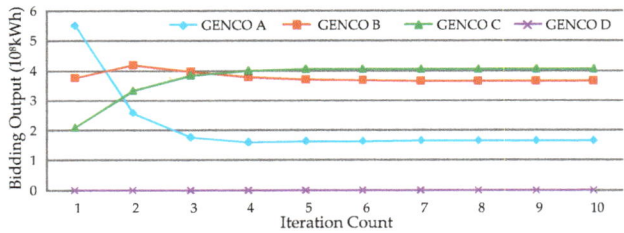

Figure 7. Evolution of bidding outputs in APP submarket of GENCOs.

Table 2. Equilibrium outputs, market clearing prices and profits of GENCOs.

GENCO	Bid in POOL (10^8 kWh)	Bid in APP (10^8 kWh)	MCP in POOL (US$/MWh)	MCP in APP (US$/MWh)	Profit (Million US$)
A	3.861	1.639			31.09
B	2.847	3.653			32.63
C	2.64	4.06	35.85	23.69	19.88
D	2.9	0			9.68

We now calculate the equilibrium results in an entire year. As discussed in Section 2.1, time coupling issues of hydropower are considered by YNPG through EOEC contracts and generation capacity caps of GENCOs according to hydropower situations in different months. So this information is important for GENCOs to make bidding strategies. Table 3 shows EOECs and generation capacity caps in different months in an entire year.

Table 3. EOEC and generation capacity caps of GENCOs.

Month	GENCO A		GENCO B		GENCO C		GENCO D	
	EOEC	Caps	EOEC	Caps	EOEC	Caps	EOEC	Caps
1	7.5	13	7.0	13.5	6.3	13	4.1	7
2	7.5	13	7.0	13.5	6.3	13	4.1	7
3	6.5	12	7.0	13.5	6.3	13	5.1	8
4	6.5	12	7.0	13.5	6.3	13	5.1	8
5	7.5	13	7.0	13.5	7.3	14	3.1	6
6	7.5	13	8.0	15.5	8.3	16	2	2
7	7.5	14.5	8.3	15.5	8.6	16	2	2
8	7.5	14.5	8.3	15.5	8.6	16	2	2
9	7.5	14.5	8.3	15.5	8.6	16	2	2
10	7.5	14.5	8.3	15.5	8.6	16	2	2
11	7.5	13	8.0	15.5	8.3	16	2	2
12	7.5	13	7.0	13.5	7.3	14	3.1	6

Figure 8 plots the MCPs in POOL and APP submarkets in an entire year. Figure 9 illustrates the equilibrium profits in an entire year of GENCOs, and Figure 10 shows generation capacity caps, EOECs, bidding outputs in POOL submarket and bidding outputs in APP submarket of all four GENCOs.

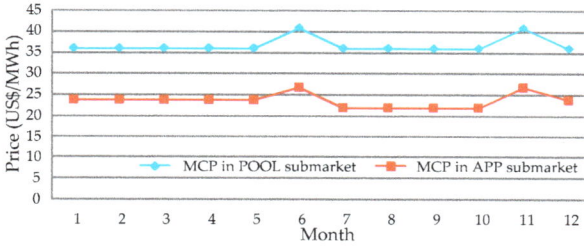

Figure 8. MCPs in POOL and APP submarkets in an entire year.

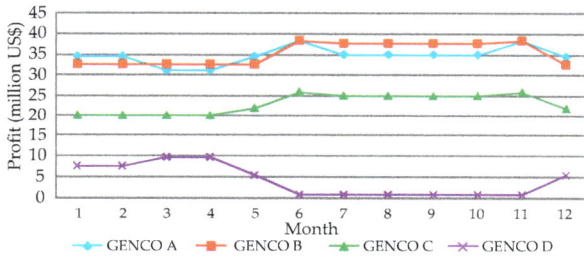

Figure 9. Profits of GENCOs in an entire year.

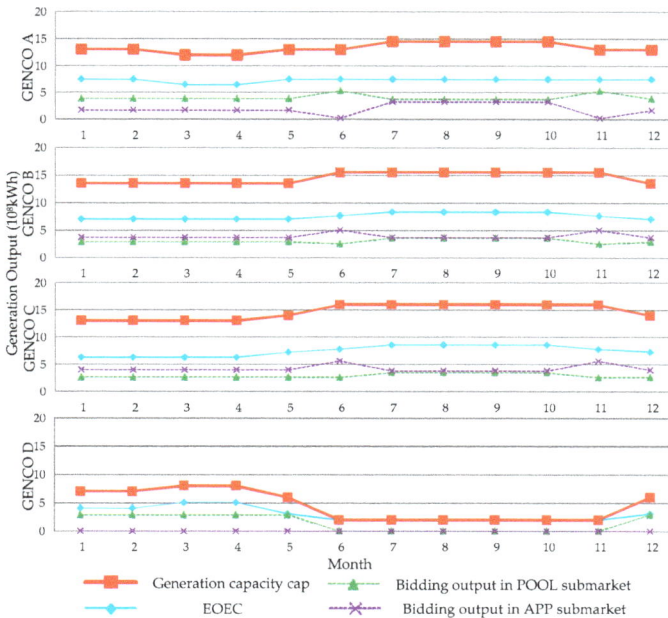

Figure 10. Bidding outputs in POOL and APP submarket of GENCOs in an entire year.

From Figures 7 and 8 we can see that hydro power plants earn more revenues in the flood season (from June to October) than the dry season (from December to April), but thermal power plants earn more revenue in the dry season. This is mainly because the generation caps for the different types of power plants in different months are different. In the flood season, demand in Yunnan can almost be entirely supplied with hydro power, so the cap for thermal power plant is relatively low in order to build a clean and green power system environment. In the dry season, the generation capacities of hydro power plants are insufficient, so the cap for thermal power plant is high enough in order to meet the demand and maintain the stability and safety of the power grid.

3. Impacts of Different Market Mechanism Parameters on Electricity Price

The previous section shows that it is feasible for GENCOs to maximize their profits by strategic bidding behaviors, so there is no reason for GENCOs to bid at relatively low prices to stimulate the demand at the cost of reducing their own profits. From the perspective of market designers, it is interesting to assess the impact of varying three market mechanism parameters in the proposed model on the electricity price and power generation in the market equilibrium state as GENCOs play strategically. We now briefly discuss the following three parameters:

Number of GENCOs: it is well known that in a perfect competitive electricity market, the number of GENCOs has no influence on the electricity price in the market. However, in an oligopolistic electricity market, each GENCO will offer a strategic bid to maximize its profit, so the number of GENCOs in the market will have an impact on the electricity price and power generation in the equilibrium state. In this paper, we measure the impact of the number of GENCOs on electricity price and power generation in the market with a hybrid pricing mechanism with contrast to perfect competition and well-known pool-based market.

EOECs: EOECs are bilateral contracts between GENCOs and YNPX, and they have great significances for market buyers, market sellers and the power grid. For market buyers and sellers, they guarantee minimum market shares for market buyers or sellers in order to maintain their market positions at early stage of Yunnan's electricity market. For power grid, because there is no real-time balancing market, ancillary service market or reserve market, the stability and security operation of the power grid are guaranteed by the regulated EOECs, which are dispatched by YNPG. So, how the existing EOEC will impact the electricity price and equilibrium generation should be assessed.

Demand elasticity: demand is usually regarded as inelastic in electricity market analysis. While the newly established Yunnan's electricity market is a monthly market, in a relatively long time span, electricity consumers can alter their production plans according to different market situations, so it is necessary to take the impact of demand elasticity into consideration. The impact of varying demand elasticity on electricity price and equilibrium generation will be analyzed and contrasted with those in perfect competition market and well-known pool-based market.

3.1. Market Equilibrium State in Different Market Mechanisms

In order to assess the impact of varying market mechanism parameters on electricity price and power generation, the analytical solutions of electricity price and power generation in a perfect competition market and a pool-based market are used for comparison. In this section, we assume generation cost parameters of all GENCOs have the same value. We first calculate the electricity price and power generation in a perfect competition market. In a perfect competition environment, the problem of maximizing profit for GENCOs is the same as Equations (10) and (11). However, no EOEC are contracted with GENCOs, and GENCOs' behaviors cannot affect the market, so they are in a market as price takers instead of price makers. Thus, the electricity price and power generation in the market can be obtained as each GENCO bid according to the price in the market:

$$p_c^c = \frac{p_{POOL} \sum_{i=1}^{n} q_{i,POOL} + p_{APP} \sum_{i=1}^{n} q_{i,APP}}{\sum_{i=1}^{n} q_{i,POOL} + \sum_{i=1}^{n} q_{i,APP}} = \frac{\beta a_0 + n\alpha b_i}{n\alpha + a_0} \tag{13}$$

$$Q_c^c = \frac{n(\beta - b_0)}{n\alpha + a_0} \tag{14}$$

where p_c^c and Q_c^c are electricity price and generation in the perfect competition market. According to the economics theory, the price and generation in the perfect competition market can be regarded as benchmarks for comparison.

In a well-known pool-based market, the clearing and pricing mechanism are just as the POOL submarket mentioned before, so the problems faced by GENCOs are:

$$\max : \pi_i = K_i q_{i,EOEC} + p_{pool} q_{i,pool} - C\left(q_{i,EOEC} + q_{i,pool}\right) (\forall i \in N) \tag{15}$$

where $q_{i,pool}$ represents the bidding output of GENCO i in a pool-based market.

The equilibrium state of the pool-based market mechanism is fully studied, so we just give out the optimal bidding strategy, electricity price and generation in the market, which are given by Equations (16)–(18), respectively:

$$q_{i,pool}^* = \frac{\left(\beta - \alpha \sum_{i=1}^{n} q_{i,EOEC} - a_0 q_{i,EOEC} - b_0\right)}{(n+1)\alpha} \quad (\forall i \in N) \tag{16}$$

$$p_{pool}^c = \frac{p_{pool} \sum_{i=1}^{n} q_{i,pool}^* + \sum_{i=1}^{n} K_i q_{i,EOEC}}{\sum_{i=1}^{n} q_{i,pool}^* + \sum_{i=1}^{n} q_{i,EOEC}} \tag{17}$$

$$Q_{pool}^c = \sum_{i=1}^{n} q_{i,pool}^* + \sum_{i=1}^{n} q_{i,EOEC} = \frac{n\beta + \alpha \sum_{i=1}^{n} q_{i,EOEC} - nb_0}{(n+1)\alpha + a_0} \tag{18}$$

where $q_{i,pool}^*$ represents the optimal bidding output of GENCO i in the pool-based market, p_{pool}^c and Q_{pool}^c represent the electricity price and power generation in the equilibrium state of a pool-based market.

In Yunnan's electricity market with a hybrid pricing mechanism, the market equilibrium state can be obtained by calculating the two-dimensional Nash equilibrium of the model. We first calculate the first-order partial derivative of the profit function *versus* $q_{i,POOL}$ and $q_{i,APP}$ of the profit function to obtain the response function of GENCO i to other GENCOs:

$$\begin{pmatrix} q_{i,POOL} \\ q_{i,APP} \end{pmatrix} = \begin{pmatrix} \partial \pi_i / \partial q_{i,POOL} \\ \partial \pi_i / \partial q_{i,APP} \end{pmatrix} \tag{19}$$

The optimal two-part bidding strategy for GENCO i is obtained when its response function equals to zero. Thus, the two-dimensional Nash equilibrium is obtained when each GENCO's optimal two-part bidding strategy is chosen, corresponding to the solution of Equation (20):

$$\begin{bmatrix} \left(\partial \pi_1 / \partial q_{1,POOL}, \partial \pi_1 / \partial q_{1,APP}\right) \\ \left(\partial \pi_2 / \partial q_{2,POOL}, \partial \pi_2 / \partial q_{2,APP}\right) \\ \vdots \\ \vdots \\ \left(\partial \pi_n / \partial q_{n,POOL}, \partial \pi_n / \partial q_{n,APP}\right) \end{bmatrix} = \begin{bmatrix} 0 \\ 0 \\ \vdots \\ \vdots \\ 0 \end{bmatrix} \tag{20}$$

Substituting Equations (5), (7) and (10) into (20) yields the two-dimensional Nash equilibrium and the two-part bidding strategy for each GENCO is shown in Equation (21):

$$\begin{pmatrix} q_{i,POOL} \\ q_{i,APP} \end{pmatrix}^* = \begin{pmatrix} \dfrac{(n-1)\left(\beta - \alpha \sum\limits_{i=1}^{n} q_{i,EOEC} - a_0 q_{i,EOEC} - b_0\right)}{(n^2+1)\alpha + (n+1)a_0} \\ \dfrac{(n-1)\left(\beta - \alpha \sum\limits_{i=1}^{n} q_{i,EOEC} - a_0 q_{i,EOEC} - b_0\right)}{(n^2+1)\alpha + (n+1)a_0} \end{pmatrix} \quad (\forall i \in N) \tag{21}$$

where the symbol * denotes the value in the equilibrium state.

Therefore, the average electricity price and power generation in Yunnan's electricity market are calculated as:

$$p_{hybrid}^c = \frac{p_{POOL} \sum\limits_{i=1}^{n} q_{i,POOL}^* + p_{APP} \sum\limits_{i=1}^{n} q_{i,APP}^* + \sum\limits_{i=1}^{n} K_i q_{i,EOEC}}{\sum\limits_{i=1}^{n} q_{i,POOL}^* + \sum\limits_{i=1}^{n} q_{i,APP}^* + \sum\limits_{i=1}^{n} q_{i,EOEC}} \tag{22}$$

$$Q_{hybrid}^c = \sum\limits_{i=1}^{n} q_{i,POOL}^* + \sum\limits_{i=1}^{n} q_{i,APP}^* + \sum\limits_{i=1}^{n} q_{i,EOEC} = \frac{(n^2+n)(\beta - b_0) - (n-1)\alpha \sum\limits_{i=1}^{n} q_{i,EOEC}}{(n^2+1)\alpha + (n+1)a_0} \tag{23}$$

where p_{hybrid}^c and Q_{hybrid}^c represent the electricity price and power generation in the equilibrium state of the market with a hybrid mechanism.

In order to assess the impact of different market mechanism parameters on electricity price, market power is used in following sections as an index. In an oligopolistic electricity market, oligopolistic GENCOs always have market power to some extent, so the electricity price is higher than in a perfect competition market. Market power can be expressed as the ratio of price in oligopolistic market to perfect competition market, which can be expressed as $\eta_{pool} = \dfrac{p_{pool}^c}{p_c^c}$ and $\eta_{hybrid} = \dfrac{p_{hybrid}^c}{p_c^c}$.

3.2. Impact of the Number of GENCOs

Figure 11 demonstrates that the electricity price in a market with hybrid mechanism is always lower than in a pool-based market. With the number of GENCOs increases, the price gradually falls in both markets. As the number of GENCOs participating in the market continues to increase, the market is closer to perfect competition, so the difference between electricity prices in different markets are getting closer. Still, as long as EOECs are contracted, the electricity price will remain higher than in perfect competition. The reason for this phenomenon will be discussed in Section 3.3.

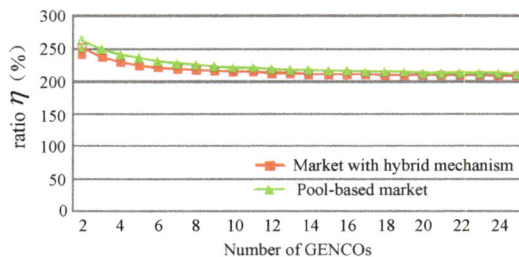

Figure 11. Impact of the number of GENCOs on electricity price.

Moreover, from Figure 12 we can see that when there are few GENCOs in the market, the market power is serious. GENCOs will increase their bidding outputs in the market with hybrid pricing mechanism and reduce their bidding outputs in the pool-based market in order to maximize their profits. Obviously, among different mechanisms, the hybrid pricing mechanism results in more generation than a perfect competition mechanism, and the well-known pool-based market results in less generation. Still, with more and more GENCOs participating in the market, there is less market power in each market, and the gaps in generation between different markets are reduced. The market power and gaps in generation will disappear until the number of GENCOs is so large to turn the market into a perfect competition situation.

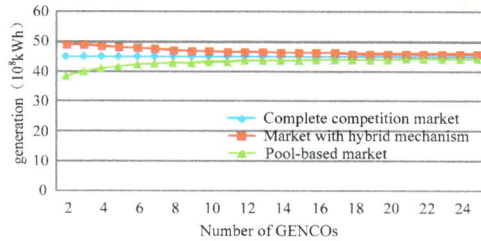

Figure 12. Impact of the number of GENCOs on generation.

3.3. Impact of Demand Elasticity

Figure 13 shows the numerical results of impacts on electricity price when demand elasticity varies in markets with different mechanisms. It is obvious that the electricity price in a well-known pool-based market is always higher than in a market with hybrid pricing mechanism. As we can see, electricity prices in both markets reduce as the demand elasticity increases at first, and then rise as the demand elasticity continues to increase. Figure 14 illustrates the trends in power generation as demand elasticity varies in markets with different mechanisms. The straightforward conclusion is that GENCOs in the market with the hybrid pricing mechanism will generate more electricity than in a perfect competition market. Moreover, the gaps of power generation between different markets narrow as demand elasticity increases.

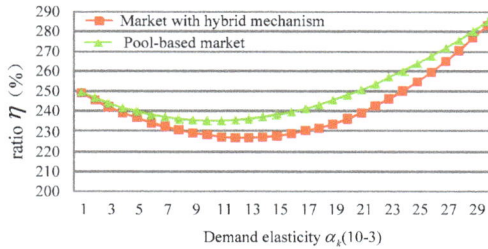

Figure 13. Impact of demand elasticity on electricity price.

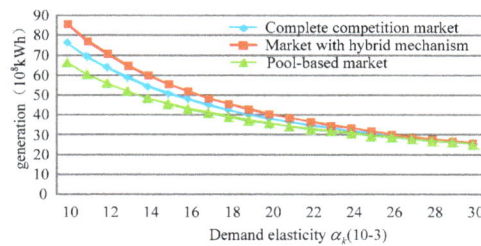

Figure 14. Impact of demand elasticity on generation.

However, the results also reveal that when the demand elasticity increases to a certain extent or a certain value, the gaps in price and power generation disappear. So far, it is still unclear whether the price in a pool-based market will be lower than in a market with hybrid pricing mechanism, or whether the power generation in a pool-based market will be more than in a market with hybrid pricing mechanism. To explain this question, the feature of EOECs is first analyzed. As introduced before, EOECs are bilateral contracts between GENCOs and the power grid on the supply side. On the demand side, the electricity contracted in EOECs must correspond to actual consumption. So the demand elasticity cannot increase indefinitely, in market with any mechanism, the electricity contracted by EOEC must be met on the demand side, that is:

$$Q^c_{pool} = \frac{n\beta + \alpha \sum_{i=1}^{n} q_{i,EOEC} - nb_0}{(n+1)\,\alpha + a_0} \geqslant \sum_{i=1}^{n} q_{i,EOEC} \tag{24}$$

$$Q^c_{hybrid} = \frac{(n^2+n)\,(\beta - b_0) - (n-1)\,\alpha \sum_{i=1}^{n} q_{i,EOEC}}{(n^2+1)\,\alpha + (n+1)\,a_0} \geqslant \sum_{i=1}^{n} q_{i,EOEC} \tag{25}$$

Calculating the expressions in Formulas (24) and (25) leads to:

$$\alpha_{pool} \leqslant \frac{n\beta - nb_0 - a_0 \sum_{i=1}^{n} q_{i,EOEC}}{n \sum_{i=1}^{n} q_{i,EOEC}} \tag{26}$$

$$\alpha_{hybrid} \leqslant \frac{n\beta - nb_0 - a_0 \sum_{i=1}^{n} q_{i,EOEC}}{n \sum_{i=1}^{n} q_{i,EOEC}} \tag{27}$$

where α_{pool} and α_{hybrid} represent possible values of the demand elasticity in a pool-based market and a market with hybrid pricing mechanism. Thus, in a market with EOECs contracted, the demand elasticity cannot increase indefinitely, and the upper limit of demand elasticity is a function of the total electricity contracted between GENCOs and the power grid.

3.4. Impact of EOEC

Figure 15 shows the impact of EOEC on the electricity price in markets with different mechanisms. When no EOEC is contracted, electricity prices in both markets are the same. Still, when there are EOECs contracted in the market, the electricity price in a pool-based market is higher than in a market with hybrid pricing mechanism. With the quantity of EOEC increasing, prices in both markets rise and the gap between the two prices widens. EOEC, as introduced before, is a kind of bilateral contract between GENCOs and the power grid. However, compared the numerical results with data in reference [30], an interesting phenomenon can be found out. In reference [30], bilateral contracts are regarded as a method of mitigating the market power. While in this paper, the electricity price rises as more EOECs contracted, resulting in more serious market power in the market. That is mainly because the price of bilateral contracts is relatively low in reference [30], while the price of EOECs is the on-grid price of each GENCO decided by NDRC, which is much higher than market clearing price in the market [31], so more contracted EOECs means GENCOs can generate at a higher price, which increases the market power of GENCOs in the market. The question raised in Section 3.2, asking why the electricity price in markets with EOEC contracted will always be higher than in a perfect competition market can also be explained.

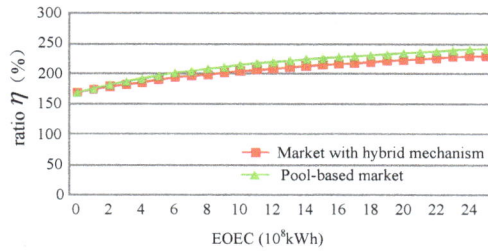

Figure 15. Impact of EOECs on electricity price.

We can tell from Figure 16 that no matter how many EOECs are contracted, GENCOs in a market with hybrid pricing mechanism generate the most electricity, and GENCOs in a pool-based market generate the least electricity.

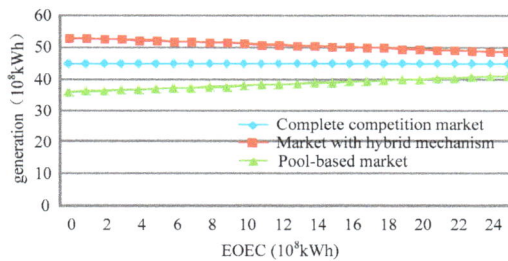

Figure 16. Impact of EOECs on generation.

With more EOECs contracted in the market, the difference between different markets in power generation becomes smaller. When all possible generation in a market with certain demand elasticity are all contracted as EOECs, power generation in markets with different mechanisms will be the same. As discussed in Section 3.3, in a market with certain demand elasticity, there is an upper limit of EOECs, expressed by the formula transformation of Equations (26) and (27):

$$\sum_{i=1}^{n} q_{i,base} \leqslant \frac{n\beta - nb_0}{n\alpha_{pool} + a_0} \tag{28}$$

$$\sum_{i=1}^{n} q_{i,base} \leqslant \frac{n\beta - nb_0}{n\alpha_{hybrid} + a_0} \tag{29}$$

4. Conclusions

According to the hybrid pricing mechanism adopted by the newly established Yunnan's electricity market, a two-dimensional Cournot model is proposed to study the strategic bidding behaviors of GENCOs, since it is not a perfect competitive market. The market equilibrium is obtained by iteratively solving each GENCO's profit maximization problem and finding their optimal bidding outputs. Numerical results demonstrate the feasibility of the proposed model. Moreover, impacts of the number of GENCOs, the demand elasticity and the quantity of EOEC on the electricity price and the power generation in market equilibrium are fully assessed. The analysis results show that all these market mechanism parameters have notable effects on the electricity price and the power generation. We may safely learn the following from the analysis results:

(1)   The electricity price in a market with hybrid pricing mechanism is always lower than those in the other two markets; and the power generated in a market with hybrid pricing mechanism is higher than those in the other two markets.

(2)   A larger number of GENCOs or fewer EOECs contracted with GENCOs will have positive effects on reducing the price. While, with the increase of demand elasticity, the price falls first and then rises because the existence of EOEC contracts.

(3)   There is a restrictive relation between demand elasticity and EOECs in markets.

These numerical results of bidding strategies and analysis results of impacts of market mechanism parameters should be of practical interest to market participants in developing bidding strategies or to market designers in setting market mechanism parameters in Yunnan' electricity market or other similar markets. The proposed model could be used for further study, which may be another subject.

Acknowledgments: This study is supported by the National Natural Science Foundation of China (No. 91547201), the Major International Joint Research Project from the National Nature Science Foundation of China (51210014) and The Fundamental Research Funds for the Central Universities (DUT16QY14).

Author Contributions: Chuntian Cheng and Fu Chen developed and solved the proposed model and carried out this analysis. Fu Chen and Gang Li wrote the manuscript and contributed to the revisions. Qiyu Tu provided the policy documents and the data used in this paper.

Conflicts of Interest: The authors declare no conflict of interest.

References

1.   The Central People's Government of PRC. *Relative Policies on Deepening the Reform of Power Industry.* Available online: http://www.ne21.com/news/show-64828.html (accessed on 24 March 2015).
2.   Cheng, C.; Liu, B.; Chau, K.; Li, G.; Liao, S. China's small hydropower and its dispatching management. *Renew. Sustain. Energy Rev.* **2015**, *42*, 43–55. [CrossRef]
3.   Hennig, T.; Wang, W.; Feng, Y.; Ou, X.; He, D. Review of Yunnan's hydropower development. Comparing small and large hydropower projects regarding their environmental implications and socio-economic consequences. *Renew. Sustain. Energy Rev.* **2013**, *27*, 585–595. [CrossRef]
4.   Yunnan Provincial Industry & Information Technology Commission. *Working Program of Yunnan's 2015 Electricity Market*; Available online: http://www.ynetc.gov.cn/Item/11399.aspx (accessed on 29 December 2014).
5.   Woo, C.; Lloyd, D.; Tishler, A. Electricity market reform failures: UK, Norway, Alberta and California. *Energy Policy* **2003**, *31*, 1103–1115. [CrossRef]
6.   Woo, C.K. What went wrong in California's electricity market? *Energy* **2001**, *26*, 747–758. [CrossRef]
7.   Yunnan Provincial Industry & Information Technology Commission. *Detailed Rules of Yunnan's 2015 Electricity Market*; Available online: http://www.ynetc.gov.cn/Item/11399.aspx (accessed on 29 December 2014).
8.   David, A.K. Competitive Bidding in Electricity Supply. *IEE Proc. C* **1993**, *140*, 421–426. [CrossRef]
9.   Conejo, A.J.; Nogales, F.J.; Arroyo, J.M. Price-taker bidding strategy under price uncertainty. *IEEE Trans. Power Syst.* **2002**, *17*, 1081–1088. [CrossRef]
10. Song, H.L.; Liu, C.C.; Lawarree, J.; Dahlgren, R.W. Optimal electricity supply bidding by Markov decision process. *IEEE Trans. Power Syst.* **2000**, *15*, 618–624. [CrossRef]
11. Krause, T.; Beck, E.V.; Cherkaoui, R.; Germond, A.; Andersson, G.; Ernst, D. A comparison of Nash equilibria analysis and agent-based modelling for power markets. *Int. J. Electr. Power* **2006**, *28*, 599–607. [CrossRef]
12. Kang, D.; Kim, B.H.; Hur, D. Supplier bidding strategy based on non-cooperative game theory concepts in single auction power pools. *Electr. Power Syst. Res.* **2007**, *77*, 630–636. [CrossRef]
13. Zhang, H.; Gao, F.; Wu, J.; Liu, K.; Liu, X. Optimal Bidding Strategies for Wind Power Producers in the Day-ahead Electricity Market. *Energies* **2012**, *5*, 4804–4823. [CrossRef]
14. Contreras, J.; Candiles, O.; de la Fuente, J.I.; Gomez, T. A cobweb bidding model for competitive electricity markets. *IEEE Trans. Power Syst.* **2002**, *17*, 148–153. [CrossRef]
15. Song, Y.; Ni, Y.; Wen, F.; Hou, Z.; Wu, F.F. Conjectural variation based bidding strategy in spot markets: fundamentals and comparison with classical game theoretical bidding strategies. *Electr. Power Syst. Res.* **2003**, *67*, 45–51. [CrossRef]

16. Song, Y.Q.; Ni, Y.X.; Wen, F.S.; Wu, F.F. Analysis of strategic interactions among generation companies using conjectured supply function equilibrium model. In Proceedings of the 2003 IEEE Power Engineering Society General Meeting, Toronto, ON, Canada, 13–17 July 2003; pp. 849–853.

17. Day, C.J.; Hobbs, B.F.; Pang, J.S. Oligopolistic competition in power networks: A conjectured supply function approach. *IEEE Trans. Power Syst.* **2002**, *17*, 597–607. [CrossRef]

18. Rahimiyan, M.; Rajabi Mashhadi, H. Supplier's optimal bidding strategy in electricity pay-as-bid auction: Comparison of the Q-learning and a model-based approach. *Electr. Power Syst. Res.* **2008**, *78*, 165–175. [CrossRef]

19. Gountis, V.P.; Bakirtzis, A.G. Bidding Strategies for Electricity Producers in a Competitive Electricity Marketplace. *IEEE Trans. Power Syst.* **2004**, *19*, 356–365. [CrossRef]

20. Song, H.L.; Liu, C.C.; Lawarree, J. Nash equilibrium bidding strategies in a bilateral electricity market. *IEEE Trans. Power Syst.* **2002**, *17*, 73–79. [CrossRef]

21. Hobbs, B.F. Linear complementary models of Nash-Cournot competition in bilateral and POOLCO power markets. *IEEE Trans. Power Syst.* **2001**, *16*, 194–202. [CrossRef]

22. Bathurst, G.N.; Weatherill, J.; Strbac, G. Trading wind generation in short term energy markets. *IEEE Trans. Power Syst.* **2002**, *17*, 782–789. [CrossRef]

23. Fujii, Y.; Okamura, T.; Inagaki, K.; Yamaji, K. Basic analysis of the pricing processes in modeled electricity markets with multi-agent simulation. In Proceedings of the 2004 IEEE International Conference on Electric Utility Deregulation, Restructuring and Power Technologies, Hong Kong, China, 5–8 April 2004.

24. Steeger, G.; Barroso, L.A.; Rebennack, S. Optimal Bidding Strategies for Hydro-Electric Producers: A Literature Survey. *IEEE Trans. Power Syst.* **2014**, *29*, 1758–1766. [CrossRef]

25. Li, G.; Shi, J.; Qu, X. Modeling methods for GenCo bidding strategy optimization in the liberalized electricity spot market—A state-of-the-art review. *Energy* **2011**, *36*, 4686–4700. [CrossRef]

26. David, A.K.; Wen, F.S. Strategic bidding in competitive electricity markets: A literature survey. In Proceedings of the 2000 IEEE Power Engineering Society Summer Meeting, Seattle, WA, USA, 16–20 July 2000.

27. Díaz, C.A.; Villar, J.; Campos, F.A.; Reneses, J. Electricity market equilibrium based on conjectural variations. *Electr. Power Syst. Res.* **2010**, *80*, 1572–1579. [CrossRef]

28. Tan, D.; Hu, P. Model of multidimensional game between two Oligarchs and its analysis on output strategies. *J. Ind. Eng. Eng. Manag.* **2004**, *1*, 123–125.

29. Zhang, H.; Gao, F.; Wu, J.; Liu, K.; Zhai, Q. A stochastic Cournot bidding model for wind power producers. In Proceedings of the 2011 IEEE International Conference on Automation and Logistics (ICAL), Chongqing, China, 15–16 August 2011.

30. Kelman, R.; Barroso, L.; Pereira, M. Market power assessment and mitigation in hydrothermal systems. *IEEE Trans. Power Syst.* **2001**, *16*, 354–359. [CrossRef]

31. Zeng, M.; Yang, Y.; Wang, L.; Sun, J. The power industry reform in China 2015: Policies, evaluations and solutions. *Renew. Sustain. Energy Rev.* **2016**, *57*, 94–110. [CrossRef]

Article

Parametric Density Recalibration of a Fundamental Market Model to Forecast Electricity Prices

Antonio Bello [1], Derek Bunn [2,*], Javier Reneses [1] and Antonio Muñoz [1]

[1] Institute for Research in Technology, Technical School of Engineering (ICAI), Universidad Pontificia Comillas, 28015 Madrid, Spain; antonio.bello@comillas.edu (A.B.); javier.reneses@comillas.edu (J.R.); antonio.munoz@comillas.edu (A.M.)
[2] London Business School, London NW1 4SA, UK
* Correspondence: dbunn@london.edu; Tel.: +44-020-7000-8827

Academic Editors: Javier Contreras and John Ringwood
Received: 18 August 2016; Accepted: 11 November 2016; Published: 17 November 2016

Abstract: This paper proposes a new approach to hybrid forecasting methodology, characterized as the statistical recalibration of forecasts from fundamental market price formation models. Such hybrid methods based upon fundamentals are particularly appropriate to medium term forecasting and in this paper the application is to month-ahead, hourly prediction of electricity wholesale prices in Spain. The recalibration methodology is innovative in seeking to perform the recalibration into parametrically defined density functions. The density estimation method selects from a wide diversity of general four-parameter distributions to fit hourly spot prices, in which the first four moments are dynamically estimated as latent functions of the outputs from the fundamental model and several other plausible exogenous drivers. The proposed approach demonstrated its effectiveness against benchmark methods across the full range of percentiles of the price distribution and performed particularly well in the tails.

Keywords: electricity; prices; forecasting; fundamentals; hybrid; densities

---

1. Introduction

In contrast to the extensive research on methods of forecasting electricity spot price expectations, the full predictive specification of price density functions has received much less attention. Point forecasts, unlike probabilistic forecasts, even when they are very accurate, provide no information on the range of risks. Yet the risks of extreme price excursions and episodes of high volatility have important managerial implications for trading, operations, and revenue planning, and this is becoming increasingly so as the technology mix shifts towards intermittent renewable power. Whilst high spiking prices in scarcity conditions have been well-documented, low price events, at times of high wind, solar, or hydro outputs, are increasingly adding to the complexity of risk management and operational control. Thus, to undertake thorough risk simulations or stochastic optimizations of alternative decisions under conditions of price risk, inputs of the full price density functions will generally be required. Furthermore, it is not the case that price densities can be reliably estimated as stable residual distributions around well-specified models for price expectations. The shape of the price density distribution often changes distinctly over the separate intraday trading periods as well as seasonally and with structural changes in the market. These distinctive densities are a function of the fundamental demand/supply drivers and market conduct together with the stochastic persistence of shocks. The task of predicting the price densities is particularly challenging, therefore in practice, the starting point is often a deterministic fundamental model of the supply stack and the price formation process. Such market models do not easily accommodate time series specifications and are generally recalibrated ex post to actual data (two exceptions are [1,2]). This two-stage process is often referred to as a "hybrid" adjustment to merge the statistical characteristics more fully with the market fundamental model in order to improve forecast accuracy. This is because by combining fundamental and statistical models, it is possible to incorporate the impact of both the projected fundamental changes in the market (such as generation expansion, mothballing of generation units, subsidies, or drops in energy demand) and the empirically revealed behavioral aspects (such as strategic and speculative behavior). However, such hybrid processes have typically focused only upon adjusting the mean bias and not the full density specifications. It is with this motivation, therefore, that we investigate the forecasting ability of a novel, fully parametric model for hourly price densities, which is sufficiently flexible in specification to accurately recalibrate the forecasts from a fundamental market model.

In a wide-ranging review on electricity price forecasting, [3] notes the paucity of research on density methods and moreover observes that even within the limited focus on predictive distributions, most of the work has been upon interval estimates or estimating specific quantiles rather than on fully parametric specifications of the density functions themselves. Quantile regression in particular has become effective in estimating specific percentiles of the predictive distribution as functions of fundamental exogenous variables (e.g., fuel prices, demand, reserve margin, etc.). Effective applications of this approach to electricity prices include [4–6], but as [7] notes, a focus upon distinct quantiles has limitations compared to a fully parametric specification and in particular, as observed by [8], the quantile regression estimates in the tails of the distribution tend to be less reliable than those generated by parametric methods. Most importantly, perhaps, is the absence of a closed form of analytic representation of the predictive distribution, as might be required for example in asset pricing, options valuation, or portfolio optimization [9].

As an alternative to quantile regression, we therefore investigate several fully parametric specifications for hourly electricity prices (from the Spanish market) to find a distributional form that not only fits all observed density shapes acceptably well but is also expressible in terms of its first four moments. These four moments can, under an appropriately specified distribution, in turn be capable of being estimated as dynamic latent variables from a Linear Additive Model (following [10]) linking them to one or more exogenous variables (e.g., the forecasts from a fundamental market model). In this respect, this is an extension to the approach taken by [11], who used a Johnson's U distribution with time varying means and variances, but constant skewness and kurtosis, to predict short-term electricity prices in the California Power Exchange and the Italian Power Exchange. However, by focusing upon the first four moments, we seek to additionally capture the changing skewness and fat tails that have added to the riskiness of power prices and furthermore to calibrate all four moments to the forecasts from a fundamental model. Whilst the dynamic estimation of the first four moments is primarily being sought to facilitate the full specification of a predictive density function, we should observe that these moment estimates are valuable in their own right. Thus, [12], for example, demonstrated that forward prices can be expressed as a Taylor expansion involving the moments of the spot price distribution.

We develop this methodology in the context of medium term electricity price forecasting, which is also relatively under-researched compared to the short term, as noted by [13]. By medium term, we consider horizons of weeks and months, over which substantial operational planning, fuel procurement, sales, and financial modelling need to be supported by forecasts and risk management. Regarding density forecasting in the medium term, the authors are only aware of the methodologies proposed in [4,14–17]. In [17], the focus is only on predicting the probability of extremely low prices given a threshold, and not on the full density function The analysis of [14] is limited to a 95% prediction interval, computed using a seasonal dynamic factor analysis, without accounting for exogenous variables, and focusing upon a short period of one week during which there were no structural changes. Furthermore, in [16], the lead time is for up to four weeks (which is a short- to medium-term horizon) and only 90% and 99% prediction intervals are given. In our work, we extend the fundamental market modelling of [4,15], and the non-parametric hybrid approach of [2], as a basis for a hybrid formulation leading to a fully parametric density specification with dynamic latent moment estimates for mean, variance, skewness, and kurtosis. Therefore, to the best of the authors' knowledge, the work presented here is the first study that evaluates real out-of-sample density forecasts in the medium term from a wide diversity of parametric models with time-varying higher moments. In addition, none of the existing works focus on the medium term and use a hybrid framework in which not only probabilistic fundamental information from a market equilibrium model is incorporated, but also information coming from statistical techniques. As a pragmatic extension, this work is furthermore innovative in combining the probabilistic forecasts from several competing parametric distributions.

The paper is structured as follows. In Section 2, an overview of the hybrid method is described, as well as theoretical details and empirical applications are presented. In Section 3, forecast combination techniques are examined to investigate increased accuracy. In Section 4, we conclude with a summary and some critical comments.

2. The Hybrid Recalibration Process

2.1. Overview of the Methodology

Following [4], we use a market equilibrium model ("MEQ") for hourly wholesale electricity prices in Spain as the starting point for the recalibration process. We do not describe this model in detail since it is well documented elsewhere (see [18,19]) and the research focus of this paper is upon the recalibration process. Briefly, this is a detailed optimization formulation in which each generating company tries to maximize its own profit ([19] subject to conjectural variations on the strategic behavior of the other market agents. In addition, therefore to the fundamentals of demand, supply, fuel costs, and plant technical characteristics it also estimates market conduct. Thus, this model is able to adequately represent the operation and behavior of the Spanish electric power system. Monte Carlo simulations based upon input distributions for the uncertain variables then provide probabilistic hourly predictions of the hourly prices. The means and selected percentiles of these MEQ derived prices are then used, alongside other exogenous variables including the production of the generating units, the international exchanges and the net demand, as regressors in the recalibration model.

The novel approach presented here comprises different blocks. On the one hand, the hybrid approach uses the probabilistic forecasts obtained as the outputs of the fundamental model (MEQ) as inputs to several general four-parameter distributions for hourly prices. The first four moments of these distributions are dynamically estimated as latent state variables and furthermore modeled as functions of several exogenous drivers. We refer to this as the two-stage approach.

Beyond this, a three-stage approach is proposed in which the percentiles of the price cumulative distribution functions of the fundamental model are firstly recalibrated with quantile regression (as suggested in [4]), and in a subsequent stage, incorporated into the two-stage forecasting approach. Finally, we investigate if, in the presence of multiple probabilistic forecasts of the same variable, it is better to combine the forecasts than to attempt to identify the single best forecasting model (Section 3). A general overview of the hybrid framework is indicated by the flowchart represented in Figure 1.

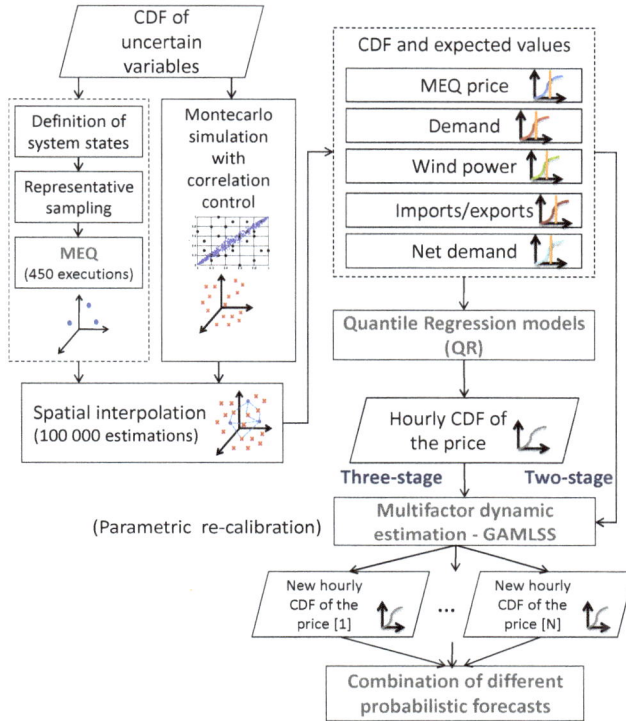

Figure 1. Overview of the methodology.

2.2. Implementation of the Proposed Methodology

The case studies in this paper use a data set that has been constructed from executions of the market equilibrium model for the Spanish day-ahead market for the period ranging between 1 May 2013 and 30 June 2014. It is sometimes observed that that the Spanish market is one of the most difficult to predict (see [20]) and certainly in this period of time it is particularly challenging with various structural and regulatory interventions.

With the objective of realistic medium-term predictions, 14 executions of the fundamental model (MEQ) have been accomplished, one per month. The forecasting horizon varies from one to two months. More precisely, for hourly predictions for month m, each execution of the fundamental model is carried out in a single step in the first hour of month m-1.

It should be noted that real ex ante forecasts of the probability distributions for all the exogenous risk factors such as the demand, wind generation, the unplanned unavailabilities of the thermal power units or fuel prices have been carried out using historical data. This set of distributions was generated in cooperation with the risk management team of a major energy utility active in the Spanish market at that time. In the case of fuel costs, the distribution functions were centered on the forward market expectations. The data corresponding to the Spanish market are available from the Iberian Energy Market Operator ([21]).

As for implementing the density recalibration model, in the second stage, the data set ranging from 1 May 2013 to 30 November 2013 that has been constructed with real ex ante predictions of the fundamental model was used for in-sample estimation of parameters, with the out-of-sample forecast evaluations being taken from 1 January 2014 to 30 June 2014. In order to thoroughly compare the forecasting capabilities, a series of multi-step forecasts with re-estimation of model parameters in an expanding window of one month was undertaken. In order to check the appropriateness of the specification of window length, we also experimented with rolling windows from 7 to 12 months. Thus, for making predictions for January 2014, the data set from 1 May 2013 to 30 November 2013 was used for estimation. Thereafter, in order to make forecasts for February 2014, the models were estimated from 1 May 2013 to 31 December 2013 and so on until June 2014.

The criterion used to evaluate the overall quality of the probabilistic forecasts is based on counting the number of observations that exceeds in each period of the out-of-sample data set the defined target percentiles: 1%, 5%, 30%, 50%, 70%, 95%, and 99%. The density recalibration model is essentially a multifactor adaptation of the Generalized Additive Model for Location, Scale, and Shape ("GAMLSS").

2.3. Generalized Additive Model for Location, Scale, and Shape

GAMLSS is a general framework that was proposed by [10] to overcome some relevant limitations of the well-known Generalized Linear Models (GLM, as proposed by [22]) and Generalized Additive Models (GAM, as introduced in [23]). The highly flexible GAMLSS models assume that the response variable presents a general parametric distribution (a GAMLSS model is parametric in the sense that it requires a parametric distribution assumption for the response variable, but this does not mean that the functions of explanatory variables cannot involve non-parametric smoothing functions such as splines). $F(\mu, \sigma, \upsilon, \tau)$ (these distributions will be explained in detail in Section 2.4) in which μ and σ are location and scale parameters and υ and τ represent the shape parameters. These parameters can be characterized by means of a wide number of functional forms and can change over time as a function of several covariates. From a mathematical point of view, let Y^S be the vector of y_h independent observations of the response variable for the hour $h = 1, ..., S$, with distribution function $F_Y(y_h; \theta_{kh})$, where θ_{kh} are the distribution parameters to predictors η_{kh} for $k = 1, 2, 3, 4$. Let g_{kh} be a known monotone link function relating the distribution parameters to explanatory variables and stochastic variables (random effects to deal with extra variability that cannot be explained by these explanatory variables) through

$$g_{kh}\left(\theta_{kh}^s\right) = \eta_{kh}^s = X_{kh}\beta_k + \sum_{j=1}^{J_k} Z_{jk}\gamma_{jk} \tag{1}$$

where θ_{kh}^S and η_{kh}^S are vectors of length S; X_{kh} is a known matrix of regressors of order $S \times J_k$; β_k is a vector of coefficients of length J_k; Z_{jk} is a fixed known $S \times q_{jk}$ design matrix and γ_{ik} is a q_{jk} dimensional random variable. The first term represents a linear function of explanatory variables and the second one represents random effects. It should be noted that Equation (1) can be equally extended to nonlinear functional terms. In order to restrict the possible combinations of structures of the general formulation, Equation (1), the functional relationship between the moments of the distributions and the covariates used in the proposed models (described later) was assumed linear, without random effects.

The estimation method for the vector of coefficients β_k and the random effects γ_{jk} is based on the maximum likelihood principle through a generalization of the algorithm presented in [24], which uses the first and (expected or approximated) second and cross derivatives of the likelihood function with respect to the distribution parameters, $\theta^S(\mu, \sigma, \upsilon, \tau)$. Due to the fact that computation of cross derivatives is sometimes problematic when the parameters $\theta^S(\mu, \sigma, \upsilon, \tau)$ are orthogonal (this is to say, the expected values of the cross derivatives in the likelihood function are null), a generalization of the algorithm developed in [25,26] has been used. This algorithm, which does not compute the expected values of the cross derivatives, is more stable (especially in the first iterations) and faster than the one developed by [24].

Thus, if in Equation (1) a distribution function of four parameters for the price $Y_i \sim D(y_i | \mu_i, \sigma_i, \upsilon_i, \tau_i)$ without random effects is considered, the likelihood to be maximized with respect to the β_k coefficients is represented as:

$$L\left(\beta_1, \beta_2, \beta_3, \beta_4\right) = \prod_{i=1}^{S} f\left(y_i | \beta_1, \beta_2, \beta_3, \beta_4\right) \tag{2}$$

where S is the number of observations. The likelihood is then maximized with an iterative algorithm that has both an outer and an inner cycle (the former one calls repeatedly the last one). The outer cycle is in charge of fitting the model of each distribution parameter θ^S $(\mu, \sigma, \upsilon, \tau)$ while the rest of the distribution parameters are fixed at their latest estimates values. Then, for each fitting of a distribution parameter θ^S $(\mu, \sigma, \upsilon, \tau)$, the inner cycle ckecks the maximization of the whole likelihood with respect to the β_k coefficients, for $k = 1, 2, 3, 4$. The outer cycle is continued until the change in the likelihood is sufficiently small. It should be noted that this algorithm requires initialization of the distribution parameter θ_0^S $(\mu_0, \sigma_0, \upsilon_0, \tau_0)$, but does not need initial values for the β_k parameters.

Following this approach, very occasional difficulties (less than 1% of the time) have arised regarding algorithm convergence. Mainly, these problems have occurred when: (i) the parametric distribution function for the electricity price is not adequate or not flexible enough; (ii) the starting values of the variables are not correctly defined; (iii) the structure of the functional form chosen is overspecified and too complex, particularly when trying to fit the higher moments υ and τ (which were the most challenging); and (iv) the step length in the Fisher's scoring algorithm is too wide. Some of these problems can be easily solved by fitting a series of models of increasing complexity. Thus, for instance, simpler models can provide starting values for the more complicated ones, as proposed in [10]. Moreover, the absence of multiple maxima has been also guaranteed by using several widely varying starting values. This point is particularly critical when the data set is small. Overall, the algorithm has been found to be fast and stable, especially when explicit derivatives are used (note that numerical derivatives can be used instead, but this results in higher computational time), which is coherent with [10].

2.4. Density Selection

Selecting the appropriate density function is a crucial aspect of the recalibration model. In this study, 32 continuous density functions with two, three, and four parameters among those listed by [27] were considered. Some distributions which have been widely used in the econometric literature were intially discarded.as being inappropriate. This is particularly the case with symmetric distributions (i.e., the t-distribution or the power exponential) since the flexible representation of skewness is important in this application. This is also the case with other distributions with three parameters (such as the skew normal), that do not show enough flexibility.

Overall, as expected, four-parameter distributions, which are able to model both skewness and kurtosis in addition to the location and scale parameters, demonstrated the best fit in terms of the global deviance criterion (which is defined as the negative of twice the fitted log likelihood function). As a basis for comparison in the later out of sample forecast validation, we retained the best four fitting densities, namely: the Box-Cox power exponential (BCPE), the Skew t type 3 (ST3) and the Skew exponential power types 2 and 3 (SEP2 and SEP3). These are all descrided in detail in [28]. For the sake of clarity, the basics of these density functions are presented in the Appendix.

2.5. Selection of the Regressors

One of the beneficial features of this approach is that the exogenous variables can be different for each of the four latent moment estimations. For the regressors, we considered both the expected values and the percentiles of the cumulative distribution function of the variables which included MEQ price, Demand, Net Demand (defined as demand less renewable production, which is a measure of the demand on the thermal price-setting generators), Wind generation, Exports, and Imports. The conventional stepwise procedure of moving from a more general model to a more specific one was followed (e.g., [29,30]). With backward elimination, we retained only statistically significant regressors at a 5% significance level, using the standard Chi-squared test, which compares the deviance change when the parameter is set to zero with a $\chi^2_{0.05}$ critical value.

The signs of the significant variables are summarized in Tables 1 and 2, for two representative models. These results present in general a coherent interpretation, mostly consistent with plausible expectations. In particular, it is noteworthy that a direct relationship between the percentiles of the MEQ price and the moments of the parametric distributions was evident across all the tested windows. In addition, Net Demand, Exports, and Wind generation output were also significant, out testing from a large range of exogenous variables. It is interesting that the kurtosis measure, which in risk measurement practice is usually taken to be an indicator of the fatness of the tails of the distribution, in the case of BCPE distribution is dependent on the 1st and 99th percentiles of the fundamental price. This is coherent and very similar to the interpretations in many quantile-based measures of peakedness [31–33] (for instance, in [33] the coefficient of kurtosis is given by: $KR = \frac{q_4 - q_0}{q_3 - q_1} - 2.91$ where $q_0 = F^{-1}(0.025)$, $q_1 = F^{-1}(0.25)$, $q_3 = F^{-1}(0.75)$, $q_4 = F^{-1}(0.975)$ and $F(y) \equiv P_0[Y_t < y]$ is the unconditional cumulative distribution function CDF of Y_t). Note also how the 99th percentile of the Net Demand, which is associated to demand shocks, has a negative impact in the kurtosis of SEP2 distribution. Regarding skewness, which conceptually describes which side of the distribution has a longer tail, it is also appealing to observe its direct relationship with the left tail of the MEQ distribution. This fact is related to a certain extent with some centile based measures of skewness proposed in the literature (for example, the Hinkley's measure of skewness [34], which is independent of position and scale and is given by: $S = \frac{(q_\alpha + q_{1-\alpha})/2 - q_{0.5}}{(q_\alpha - q_{1-\alpha})/2}$ where $q_\alpha = F^{-1}(\alpha)$ and $F(y) \equiv P_0[Y_t < y]$ is the unconditional CDF of Y_t. Note that a common value widely used for the α is 0.75). In addition, it seems that the expected wind production, which is a significant variable in the SEP2 distribution for increasing skewness, helps to justify changes in the skewness between different hours. Finally, it can be seen that the MEQ price and the net demand variables (both mean and median) are positive on price levels, indicating adaptive behavior.

Table 1. Summary of significant signs from the proposed multifactor model-BCPE.

Variable	μ	σ	v	τ
MEQ Price P50	+			
Net Demand P50	+			
MEQ Price P1		-	+	-
MEQ Price Mean		+		
MEQ Price P30		-		
MEQ Price P99				+
Intercept	-	-	+	+

Notes: +/- means that the variable has a positive/negative influence on the distribution parameter.

Table 2. Summary of significant signs from the proposed multifactor model-SEP2.

Variable	μ	σ	v	τ
MEQ Price Mean	+			
Net Demand Mean	+			
Exports Mean	-			
MEQ Price P1		-	+	
MEQ Price Mean		+		
Wind Mean		+		
MEQ Price P30			+	
Net Demand P99				-
Intercept	+	+	-	+

Notes: +/- means that the variable has a positive/negative influence on the distribution parameter.

2.6. Performance Analysis

In this subsection, the accuracies of the best four hybrid models are empirically evaluated when making ex ante forecasts for each hour of the validation period (1 January 2014 to 30 June 2014).

First, we illustrate the flexibility of these models in capturing a wide diversity of shapes of the probability density function (PDF). For example, Figure 2 presents the predicted hourly PDF for the particular case of the BCPE model during one week of June 2014. As can be seen, this approach is able to encompass, among other characteristics, the features of asymmetry and fat tails.

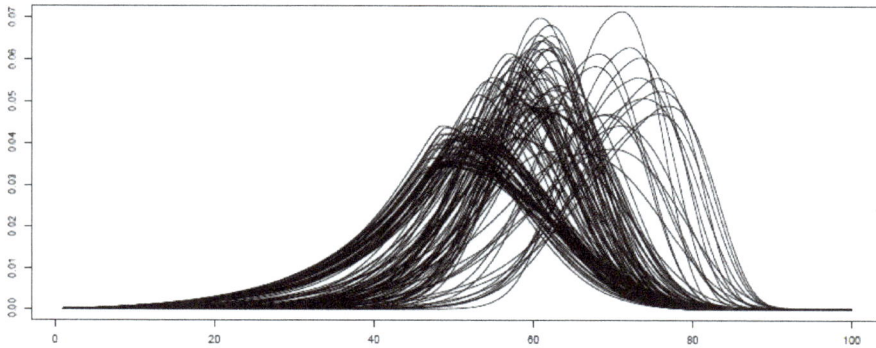

Figure 2. Predicted probability density function for a representative week-BCPE.

In addition to this, Figure 3 shows the PDFs that have been estimated on an hourly basis with the SEP2 model. The color scale ranges from white (null probability) to blue (the highest predicted probability). It is evident that the actual price always falls within the range of predicted prices. Moreover, the actual price for most days is near the mode value, and the model assigns a high probability of occurrence to it.

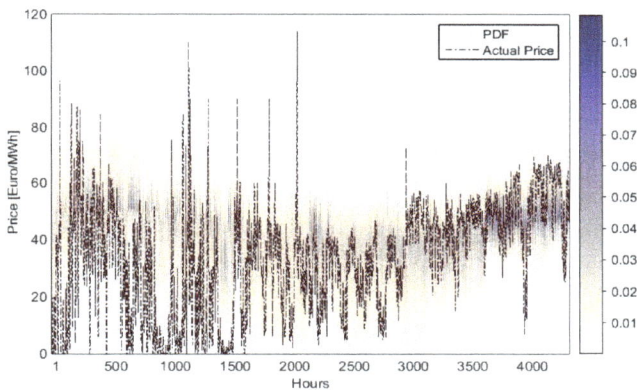

Figure 3. Comparison between the predicted PDF using SEP2 model and the actual price during the period from 1 January 2014 to 30 June 2014.

Table 3 shows the exceedence rate for the seven target percentiles throughout the validation sample for each of the models. The proposed recalibration process is benchmarked against the uncalibrated market equilibrium model MEQ percentiles and the hybrid approach based upon recalibrations using conventional quantile regression (denoted as MEQ-QR), which performed best in [4]. Overall, as seen in Table 3, the proposed methodology is not outperformed by the nonparametric MEQ-QR benchmark, which furthermore does not offer analytical solutions, but would have been expected to estimate the percentiles more precisely. Moreover, the extreme quantiles appear to be better calibrated with the proposed parametric approach, which is crucial in risk analysis. This is consistent with the well-known underperformance of quantile regression in the extreme tails because of the scarcity of data points. Regarding the purely fundamental model (MEQ), it is evident that it does not fully capture all of the important characteristics of the electricity price series, particularly in the tails of the price density function where behavioral factors are apparently more relevant than basic fundamentals. Overall, these results validate the proposed parametric recalibration process as an effective methodology to capture not only the projected fundamental changes in the market, but also behavioral aspects.

Table 3. Forecasting results for the target percentiles in terms of the exceedance rate.

(%)	P1	P5	P30	P50	P70	P95	P99
BCPE	93.43	81.95	55.50	34.63	19.86	0.18	0.02
SEP2	**96.20**	**90.27**	56.29	38.48	11.61	0.42	0.05
SEP3	91.82	85.84	56.64	39.00	22.38	3.34	0.11
ST3	91.93	84.71	58.17	**40.57**	22.77	4.02	**0.16**
MEQ	77.97	70.06	48.20	37.34	28.01	12.34	7.65
MEQ-QR	93.14	80.43	**61.81**	36.76	**30.92**	8.20	2.39

Note: The best value obtained in each target percentile is highlighted in bold.

3. Combinations for Increased Accuracy

In this section, we investigate if further modelling combinations can increase accuracy. We first observe that the MEQ-QR values could be taken as inputs to the GAMLSS recalibration process instead of the basic MEQ values. This approach has been indicated previously as the three-stage hybrid approach. Table 4 reports a summary of the exceedance rates across the target quantiles. These findings demonstrate, perhaps not surprisingly, that the effectiveness of the hybrid recalibration technique is apparently increased if both quantile regression and GAMLSS recalibrations are used in sequence for additional refinement in model specification. Note that this three-stage hybridization scheme allows a relaxation of the constraints imposed to avoid the crossing between the quantile curves (since the most common approach to estimate quantile regression curves is to fit a function for each target percentile individually, the quantile curves can cross when multiple percentiles are estimated. This can lead to a lack of monotonicity and therefore, to an inconsistent distribution for the response) as these restrictions are naturally forced in the parametric distributions. Consequently, this fact leads to a major flexibility and a reduction in possible bias.

Table 4. Forecasting results for the target percentiles in terms of the exceedence rates.

(%)	P1	P5	P30	P50	P70	P95	P99
MEQ-QR-BCPE	94.54	82.19	56.22	32.18	21.93	4.19	0.53
MEQ-QR-SEP2	95.30	93.04	**64.85**	**47.06**	28.91	6.84	2.57
MEQ-QR-ST3	95.19	93.29	77.19	58.45	31.65	**4.72**	**0.91**
MEQ-QR-SEP3	**95.59**	**94.74**	79.23	58.61	32.45	3.06	0.27
BCPE	93.43	81.95	55.50	34.63	19.86	0.18	0.02
SEP2	96.20	90.27	56.29	38.48	11.61	0.42	0.05
SEP3	91.82	85.84	56.64	39.00	22.38	3.34	0.11
ST3	91.93	84.71	58.17	40.57	22.77	4.02	0.16
MEQ-QR	93.14	80.43	61.81	36.76	**30.92**	8.20	2.39

Note: The best value obtained in each target percentile is highlighted in bold.

Also perhaps not surprising is that there is no clear winner amongst the different density models presented. It is appealing to observe that the three-stage approach with SEP2 distribution improves the forecasting accuracy in the center of the distribution, but loses precission at the tails in comparison to the two-stage approach. It is also of interest to note that a distribution such as ST3, that is not able to model platykurtosis in certain hours, gains accuracy in the upper tail of the distribution. This observation naturally leads to a proposal for combining the probabilistic forecasts.

It is well known for both point [35] and density [36–38] forecasting that combinations of forecasts may offer diversification gains and can provide insurance against possible model misspecification, data sets that are not sufficiently informative and structural changes (see [39]). Although the idea of combining forecasts in itself is not new, it has been barely touched upon in the context of electricity spot prices (see the discussion of [3]). Thus, we test different combination schemes on the methods (MEQ-QR-SEP2, MEQ-QR-SEP3 and MEQ-QR-ST3). Note that including MEQ-QR-BCPE, which is the worst model, in the combinations leads to poor forecasting performance. This is consistent, as expected, with [40] and others who have observed that it is advisable to restrict combinations to a few good models.

Unlike [41,42], we used the probabilistic forecasts instead of the point predictions. As stated in [35], combinations of probability density forecasts impose extra requirements beyond those that have been highlighted for combinations of point forecasts. The fundamental requirement is that the combination must be convex with weights restricted to the zero-one interval so that the probability forecast never becomes negative and always sums to one. Consistently with this prerequisite, we test equal weighting, as that is generally advocated as the most robust, and one performance-based weighting. For the latter, we use weights which are inversely proportional to the least absolute deviation (LAD).

The most natural approach to forecast averaging is the use of the arithmetic mean of all forecasts produced by the different models. This scheme, which is denoted as Equally Weighted Combination, EWC, is robust and is widely recommended in forecast combinations (see e.g., [43–45]). In Equation (3) the EWC scheme is represented, where W is the number of methods used (three in the case study here presented: MEQ-QR-SEP2, MEQ-QR-SEP3, and MEQ-QR-ST3), $\hat{y}_{h,w}^{\alpha}$ is the forecast of the price in hour h for the percentile α from parametric method w (i.e., $\hat{y}_{h,w}^{\alpha} = F_{h,w}^{-1}(\alpha)$, in which $F_{h,w}$ is the unconditional CDF) and $\hat{y}_{h,EWC}^{\alpha}$ is the final weighted prediction for the percentile α and hour h.

$$\hat{y}_{h,EWC}^{\alpha} = \frac{1}{W} \sum_{w=1}^{W} \hat{y}_{h,w}^{\alpha} \tag{3}$$

The other method for combining distributions we considered is based on the LAD, which hereinafter will be referred to as quantile regression averaging (QRA), and proceeds as follows:

(1) As in the previous combination approach, for each proposed parametric distribution function w (MEQ-QR-SEP2, MEQ-QR-SEP3, and MEQ-QR-ST3), the corresponding quantile functions $\hat{y}_{h,w}^{\alpha}$ are derived for the usual target percentiles, so that $\hat{y}_{h,w}^{\alpha} = F_{h,w}^{-1}(\alpha)$. Note as the distribution functions vary with time (i.e., as the information set of explanatory variables evolves) so will the quantile functions.

(2) The predicted quantile functions $\hat{y}_{h,w}^{\alpha}$ are combined for each percentile α using the asymmetric absolute loss function to yield the LAD regression. The LAD regression may be viewed as a particular case of quantile regression and intuitively, this method assigns specific weights for each percentile depending on the inverse of the absolute deviation error, so that larger weights are given to models that show smaller deviation error during the in-sample data set. Recall that the weights are sequentially updated after each additional moving window. As constructed quantile functions are sample unbiased, then we might expect that the weights' sum to unity and there is strong intuitive appeal for omitting the constant (see [46]).

Table 5 reports the accuracy metrics for the combinations. Again, there is not a clear winner in terms of forecasting performance. However, it seems that model averaging, following the discussion of [47] and the findings of [48] (in this reference, it is shown that simple combination schemes perform better than more sophisticated rules relying on estimating optimal weights that depend on the full variance-covariance matrix of forecast errors) and [35] can capture different aspects of market conditions (particularly when the different approaches contain distinct information), provide a model diversification strategy that can improve forecasting robustness. As might be expected, this improvement is more apparent in the central percentiles than in the tails. Nevertheless, the improvements from these more elaborate combining methods are not huge and if the focus is upon tail risks it is better to find the best method instead of a combination. In these results, the indications are that ST3 is the most accurate for the high tails, and SEP3 for the low tails. All of this further vindicates the basic recalibration process based upon GAMLSS methodology.

Table 5. Forecasting results for the target percentiles in terms of the excedance rates.

(%)	P1	P5	P30	P50	P70	P95	P99
QRA	95.14	90.46	74.52	**52.38**	31.80	3.53	1.97
EWC	95.36	93.61	**72.94**	54.57	**30.73**	5.80	1.59
MEQ-QR-SEP2	95.30	93.04	64.85	47.06	28.91	6.84	2.57
MEQ-QR-ST3	95.19	93.29	77.19	58.45	31.65	**4.72**	**0.91**
MEQ-QR-SEP3	**95.59**	**94.74**	79.23	58.61	32.45	3.06	0.27

Note: The best value obtained in each target percentile is highlighted in bold.

4. Conclusions

The parametric recalibration methodology, as presented in this paper, offers a potentially valuable technique for the practical use of fundamental market models and their translation into well-calibrated density forecasts. The recalibration process improves accuracy compared to state of the art baseline techniques and is not substantially outperformed by more elaborate combining methods. Especially for the tail percentiles, where risk management is most crucial, the GAMLSS recalibration procedure—allied to the selection of an appropriate four parameter density function—would appear to offer the most accurate and analytically attractive approach to price density forecasting. In addition, the dynamic estimation of the first four moments is potentially beneficial in many analytical models of derivative pricing and portfolio optimization.

We did not discuss in detail the underlying fundamental model that was being recalibrated. In principle, the density recalibration process as presented, is applicacable to outputs from any fundamental model, whether such a model is a simple supply function stack or a computationally-intensive market equilibrium model. One would expect, however, that with simpler fundamental predictors, the scope for recalibration would be greater and that more exogenous variables would be significant in the dynamic estimation of the regressor coefficients for the latent moments. Nevertheless, selecting the appropriate regressors is a delicate process and overfitting should be a crucial concern in applying this methodology. Extensive out-of-sample validation testing is clearly required.

Author Contributions: Antonio Bello conducted the research, Derek Bunn proposed the density recalibration approach, whilst Javier Reneses and Antonio Muñoz supervised this paper. In addition, Antonio Bello and Derek Bunn wrote the manuscript.

Conflicts of Interest: The authors declare no conflict of interest.

Appendix A. Density Functions

Appendix A.1. Box-Cox Power Exponential

The Box-Cox power exponential family distribution, which is denoted by $\text{BCPE}(\mu, \sigma, \upsilon, \tau)$, was introduced by [49]. It provides a model for a response variable exhibiting both skewness (positive or negative) and kurtosis (leptokurtosis or platykurtosis). The four distribution parameters define the shape of the curve. For the sake of clarity, Figure A1 plots the BCPE($\mu, \sigma, \upsilon, \tau$) distribution for different values of the parameters.

An identity link function has been assumed for $g_1(\cdot)$ and $g_3(\cdot)$, whereas logarithmic link functions have been assumed for $g_2(\cdot)$ and $g_4(\cdot)$ to ensure positivity for the parameters σ and τ of hourly prices. Remember that the terms $g_k(\cdot)$ for the k moments were previously defined in Equation (1).

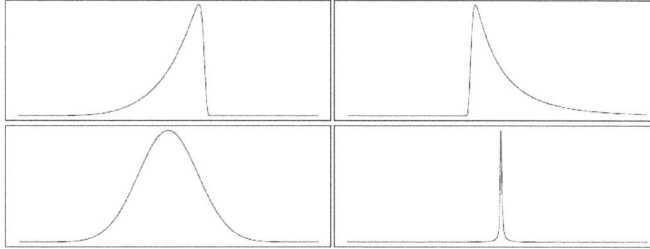

Figure A1. The PDF from a BCPE distribution for specific parameter values. Upper-left: BCPE(5, 0.2, 6, 10), upper-right: BCPE(5, 0.2, −4, 10), lower-left: BCPE(5, 0.2, 1, 2) and lower-right: BCPE(5, 0.2, 6, 0.1).

Appendix A.2. Skew Exponential Power Type 2

This distribution, which is flexible enough to incorporate a wide range of shapes, was introduced by [50] as his type 2 distribution and was further developed by [51]. Figure A2 shows the flexibility of this distribution for representative values of each one of the four parameters.

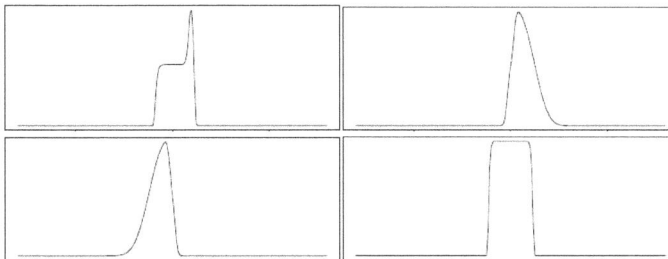

Figure A2. The PDF from a SEP2 distribution for specific parameter values. Upper-left: SEP2(0, 1, 10, 3), upper-right: SEP2(0, 1, 10, 1), lower-left: SEP2(0, 1, −10, 1), and lower-right: SEP2(0, 1, 0, 3).

An identity link function has been assumed for $g_1(\cdot)$ and $g_3(\cdot)$, whereas logarithmic link functions have been used for $g_2(\cdot)$ and $g_4(\cdot)$ to ensure positivity for the σ and τ of hourly electricity prices.

Appendix A.3. Skew Exponential Power Type 3

This four-parameter distribution is a "spliced-scale" distribution (see [28]) with a PDF that is denoted by SEP3($\mu, \sigma, \upsilon, \tau$).

An identity link function for $g_1(\cdot)$ has been used, whereas logarithmic link functions have been assumed for $g_2(\cdot)$, $g_3(\cdot)$, and $g_4(\cdot)$ to ensure positivity for the σ, υ, and τ of hourly electricity prices. Figure A3 reflects some representative shapes of SEP3 distribution for different parameter values.

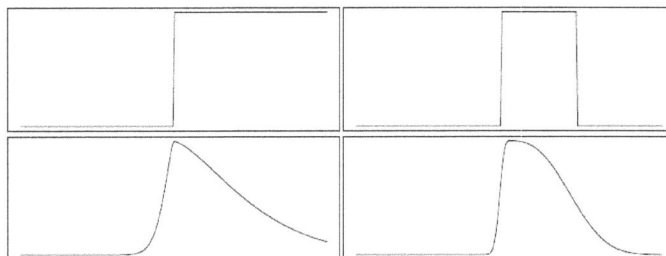

Figure A3. The PDF from a SEP3 distribution for specific parameter values. Upper-left: SEP3(0, 1, 7, 7), upper-right: SEP3(0, 1, 1, 7), lower-left: SEP3(0, 1, 1, 0.1), and lower-right: SEP3(0, 1, 1, 1).

Appendix A.4. Skew t Type 3

This four-parameter distribution (denoted by $ST3(\mu, \sigma, \upsilon, \tau)$), unlike the previous ones, is able to model only leptokurtosis. This can be seen more easily in Figure A4 for specific parameter values. It should be noted that an identity link function for $g_1(\cdot)$ has been assumed, whereas logarithmic link functions have been used for $g_2(\cdot)$, $g_3(\cdot)$, and $g_4(\cdot)$ to ensure positivity for the σ, υ, and τ of hourly electricity prices.

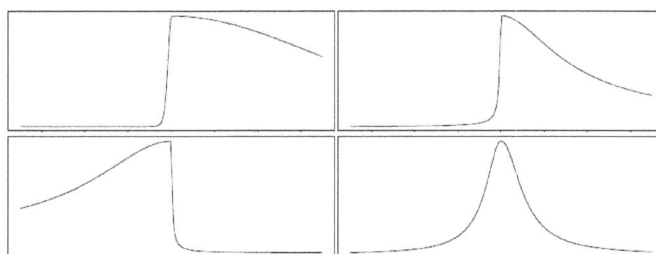

Figure A4. The PDF from a ST3 distribution for specific parameter values. Upper-left: ST3(0, 1, 7, 7), upper-right: ST3(0, 1, 7, 0.1), lower-left: ST3(0, 1, 0.1, 0.1), and lower-right: ST3(0, 1, 1, 1).

References

1. Coulon, M.; Jacobsson, C.; Ströjby, J. Hourly resolution forward curves for power: Statistical modeling meets market fundamentals. In *Energy Pricing Models: Recent Advances, Methods and Tools*; Prokopczuk, M., Ed.; Palgrave Macmillan: Basingstoke, UK, 2014.
2. Ziel, F.; Steinert, R. Electricity Price Forecasting using Sale and Purchase Curves: The X-Model. *Quant. Financ.* **2015**, *53*, 435–454. [CrossRef]
3. Weron, R. Electricity price forecasting: A review of the state-of-the-art with a look into the future. *Int. J. Forcast.* **2014**, *30*, 1030–1081. [CrossRef]
4. Bello, A.; Bunn, D.W.; Reneses, J.; Muñoz, A. Medium-Term Probabilistic Forecasting of Electricity Prices: A hybrid approach. *IEEE Trans. Power Syst.* **2016**. [CrossRef]
5. Westgaard, D.B.; Andresen, A.; Chen, S.D. Analysis and Forecasting of Electricty Price Risks with Quantile Factor Models. *Energy J.* **2016**, *37*, 1.
6. Hagfors, L.I.; Bunn, D.; Kristoffersen, E.; Staver, T.T.; Westgaard, S. Modeling the UK electricity price distributions using quantile regression. *Energy* **2016**, *102*, 231–243. [CrossRef]
7. Rigby, R.A.; Stasinopoulos, M.D.; Voudouris, V. Discussion: A comparison of GAMLSS with quantile regression. *Stat. Model.* **2013**, *13*, 335–348. [CrossRef]
8. Van Buuren, S. Worm plot to diagnose fit in quantile regression. *Stat. Model.* **2007**, *7*, 363–376. [CrossRef]

9. Nicolau, J. Nonparametric density forecast based on time- and state-domain. *J. Forecast.* **2011**, *30*, 706–720. [CrossRef]
10. Rigby, R.A.; Stasinopoulos, D.M. Generalized Additive Models for Location, Scale and Shape, (with discussion). *Appl. Stat.* **2005**, *54*, 507–554. [CrossRef]
11. Serinaldi, F. Distributional modeling and short-term forecasting of electricity prices by Generalized Additive Models for Location, Scale and Shape. *Energy Econ.* **2011**, *33*, 1216–1226. [CrossRef]
12. Bessembinder, H.; Lemmon, M.L. Equilibrium Pricing and Optimal Hedging in Electricity Forward Markets. *J. Financ.* **2002**, *57*, 1347–1382. [CrossRef]
13. Torbaghan, S.S.; Motamedi, A.; Zareipour, H.; Tuan, L.A. Medium-term electricity price forecasting. In Proceedings of the 2012 North American Power Symposium (NAPS), Champaign, IL, USA, 9–11 September 2012; pp. 1–8.
14. Alonso, A.M.; García, C.; Rodríguez, J.; Sánchez, M.J. Seasonal Dynamic Factor Analysis and Bootstrap Inference: Application to Electricity Market Forecasting. *Technometrics* **2011**, *53*, 137–151. [CrossRef]
15. Bello, A.; Reneses, J.; Muñoz, A.; Delgadillo, A. Probabilistic forecasting of hourly electricity prices in the medium-term using spatial interpolation techniques. *Int. J. Forecast.* **2016**, *32*, 966–980. [CrossRef]
16. Ziel, F.; Steinert, R.; Husmann, S. Efficient modeling and forecasting of electricity spot prices. *Energy Econ.* **2015**, *47*, 98–111. [CrossRef]
17. Bello, A.; Reneses, J.; Muñoz, A. Medium-term probabilistic forecasting of extremely low prices in electricity markets: Application to the Spanish case. *Energies* **2016**, *9*, 193. [CrossRef]
18. Barquín, J.; Centeno, E.; Reneses, J. Stochastic market equilibrium model for generation planning. *Probab. Eng. Inf. Sci.* **2005**, *19*, 533–546. [CrossRef]
19. Centeno, E.; Reneses, J.; Reneses, B. Strategic Analysis of Electricity Markets under Uncertainty: A Conjectured-Price-Response Approach. *IEEE Trans. Power Syst.* **2007**, *22*, 423–432. [CrossRef]
20. Jin, C.H.; Pok, G.; Lee, Y.; Park, H.-W.; Kim, K.D.; Yun, U.; Ryu, K.H. A {SOM} clustering pattern sequence-based next symbol prediction method for day-ahead direct electricity load and price forecasting. *Energy Convers. Manag.* **2015**, *90*, 84–92. [CrossRef]
21. Iberian Energy Market Operator (OMIE). Available online: http://www.omie.es/en/inicio (accessed on 20 February 2016).
22. Nelder, J.A.; Wedderburn, R.W.M. Generalized linear models. *J. R. Stat. Soc. Ser. A Gen.* **1972**, *135*, 370–384. [CrossRef]
23. Hastie, T.J.; Tibshirani, R.J. *Generalized Additive Models*; Chapman & Hall: London, UK, 1990.
24. Cole, T.J.; Green, P.J. Smoothing reference centile curves: The lms method and penalized likelihood. *Stat. Med.* **1992**, *11*, 1305–1319. [CrossRef] [PubMed]
25. Rigby, R.A.; Stasinopoulos, D.M. A semi-parametric additive model for variance heterogeneity. *Stat. Comput.* **1996**, *6*, 57–65. [CrossRef]
26. Rigby, R.A.; Stasinopoulos, M.D. Statistical Theory and Computational Aspects of Smoothing. In Proceedings of the COMPSTAT '94 Satellite Meeting, Semmering, Austria, 27–28 August 1994.
27. Stasinopoulos, D.; Rigby, R. Generalized Additive Models for Location Scale and Shape GAMLSS in R. *J. Stat. Softw.* **2007**, *23*, 1–46. [CrossRef]
28. Rigby, B.; Stasinopoulos, M. *A Flexible Regression Approach Using GAMLSS in R*; University of Lancaster: Lancaster, UK, 2010.
29. Gilbert, C. Professor Hendry's Methodology. *Oxf. Bull. Econ. Stat.* **1986**, *48*, 283–307. [CrossRef]
30. Brooks, C. *Introductory Econometrics for Finance*; Cambridge University Press: Cambridge, UK, 2008.
31. Andrews, D.F.; Bickel, P.J.; Hampel, F.R.; Huber, P.J.; Rogers, W.H.; Tukey, J.W. *Robust Estimates of Location: Survey and Advances*; Princeton University Press: Princeton, NJ, USA, 1973.
32. Kim, T.-H.; White, H. On more robust estimation of skewness and kurtosis. *Financ. Res. Lett.* **2004**, *1*, 56–73. [CrossRef]
33. Crow, E.L.; Siddiqui, M.M. Robust Estimation of Location. *J. Am. Stat. Assoc.* **1967**, *62*, 353–389. [CrossRef]
34. Hinkley, D.V. On power transformations to symmetry. *Biometrika* **1975**, *62*, 101–111. [CrossRef]
35. Timmermann, A. *Chapter 4: Forecast Combinations*; Elliott, G., Timmermann, A., Eds.; Elsevier: Amsterdam, The Netherlands, 2006.
36. Aastveit, K.; Gerdrup, K.; Jore, A.; Thorsrud, L. Nowcasting GDP in real time: A density combination approach. *J. Bus. Econ. Stat.* **2014**, *32*, 48–68. [CrossRef]

37. Hall, S.G.; Mitchell, J. Combining density forecasts. *Int. J. Forecast.* **2007**, *23*, 1–13. [CrossRef]
38. Wallis, K.F. Forecasting with an econometric model: The 'ragged edge' problem. *J. Forecast.* **1986**, *5*, 1–13. [CrossRef]
39. Hjort, N.L.; Claeskens, G. Frequentist Model Average Estimators. *J. Am. Stat. Assoc.* **2003**, *98*, 879–899. [CrossRef]
40. Zou, H.; Yang, Y. Combining time series models for forecasting. *Int. J. Forecast.* **2004**, *20*, 69–84. [CrossRef]
41. Nowotarski, J.; Weron, R. Computing electricity spot price prediction intervals using quantile regression and forecast averaging. *Comput. Stat.* **2015**, *30*, 791–803. [CrossRef]
42. Maciejowska, K.; Nowotarski, J.; Weron, R. Probabilistic forecasting of electricity spot prices using Factor Quantile Regression Averaging. *Int. J. Forecast.* **2015**, *32*, 957–965. [CrossRef]
43. Raviv, E.; Bouwman, K.E.; van Dijk, D.J.C. Forecasting day-ahead electricity prices: Utilizing hourly prices. *Energy Econ.* **2015**, *50*, 227–239. [CrossRef]
44. Genre, V.; Kenny, G.; Meyler, A.; Timmermann, A. Combining expert forecasts: Can anything beat the simple average? *Int. J. Forecast.* **2013**, *29*, 108–121. [CrossRef]
45. Stock, J.H.; Watson, M.W. Combination forecasts of output growth in a seven country data set. *J. Forecast.* **2004**, *23*, 405–430. [CrossRef]
46. Taylor, J.W.; Bunn, D.W. Combining forecast quantiles using quantile regression: Investigating the derived weights, estimator bias and imposing constraints. *J. Appl. Stat.* **1998**, *25*, 193–206. [CrossRef]
47. Bordignon, S.; Bunn, D.W.; Lisi, F.; Nan, F. Combining day-ahead forecasts for British electricity prices. *Energy Econ.* **2013**, *35*, 88–103. [CrossRef]
48. Bunn, D.W. Statistical efficiency in the linear combination of forecasts. *Int. J. Forecast.* **1985**, *1*, 151–163. [CrossRef]
49. Rigby, R.A.; Stasinopoulos, D.M. Smooth centile curves for skew and kurtotic data modelled using the Box-Cox power exponential distribution. *Stat. Med.* **2004**, *23*, 3053–3076. [CrossRef] [PubMed]
50. Azzalini, A. Further results on a class of distributions which includes the normal ones. *Statistica* **1986**, *46*, 199–208.
51. DiCiccio, T.J.; Monti, A.C. Inferential Aspects of the Skew Exponential Power Distribution. *J. Am. Stat. Assoc.* **2004**, *99*, 439–450. [CrossRef]

Chapter 5:
Ensemble and Portfolio Decision Models

energies

MDPI

Article

Portfolio Decision of Short-Term Electricity Forecasted Prices through Stochastic Programming

Agustín A. Sánchez de la Nieta *, Virginia González and Javier Contreras

E. T. S. de Ingenieros Industriales, University of Castilla-La Mancha, UCLM, 13071 Ciudad Real, Spain; virginia.glez.lopez@gmail.com (V.G.); Javier.Contreras@uclm.es (J.C.)
* Correspondence: agustinsnl@gmail.com; Tel.: +34-926-29-5300

Academic Editor: Tapas Mallick
Received: 29 August 2016; Accepted: 11 December 2016; Published: 16 December 2016

Abstract: Deregulated electricity markets encourage firms to compete, making the development of renewable energy easier. An ordinary parameter of electricity markets is the electricity market price, mainly the day-ahead electricity market price. This paper describes a new approach to forecast day-ahead electricity market prices, whose methodology is divided into two parts as: (i) forecasting of the electricity price through autoregressive integrated moving average (ARIMA) models; and (ii) construction of a portfolio of ARIMA models per hour using stochastic programming. A stochastic programming model is used to forecast, allowing many input data, where filtering is needed. A case study to evaluate forecasts for the next 24 h and the portfolio generated by way of stochastic programming are presented for a specific day-ahead electricity market. The case study spans four weeks of each one of the years 2014, 2015 and 2016 using a specific pre-treatment of input data of the stochastic programming (SP) model. In addition, the results are discussed, and the conclusions are drawn.

Keywords: ARIMA models; day-ahead electricity market price; forecasting portfolio; stochastic programming

1. Introduction

Electric energy systems have started being controlled by governments; since then, a constant search for improving such systems towards deregulation, seeking a competitive structure and attaining growth by satisfying the needs of society, has been pursued.

Electricity market prices have achieved a relevant importance as a consequence of the deregulation of the electricity sector. The importance of prices for the generation sector has been increasing in the last few years due to the high penetration of renewable energy sources, whose revenues come from selling the energy generated at market prices. Hence, price forecasting is still an active field of research, especially due to the incorporation of new technologies, such as wind and photovoltaic energy.

The high amount of renewable energies in the markets has decreased electricity market prices due to the fact that some renewable energies are offered in the market at zero price owing to their close-to-zero marginal costs. However, other effects in market prices can be observed since some generators could be working at a higher marginal cost to be used as reserves.

Literature Review and Contributions

Deregulation of electric energy systems started with the growth of the industry and the new demands for electricity [1,2]. After that, several research lines were created to reduce generation uncertainties, as presented in [3–6].

Thus, electricity market price forecasting has a high importance for generators, and electricity price forecasting is performed through different approaches [7], such as multi-agent, fundamental, reduced-form, statistical and computational intelligence methods.

Neural networks are presented in [8]. Another approach is based on a combinatorial neural network [9]. In addition, different models used for the Pennsylvania-New Jersey-Maryland (PJM) and Spanish electricity markets are compared in [10]. An artificial neural network with the preparation of input data through cluster algorithms is developed in [11]. The work in [12] combines an artificial neural network with a clustering algorithm. The works in [13–15] present time series analysis, forecasting and control models. Hence, [16] uses neural networks to forecast day-ahead market prices, while [17] forecasts through an autoregressive integrated moving average (ARIMA) model. Moreover, some models are based on forecasting the volatility [18] as a result of generalized autoregressive conditional heteroskedasticity (GARCH) models. In this way, forecasting trends of time series can be useful [19,20], as well as the use of filters [21].

Some forecasting methods are based on the combination or a portfolio of several models, as proposed by [22–25].

In this regard, an interesting procedure is presented in [26], proposing an enhanced hybrid approach composed of an innovative combination of wavelet transform, differential evolutionary particle swarm optimization and an adaptive neuro-fuzzy inference system to forecast electricity market price signals in the short-term through historical data.

Another work to estimate uncertainty uses a statistical approach for interval forecasting of the electricity price [27] based on a support vector machine (SVM) where some model parameters are estimated by means of maximum likelihood estimation (MLE). A possible accuracy gain from using factor models, quantile regression and forecast averaging to compute interval forecasts of electricity spot prices is evaluated in [28]. A general survey of support vector machines is shown in [29]. An ensemble method for weather conditions is described in [30].

This paper sets out a new stochastic programming model [31] combining many models whose forecasts are made by way of ARIMA models. Note that any forecast has an error because of future uncertainty.

In this paper, we propose a new stochastic programming model with many input data, which may help to reduce the error, where the combination of models comes from several forecasts. In contrast, perfect input data considering our forecast methodology could achieve a perfect forecast. However, this paper is only focused on the stochastic programming model and its features and not on the best input data for the model.

The main contributions of this paper are as follows:

1. A new stochastic programming model to create a forecasting portfolio.
2. A new combination approach using multiple input data in order to apply stochastic programming.

A description of the effects of input data on the created optimal forecasting portfolio is drawn. An application of forecasting in a real market with real-time series is also presented. A recent period is analyzed because the integration of renewable energy sources in the Spanish electricity market has produced a downward effect on the price from 2010 onwards, where the participation of renewable energy sources on some days (25 February 2015) was higher than 70%, as shown in [32].

The remainder of the paper is structured as follows: Section 2 describes the mathematical model, the case study and the input data, and the ARIMA models used are shown in Section 3; in Section 4, the results and a discussion are presented; and the conclusions are portrayed in Section 5.

2. Mathematical Model

The aim of the paper is to create a new model whose final error can be reduced by way of stochastic programming, after combining several forecasting models and their errors. The main decision made is the weight of each model per period seeking the lower error of the combination of forecasts.

A stochastic mixed integer linear programming model to combine the forecasting models and their portfolio (SP) is created. The variables of the stochastic programming model are $\Theta = \left\{ error_{p,s}, error_{p,s}^{+}, error_{p,s}^{-}, \hat{e}_{p,s}^{\pm}, \hat{e}_{p,s}^{+}, \hat{e}_{p,s}^{-}, \hat{\lambda}_{p}, \beta_{p,s}, y_{p,s}, z_{p,s} \right\}$.

$$\min_{\Theta} \quad \sum_{s} \sum_{p} \left(error_{p,s}^{+} + error_{p,s}^{-} \right); \tag{1}$$

subject to:

$$error_{p,s} = error_{p,s}^{+} - error_{p,s}^{-}; \tag{2}$$

$$error_{p,s}^{+} \leq m \cdot y_{p,s}; \tag{3}$$

$$error_{p,s}^{-} \leq m \cdot \left(1 - y_{p,s} \right); \tag{4}$$

$$error_{p,s}^{+} \geq 0; \tag{5}$$

$$error_{p,s}^{-} \geq 0; \tag{6}$$

$$\hat{e}_{p,s}^{+} = \hat{e}_{p,s}^{forecasts} \cdot z_{p,s}; \tag{7}$$

$$\hat{e}_{p,s}^{-} = -\hat{e}_{p,s}^{forecasts} \cdot \left(1 - z_{p,s} \right); \tag{8}$$

$$\hat{e}_{p,s}^{\pm} = \hat{e}_{p,s}^{+} - \hat{e}_{p,s}^{-}; \tag{9}$$

$$\hat{e}_{p,s}^{+} \geq 0; \tag{10}$$

$$\hat{e}_{p,s}^{-} \geq 0; \tag{11}$$

$$\hat{\lambda}_{p} = \sum_{s} \left(\lambda_{p,s}^{forecasts} \cdot \beta_{p,s} \right); \tag{12}$$

$$\sum_{s} \beta_{p,s} = 1; \tag{13}$$

$$error_{p,s} = \hat{\lambda}_{p} - \left(\lambda_{p,s} + \hat{e}_{p,s}^{\pm} \right); \tag{14}$$

$$\beta_{p,s} \in [0,1]; \, y_{p,s} \in \{0,1\}; \, z_{p,s} \in \{0,1\}. \tag{15}$$

where the objective is to minimize the variables related to the errors (1), positive $error_{p,s}^{+}$ or negative $error_{p,s}^{-}$, in each period p and scenario s. The error variables of (1) are both positive, as shown in (5) and (6); $error_{p,s}$ can be positive or negative, as shown in (3) and (4), which are not zero through binary variable $y_{p,s}$; depending on (5) or (6); if $y_{p,s} = 1$, $error_{p,s} = error_{p,s}^{+}$, whereas, if $y_{p,s} = 0$, $error_{p,s} = -error_{p,s}^{-}$ (negative). Constant m is a big enough value.

There are three input data, each one being a parameter, namely the forecasted errors $\hat{e}_{p,s}^{forecasts}$, the forecasted trends $\lambda_{p,s}$ and the forecasted prices $\lambda_{p,s}^{forecasts}$.

The forecasted error (parameter) is an error that corrects the forecasted trends and the forecasted prices, i.e., $\hat{e}_{p,s}^{forecasts}$. This parameter is the possible distance between the real price and the forecasted value. When the real price is lower than the forecasted price, $\hat{e}_{p,s}^{forecasts}$ is negative, and variable $\hat{e}_{p,s}^{-}$ is positive (11), i.e., $z_{p,s}$ binary variable is zero, so $\hat{e}_{p,s}^{\pm} = -\hat{e}_{p,s}^{-}$. The opposite case is $z_{p,s} = 1$ and $\hat{e}_{p,s}^{\pm} = \hat{e}_{p,s}^{+}$, where $\hat{e}_{p,s}^{+}$ is a positive variable, as shown in (10).

The forecasting portfolio is evaluated in (12), where $\lambda_{p,s}^{forecasts}$ is the parameter whose values are the forecasts made; one scenario for this parameter is a forecast per period, where variable $\hat{\lambda}_{p}$ is the final price of the portfolio created by the forecasts, $\lambda_{p,s}^{forecasts}$, and the weight that has to be decided,

$\beta_{p,s}$, whose sum, $\sum_s \beta_{p,s}$, has to be equal to one, as shown in (13). $\sum_s \beta_{p,s}$ could be different from one, even being an interval, and the model should decide what is the best value. In this paper, $\sum_s \beta_{p,s}$ is used as in (13).

Equation (14) decides the value of $\hat{\lambda}_p$ variable; this variable comes from the multiplication of $\lambda_{p,s}^{forecasts}$ and the weight $\beta_{p,s}$, being $\hat{\lambda}_p$ the final price from (12). Equation (14) reduces the error whose value is the difference between the variable $\hat{\lambda}_p$ and the parameters that are the input data of the model, such as $\lambda_{p,s}^{forecasts}$, $\lambda_{p,s}$ and $\hat{e}_{p,s}^{\pm}$. The parameter $\lambda_{p,s}$ represents the trend of the price; thus, more forecasts provide more information for the possible behavior of the real unknown price. On the other hand, the differences between $\hat{\lambda}_p$ and $\lambda_{p,s}$ can be corrected through parameter $\hat{e}_{p,s}^{\pm}$, but also, $\hat{e}_{p,s}^{\pm}$ tries to reduce the imbalance between $\hat{\lambda}_p$ and the real unknown price. Therefore, the forecasting portfolio could improve the ordinary forecasts, but always following parameter $\lambda_{p,s}^{forecasts}$.

To sum up, the stochastic programming model is composed of three kinds of input data. These input data are: (i) forecasted prices; (ii) errors of the forecasts that show the differences between the real price (unknown), forecasted price and the trend; and (iii) the trend of the prices (it could be obtained through more forecasts) in order to describe the possible evolution of the price. On the other hand, the variables of the model, $\Theta = \left\{ error_{p,s}, error_{p,s}^+, error_{p,s}^-, \hat{e}_{p,s}^{\pm}, \hat{e}_{p,s}^+, \hat{e}_{p,s}^-, \lambda_p, \beta_{p,s}, y_{p,s}, z_{p,s} \right\}$, depend on the input data and can be calculated through different techniques.

Index p represents the hour, from Hour 1 spanning the time horizon of forecasting, and index s is the scenario of the stochastic programming model; nevertheless, each s scenario of each input datum can be achieved using any technique to forecast the prices, the error and the trend.

3. Case Study

The forecasts are obtained for the Spanish electricity market prices. The prices are quite different for each year; Table 1 shows a summary of the statistics, and Table 2 presents the correlation matrix of the Spanish electricity prices from 2010–2015 [33]. The high standard deviation of prices of years 2013 and 2014 is remarkable, whose values are €20.73/MWh and €21.14/MWh, respectively. These values are two-times the standard deviation of the prices of 2011. This analysis is made using ECOTOOL (2016) [34], a forecasting toolbox in MATLAB® (R2011b) [35].

Table 1. Summary statistics of the Spanish electricity prices (€/MWh) of 2010–2015.

Statistics	2010	2011	2012	2013	2014	2015
Data points	8760	8760	8760	8760	8760	8760
Minimum	0.0	0.0	0.0	0.0	0.0	4.0
10% percentile	15.13	38.57	30.49	10.0	10.0	33.99
25% percentile	30.08	45.93	40.77	34.00	30.50	43.10
Mean	37.00	49.92	47.21	44.26	43.09	50.32
Geometric mean	-	-	-	-	-	48.24
Harmonic mean	10.85	24.70	13.99	4.86	6.48	44.54
Median	40.00	52.00	50.00	47.00	44.83	51.20
75% percentile	46.41	55.23	55.0	55.96	56.17	60.00
90% percentile	51.50	60.18	60.12	65.13	67.16	64.88
Maximum	145.00	91.01	90.13	112.00	113.92	85.05
Interquartile range	16.33	9.29	14.23	21.96	25.67	16.90
Range	36.37	21.60	29.63	55.13	57.16	30.89
Standard deviation	14.69	10.60	13.13	20.73	21.14	12.37
Variance	216.02	112.45	172.64	429.99	447.00	153.03
Mean absolute deviation	11.10	7.17	9.86	15.44	16.17	9.81
Median absolute deviation	7.77	3.99	6.52	10.32	12.34	8.50
Mean/Standard deviation	2.51	4.70	3.59	2.13	2.03	4.06
Skewness	−0.60	−1.33	−1.18	−0.29	−0.07	−0.71
Kurtosis	1.18	4.86	1.98	0.45	0.12	0.64

Table 2. Correlation matrix of the Spanish electricity prices of 2010–2015.

Years	2010	2011	2012	2013	2014	2015
2010	1	0.49	0.26	0.50	0.57	0.53
2011	0.49	1	0.30	0.31	0.40	0.28
2012	0.26	0.30	1	0.33	0.25	0.31
2013	0.50	0.31	0.33	1	0.58	0.37
2014	0.57	0.40	0.25	0.58	1	0.46
2015	0.53	0.28	0.31	0.37	0.46	1

Figure 1 portrays the scatter plots and the histograms of all of the hourly prices of each one of the six years. Red lines in the histograms indicate the shape of the normal distribution, whilst the green lines describe the real shape of the distribution that follows those data. Figure 2 depicts the sample of 2016, from 1 January–10 June 2016; the mean price is equal to €29.17/MWh, and the standard deviation of prices is €12.35/MWh. The time series of the day-ahead electricity market prices is transformed using a logarithmic transformation to make the dispersion constant.

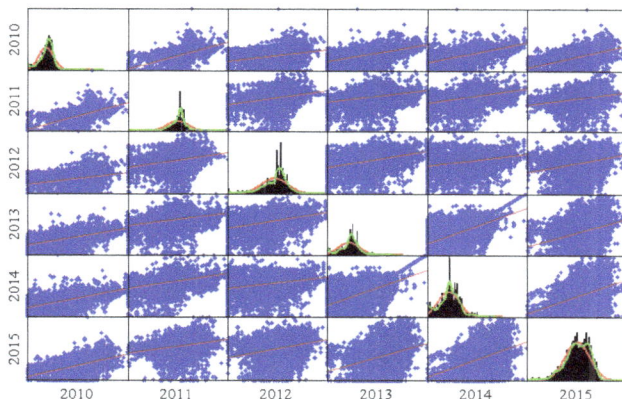

Figure 1. Scatter plot of the Spanish electricity prices of 2010, 2011, 2012, 2013, 2014 and 2015.

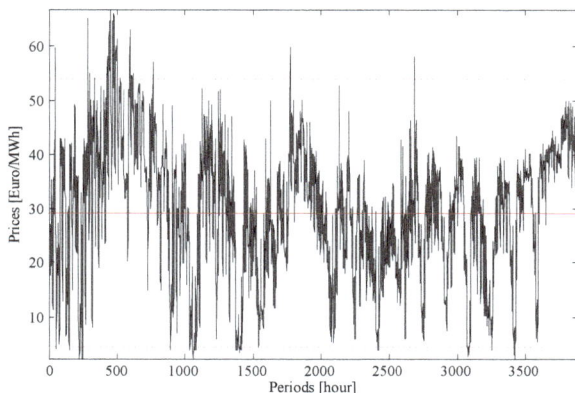

Figure 2. Prices from 1 January–10 June 2016 of the Spanish electricity prices.

3.1. Input Data for the SP Model

As presented in Section 2, there are three input data: forecasted prices, forecasted errors and forecasted trends. This section shows how these input data are calculated to test the SP model, achieving a portfolio of the forecasted prices, which are input data.

Figure 3 shows the three input data: forecasted prices, forecasted errors and forecasted trends; all of them represent the information used in the SP model. Moreover, Figure 3 shows the main output data, i.e., the final forecasted prices.

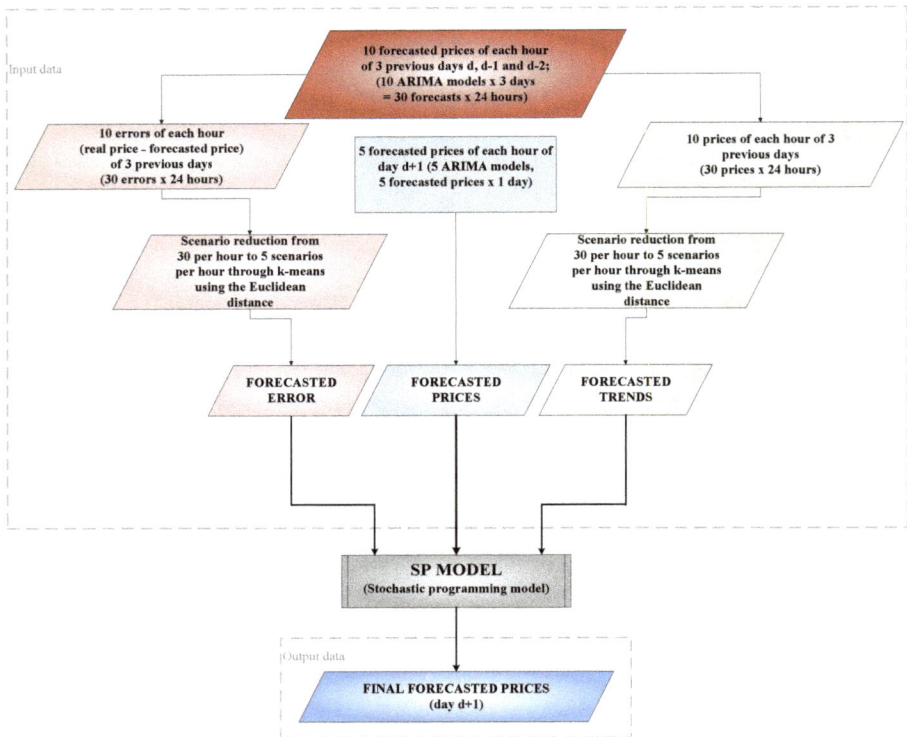

Figure 3. Diagram of the input data of the SP model.

Figure 3 presents how the different inputs are calculated. There are two forecasting processes; first, some forecasts are made for the three days $(d, d-1,$ and $d-2)$ previous to the final day $(d+1)$ in order to have more information; and second, some forecasts are made only for the final day $(d+1)$.

The first forecasting process is used to determine the behavior of the errors and trends of the three days previous to day $d+1$, where the models are similar to the models used in the second forecasting process. The three previous days are utilized because the real prices of these days are known; thus, the behavior of the errors and trends can be calculated from these forecasts. Thirty scenarios of trends and errors are attained for each hour of the day, 3 days $\times$ 10 models $\times$ 24 h. The 30 scenarios per hour are reduced to five scenarios per hour for the errors and trends. The scenario reduction from 30 scenarios per hour to five scenarios per hour is done by the k-means method using the Euclidean distance, where the centroid is the mean of the points of the cluster of the five scenarios.

The second forecasting process is applied to day $d + 1$, that is the real forecasted day, and five forecasted prices are obtained from the first five ARIMA models of Table 3. Five ARIMA models are used because the stochastic programming model presented in Section 2 is tested utilizing five scenarios.

Table 3. Terms of the ARIMA models.

Model	Seasonal Parameters $(1 - B_j^s)$	AR Parameters $\phi_{p_j}(B^{s_j})$	MA Parameters $\theta_{q_j}(B^{s_j})$
ARIMA1	1, 24	1, 2, 24	2, 24
ARIMA2	1, 24	1, 2, 7, 24	1, 2
ARIMA3	-	1, 2, 7, 9, 24	2
ARIMA4	-	1, 2, 7, 19, 24, 27	1
ARIMA5	-	1, 2, 3, 7, 19, 21, 24	1, 2
ARIMA6	-	1, 2, 3, 7, 9, 24, 27, 48	2
ARIMA7	-	1, 2, 3, 7, 19, 21, 24	1, 2
ARIMA8	1, 24	1, 24, 168	1
ARIMA9	1, 24	1, 2, 24, 168	1
ARIMA10	1, 24	1, 24, 168	2

Note that the forecasting processes previous to the SP model can be done by means of neural networks, support vector machines, ensemble methods or by a mixture of all of them, with the possibility of including other methods.

The information of the input data comes from the use of ARIMA models. The behavior of previous days for each forecast is used in order to obtain the forecasted error and the forecasted trend.

3.2. ARIMA Models

The proposed general ARIMA formulation [34] is as follows:

$$
\begin{aligned}
y_t = c + &\frac{1}{(1 - B)^{d_0}(1 - B^{s_1})^{d_1} \dots (1 - B^{s_k})^{d_k}} \\
&\times \frac{\theta_{q_0}(B)}{\phi_{p_0}(B)} \frac{\theta_{q_1}(B^{s_1})}{\phi_{p_1}(B^{s_1})} \times \dots \times \frac{\theta_{q_k}(B^{s_k})}{\phi_{p_k}(B^{s_k})} \epsilon_t.
\end{aligned}
\tag{16}
$$

where y_t is the observed time series, ϵ_t is the residual term, s_j, $j = 0, ..., k$ is a set of seasonal periods, $s_0 = 1$, $(1 - B_j^s)$, $j = 0, 1, ..., k$ are the $k + 1$ differencing operators necessary to reduce the time series and to achieve mean stationary, $\phi_{p_j}(B^{s_j})$ and $\theta_{q_j}(B^{s_j})$, $j = 0, 1, ..., k$ are the AR and MA polynomials of the back shift operator B: $B^l y_t = y_{t-l}$ of $\theta_{q_j}(B^{s_j}) = (1 + \theta_1 B^{s_j} + \theta_2 B^{2s_j} + \dots + \theta_j B^{q_j s_j})$, and c is a constant.

Previously, the time series is transformed through logarithmic transformation to stabilize the variance; after that, ARIMA models can be applied.

Following the formulation of (16), the indexes of each term for every ARIMA model are shown in Table 3. The process to select each term of each ARIMA model is based on the evaluation of each autocorrelation function (ACF) and partial autocorrelation function (PACF) of the residual component, ϵ_t, cf each ARIMA model, as shown in Figures 4 and 5.

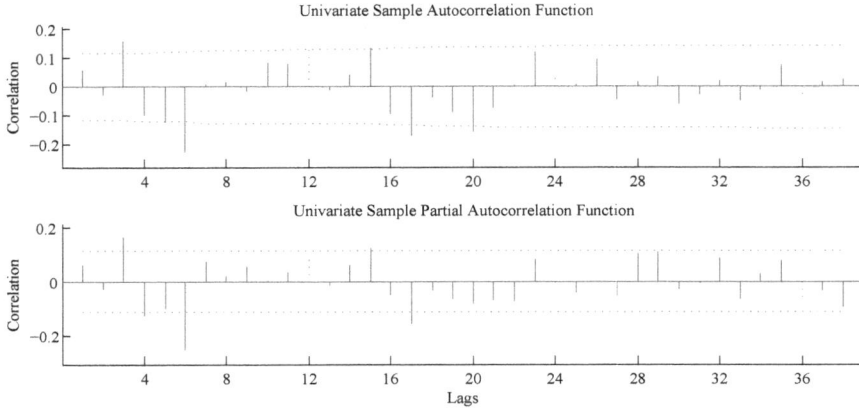

Figure 4. Autocorrelation function (ACF) and partial ACF (PACF) of the residual component for the ARIMA 2 model of 5 June 2016.

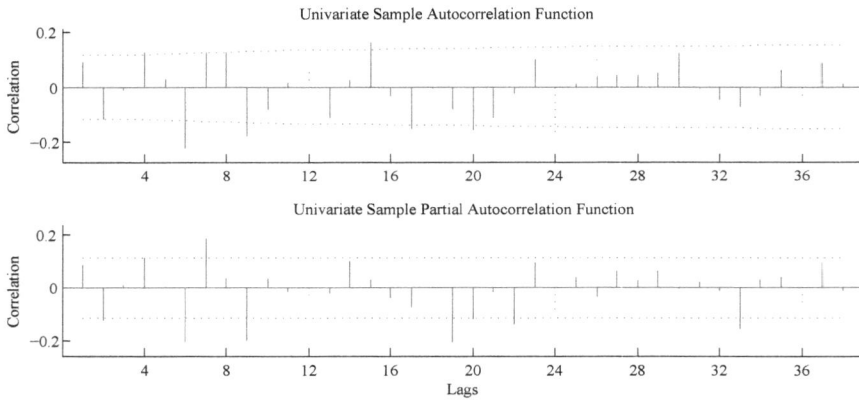

Figure 5. ACF and PACF of the residual component for the ARIMA 2 model of 6 June 2016.

Figures 4 and 5 portray the ACF and PACF of the residual terms for the ARIMA 2 model for 5 and 6 June 2016. The ARIMA model formulation of the ARIMA 2 model is presented in (17), where it is easy to identify every term.

$$y_t = c + \frac{1}{(1-B)(1-B^{24})} \frac{(1-\theta_1 B - \theta_2 B^2)}{(1-\phi_1 B - \phi_2 B^2 - \phi_7 B^7)} \frac{1}{(1-\phi_{24}B^{24})} \epsilon_t. \tag{17}$$

3.3. Forecasted Prices

Forecasted prices are obtained through ARIMA models; this case study uses five scenarios. These five scenarios for the input data of forecasted prices are ARIMA 1, ARIMA 2, ARIMA 3, ARIMA 4 and ARIMA 5, as presented in Table 3, where each ARIMA model represents one scenario. An econometric toolbox of MATLAB® [35], ECOTOOL [34], is used to obtain the forecasts. The sample used to forecast every day with each ARIMA model is 15 days, i.e., 360 h. A sample spanning 360 h has been selected because the sample changes every two or three weeks as a consequence of the price volatility. The computing time increases for a sample spanning more days.

The five scenarios of forecasted prices are depicted in Figure 6, where one week of forecasts is shown. The forecasted days span from the 4–10 June 2016. After this, the models are verified for one week of each season of 2014, 2015 and 2016, the forecasting horizon being 24 h using a 24-h rolling horizon window for the next day until every day of each week is evaluated.

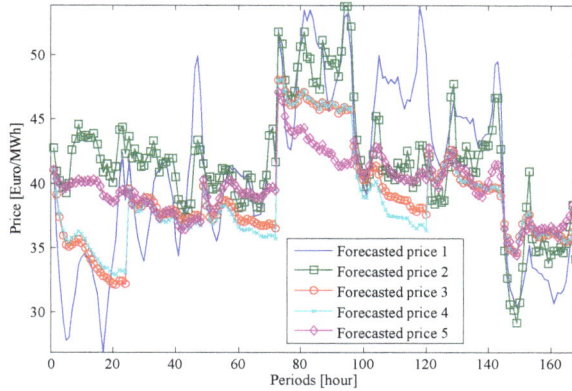

Figure 6. Input data of forecasted prices.

3.4. Forecasted Errors

Errors are the differences between the real price and the forecasted price, being positive when the real price is higher than the forecasted price and negative otherwise. It is remarkable that, if for the forecasting day the real price are unknown, then the error is also unknown. However, the error can be calculated for previous days since the real price is known. Thus, the ten ARIMA models of Table 3 are used to make forecasts of the three previous days. As a consequence of using 10 ARIMA models in these three days, the number of scenarios of forecasted errors would be 30 (10 ARIMA models multiplied by three days), but they can be reduced to five scenarios. Scenario reduction is performed through the squared Euclidean distance, and each centroid is the mean of the points in the cluster for five scenarios, reducing them from 30 down to five. The input data forecasted errors are these five scenarios.

The five scenarios obtained from the 30 scenarios are shown in Figure 7.

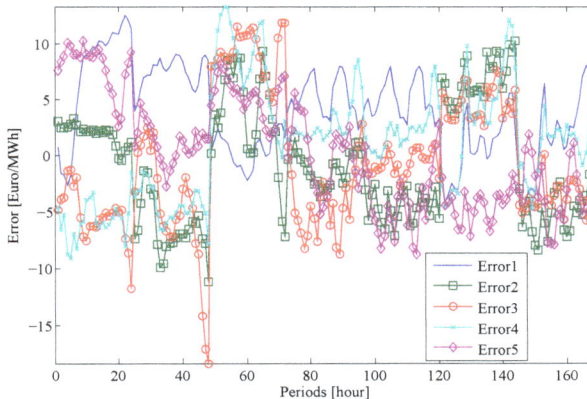

Figure 7. Input data of the forecasted errors.

3.5. Forecasted Trends

   Forecasted trends are made through the 10 ARIMA models of Table 3. The forecasted trend is made for the three previous days, trying to recover some behaviors of previous days. Therefore, as happened for the forecasted error input data, the forecasted trend has 30 scenarios, three days multiplied by 10 forecasts of the 10 ARIMA models. The scenarios are reduced through the squared Euclidean distance as done for the forecasted error. The five scenarios of forecasted trends are depicted in Figure 8.

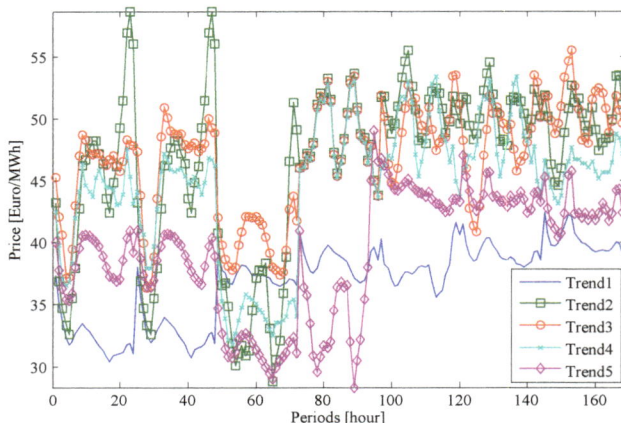

Figure 8. Input data of the forecasted trends.

4. Results and Discussion

4.1. Results

   This subsection shows the results obtained using this technique for the forecasting week. Firstly, the daily average errors (%) (DAE) of the week under study are calculated in (18), where p_t is the real price, $\hat{p}_t$ is the forecasted price and $\bar{p}$ is the average real price of the 24 hours. Secondly, the errors (18) from 1–10 June 2016 are presented in Table 4. Several input data are calculated through the forecasts of the three previous days, and due to that, Days 1, 2 and 3 of June are presented in Table 4, as well.

$$DAE(\%) = e_{day}(\%) = \frac{1}{24} \sum_{t=1}^{24} \frac{|p_t - \hat{p}_t|}{\bar{p}} \cdot 100. \tag{18}$$

Table 4. Daily average error (%) of 1–10 June 2016, for ARIMA Models 1–10.

Model/Day of June (%)	1	2	3	4	5	6	7	8	9	10
ARIMA 1	5.56	14.16	23.33	16.09	6.24	10.49	8.03	9.47	8.03	18.08
ARIMA 2	13.05	16.33	11.42	7.28	5.55	9.31	8.31	8.28	4.54	13.74
ARIMA 3	20.22	21.67	21.17	13.16	4.61	15.48	3.35	12.01	7.73	10.48
ARIMA 4	21.38	22.17	21.15	11.56	5.72	16.81	3.43	13.58	7.65	10.39
ARIMA 5	20.01	20.11	23.97	3.10	5.04	11.53	6.36	8.98	6.94	9.92
ARIMA 6	21.55	7.56	15.50	8.43	7.78	15.59	5.11	12.76	6.26	9.43
ARIMA 7	20.01	20.11	23.97	3.10	5.04	11.53	6.36	8.98	6.94	9.92
ARIMA 8	16.34	11.93	15.56	6.38	12.25	10.71	11.15	6.37	4.40	9.13
ARIMA 9	15.73	12.91	15.87	6.38	12.42	10.33	10.60	6.42	6.23	8.68
ARIMA 10	16.08	12.82	15.78	6.27	11.87	9.97	11.45	6.40	4.46	9.09

DAE for the SP model of Table 5 can be lower and higher than the average price forecasting (AVG), but this depends on the forecasted errors and trends used as input data, where the final error is lower when the forecasted trend and forecasted error follow the forecasted price. The differences between forecasted price, trend and error can increase or reduce the final error.

Table 5. Daily average error (%) of 4–10 June 2016, for the SP model, average, and ARIMA Models 1–5.

Model/Day of June (%)	4	5	6	7	8	9	10
SP	7.36	**2.91**	11.93	**2.99**	**7.55**	7.84	**9.16**
Average	7.17	3.15	12.73	3.36	7.92	5.57	12.19
ARIMA 1	16.09	6.24	10.49	8.03	9.47	8.03	18.08
ARIMA 2	7.28	5.55	9.31	8.31	8.28	4.54	13.74
ARIMA 3	13.16	4.61	15.48	3.35	12.01	7.73	10.48
ARIMA 4	11.56	5.72	16.81	3.43	13.58	7.65	10.39
ARIMA 5	3.10	5.04	11.53	6.36	8.98	6.94	9.92

Note that the numbers in bold indicate that the SP model has the lowest error compared to the average model.

The forecasted week of June of 2016 is portrayed in Figure 9, and Table 6 shows the weight of each forecasted price, $\lambda_{p,s}^{forecasts}$, introduced in the SP model for the second and third forecasted days, the best and worst forecasted day of the SP model, respectively.

For this case study, $\beta_{p,s} \in [0,1]$, but this value could be different allowing for an increase or a decrease in the forecasted price weight. The main dissimilarity between $\beta_{p,s}$ of Table 6 is for the best day; the percentage of each forecasted price of the input data is more distributed between two scenarios of the input data.

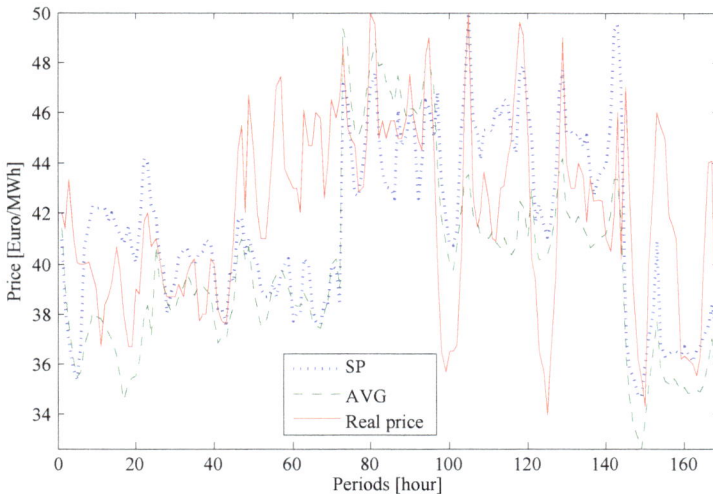

Figure 9. Prices of the SP model, average forecasted price (AVG) and real prices for the week of 4–10 June 2016.

Table 6. $\beta \in [0,1]$ values in each scenario of the forecasted price for the second and third days of the week.

Hour/Scenario	1	2	3	4	5	Hour/Scenario	1	2	3	4	5
25	0.94	0.06	0	0	0	49	0.53	0	0	0.47	0
26	0.60	0.40	0	0	0	50	0.04	0.96	0	0	0
27	0.53	0.47	0	0	0	51	0.21	0.79	0	0	0
28	0.65	0.35	0	0	0	52	0.16	0.84	0	0	0
29	0.35	0.65	0	0	0	53	0.24	0.76	0	0	0
30	0.28	0.72	0	0	0	54	0.43	0.57	0	0	0
31	0.27	0.73	0	0	0	55	0.81	0.19	0	0	0
32	0.37	0.63	0	0	0	56	0.28	0	0	0.72	0
33	0.82	0.18	0	0	0	57	0.49	0	0	0.51	0
34	0	0.54	0	0.46	0	58	0.70	0	0	0.30	0
35	0.73	0	0	0.27	0	59	0.78	0	0	0.22	0
36	0.70	0	0	0.30	0	60	0.26	0	0	0.74	0
37	0.75	0	0	0.25	0	61	0.38	0	0	0.62	0
38	0.62	0.38	0	0	0	62	0.41	0	0	0.59	0
39	0.36	0.64	0	0	0	63	0.90	0	0	0.10	0
40	0.15	0.85	0	0	0	64	0.32	0.68	0	0	0
41	0	0.75	0	0	0.25	65	0	0	0	0.28	0.72
42	0	1.00	0	0	0	66	0.89	0	0	0	0.11
43	1.00	0	0	0	0	67	1.00	0	0	0	0
44	0.80	0	0	0.20	0	68	0.93	0	0	0.07	0
45	0.47	0	0	0.53	0	69	0.59	0	0	0.41	0
46	0.40	0	0	0.60	0	70	0.51	0	0	0.49	0
47	0	0.76	0	0.24	0	71	0	0.46	0	0.54	0
48	0.35	0	0	0.65	0	72	0.44	0	0	0.56	0

After showing one week for the spring season of the current year, 2016, and in order to evaluate the forecasts per season, Tables 7–10 show several error measures, such as DAE, $FMSE$ and the error variance. The studied seasons are winter, spring, summer and autumn for the last three years, 2014, 2015 and 2016. The SP model is presented together with the average of the first five ARIMA models of Table 5 and the naïve model (see [10]), whose forecasts for Monday, Saturday and Sunday are the previous Monday, Saturday and Sunday, respectively, while for Tuesday, Wednesday, Thursday and Friday, the forecasts are their previous days, respectively.

The errors of Tables 7–10 are calculated in (18)–(21):

$$DAE = e_{day} = \frac{1}{24} \sum_{t=1}^{24} \frac{|p_t - \hat{p}_t|}{\bar{p}}; \tag{19}$$

$$e_{FMSE}^{day} = \sqrt{\frac{1}{24} \sum_{t=1}^{24} (p_t - \hat{p}_t)}; \tag{20}$$

$$\sigma_{e,day}^2 = \frac{1}{24} \sum_{t=1}^{24} \left(\frac{|p_t - \hat{p}_t|}{\bar{p}} \right)^2 - \left(e_{day}^2 \right). \tag{21}$$

Table 7. e_{day} (%), e_{FMSE}^{day} and $\sigma_{e,day}^2$ of winter of 2016, 2015 and 2014 for the SP model, average (AVG) and naïve models.

Day	Winter (January 2016)								
	SP			AVG			Naïve		
	e_{day} %	e_{FMSE}^{day}	$\sigma_{e,day}^2$	e_{day} %	e_{FMSE}^{day}	$\sigma_{e,day}^2$	e_{day} %	e_{FMSE}^{day}	$\sigma_{e,day}^2$
Saturday	19.03	7.95	0.0151	10.52	4.84	0.0080	22.24	10.74	0.0441
Sunday	13.44	6.16	0.0093	18.29	7.99	0.0126	68.60	25.78	0.0083
Monday	16.34	8.04	0.0093	20.31	9.90	0.0133	39.30	18.26	0.0311
Tuesday	20.34	11.97	0.0116	32.87	19.48	0.0322	18.82	11.18	0.0108
Wednesday	15.11	9.17	0.0071	23.33	14.45	0.0197	3.58	2.61	0.0012
Thursday	17.29	10.23	0.0102	9.67	5.95	0.0042	7.04	4.61	0.0032
Friday	6.67	3.94	0.0025	16.92	9.22	0.0095	10.38	5.69	0.0038

Day	Winter (January 2015)								
	SP			AVG			Naïve		
	e_{day} %	e_{FMSE}^{day}	$\sigma_{e,day}^2$	e_{day} %	e_{FMSE}^{day}	$\sigma_{e,day}^2$	e_{day} %	e_{FMSE}^{day}	$\sigma_{e,day}^2$
Saturday	8.85	5.60	0.0066	11.95	6.56	0.0056	21.75	11.28	0.0113
Sunday	14.60	8.02	0.0138	13.97	7.30	0.0096	18.06	8.78	0.0095
Monday	11.31	7.32	0.0069	22.39	13.66	0.0185	14.64	8.58	0.0056
Tuesday	10.31	7.87	0.0062	10.42	6.93	0.0023	13.87	9.73	0.0066
Wednesday	12.78	9.14	0.0144	14.38	9.95	0.0158	20.94	12.79	0.0164
Thursday	9.58	5.66	0.0033	12.53	7.34	0.0053	15.12	8.93	0.0082
Friday	4.40	3.14	0.0015	6.64	4.13	0.0015	6.36	4.46	0.0029

Day	Winter (January 2014)								
	SP			AVG			Naïve		
	e_{day} %	e_{FMSE}^{day}	$\sigma_{e,day}^2$	e_{day} %	e_{FMSE}^{day}	$\sigma_{e,day}^2$	e_{day} %	e_{FMSE}^{day}	$\sigma_{e,day}^2$
Saturday	19.65	10.94	0.0087	13.69	9.38	0.0161	90.07	46.07	0.0286
Sunday	27.86	11.92	0.0227	9.04	4.77	0.0079	89.13	33.98	0.0193
Monday	22.83	11.08	0.0294	23.16	11.68	0.0370	72.07	31.79	0.1515
Tuesday	43.30	16.27	0.0225	45.91	18.26	0.0536	45.27	18.13	0.0906
Wednesday	25.24	14.41	0.0199	21.99	12.16	0.0112	34.03	20.63	0.0555
Thursday	14.90	7.08	0.0066	18.42	9.00	0.0125	21.29	10.14	0.0136
Friday	24.22	11.33	0.0189	46.91	21.30	0.0539	11.94	7.05	0.0157

Table 8. e_{day} (%), e_{FMSE}^{day} and $\sigma_{e,day}^2$ of spring of 2016, 2015 and 2014 for the SP model, average and naïve models.

Day	Spring (June, 2016)								
	SP			AVG			Naïve		
	e_{day} %	e_{FMSE}^{day}	$\sigma_{e,day}^2$	e_{day} %	e_{FMSE}^{day}	$\sigma_{e,day}^2$	e_{day} %	e_{FMSE}^{day}	$\sigma_{e,day}^2$
Saturday	7.36	3.30	0.0015	7.17	3.18	0.0013	40.20	16.30	0.0068
Sunday	2.91	1.57	0.0007	3.14	1.68	0.0008	68.09	27.73	0.0232
Monday	11.93	5.63	0.0019	12.72	5.85	0.0012	20.39	9.73	0.0065
Tuesday	2.99	1.67	0.0004	3.36	1.77	0.0003	3.86	2.06	0.0005
Wednesday	7.55	3.75	0.0021	7.92	4.00	0.0026	9.01	4.73	0.0043
Thursday	7.84	3.93	0.0028	5.56	2.73	0.0012	5.05	2.90	0.0024
Friday	9.16	4.75	0.0057	12.18	5.82	0.0063	7.93	3.85	0.0030

Table 8. Cont.

Spring (June 2015)									
	SP			AVG			Naïve		
Day	e_{day} %	e_{FMSE}^{day}	$\sigma_{e,day}^2$	e_{day} %	e_{FMSE}^{day}	$\sigma_{e,day}^2$	e_{day} %	e_{FMSE}^{day}	$\sigma_{e,day}^2$
Saturday	12.78	7.72	0.0091	10.62	6.47	0.0066	8.86	5.15	0.0035
Sunday	9.98	5.95	0.0067	11.51	6.62	0.0074	7.64	4.09	0.0020
Monday	8.92	5.81	0.0028	11.35	7.20	0.0037	1.66	1.37	0.0003
Tuesday	10.00	6.57	0.0032	6.58	4.00	0.0005	2.42	1.78	0.0003
Wednesday	3.13	2.25	0.0005	4.45	3.44	0.0016	5.35	3.81	0.0016
Thursday	5.38	3.59	0.0005	5.17	3.52	0.0006	6.04	4.05	0.0008
Friday	3.52	2.75	0.0009	5.92	4.25	0.0018	5.71	3.89	0.0012
Spring (June 2014)									
	SP			AVG			Naïve		
Day	e_{day} %	e_{FMSE}^{day}	$\sigma_{e,day}^2$	e_{day} %	e_{FMSE}^{day}	$\sigma_{e,day}^2$	e_{day} %	e_{FMSE}^{day}	$\sigma_{e,day}^2$
Saturday	15.33	10.34	0.0115	10.88	7.03	0.0044	25.18	14.90	0.0093
Sunday	7.26	4.66	0.0030	8.40	5.04	0.0026	61.46	31.95	0.0123
Monday	14.12	10.28	0.0064	18.05	12.79	0.0081	35.02	24.02	0.0210
Tuesday	3.49	3.03	0.0009	5.77	3.86	0.0002	4.20	4.10	0.0023
Wednesday	11.86	8.77	0.0102	10.82	7.57	0.0063	15.56	11.41	0.0169
Thursday	2.68	2.05	0.0004	7.13	4.60	0.0007	6.76	5.41	0.0035
Friday	7.06	5.57	0.0029	5.84	4.12	0.0008	7.77	6.51	0.0047

Table 9. e_{day} (%), e_{FMSE}^{day} and $\sigma_{e,day}^2$ of summer of 2016, 2015 and 2014 for the SP model, average and naïve models.

Summer (August 2016)									
	SP			AVG			Naïve		
Day	e_{day} %	e_{FMSE}^{day}	$\sigma_{e,day}^2$	e_{day} %	e_{FMSE}^{day}	$\sigma_{e,day}^2$	e_{day} %	e_{FMSE}^{day}	$\sigma_{e,day}^2$
Saturday	12.74	6.17	0.0060	7.56	3.74	0.0024	11.98	5.40	0.0026
Sunday	8.99	4.18	0.0035	16.09	7.08	0.0073	15.70	6.46	0.0030
Monday	4.52	2.13	0.0006	4.69	2.32	0.0009	9.22	4.12	0.0015
Tuesday	9.19	5.20	0.0050	7.09	4.17	0.0036	9.20	5.63	0.0073
Wednesday	3.17	1.60	0.0003	2.12	1.32	0.0004	3.87	2.37	0.0015
Thursday	4.02	2.01	0.0005	4.27	2.11	0.0005	3.53	1.88	0.0006
Friday	3.45	1.78	0.0005	3.88	1.90	0.0005	4.64	2.39	0.0011
Summer (August 2015)									
	SP			AVG			Naïve		
Day	e_{day} %	e_{FMSE}^{day}	$\sigma_{e,day}^2$	e_{day} %	e_{FMSE}^{day}	$\sigma_{e,day}^2$	e_{day} %	e_{FMSE}^{day}	$\sigma_{e,day}^2$
Saturday	6.97	5.48	0.0051	7.89	5.53	0.0039	3.76	2.72	0.0011
Sunday	20.27	10.64	0.0114	14.25	7.64	0.0068	5.99	3.25	0.0013
Monday	6.50	5.22	0.0033	11.97	7.87	0.0027	4.38	3.65	0.0017
Tuesday	3.34	2.85	0.0009	5.91	4.06	0.0007	3.44	3.42	0.0019
Wednesday	5.11	4.50	0.0031	5.12	3.99	0.0019	4.99	4.36	0.0029
Thursday	9.22	6.29	0.0060	7.80	5.03	0.0032	13.81	8.48	0.0072
Friday	11.27	7.09	0.0038	12.91	8.46	0.0068	11.36	7.55	0.0058

Table 9. Cont.

Day	Summer (August 2014)								
	SP			AVG			Naïve		
	e_{day} %	e_{FMSE}^{day}	$\sigma_{e,day}^2$	e_{day} %	e_{FMSE}^{day}	$\sigma_{e,day}^2$	e_{day} %	e_{FMSE}^{day}	$\sigma_{e,day}^2$
Saturday	11.80	6.61	0.0048	10.67	5.71	0.0026	8.89	5.34	0.0043
Sunday	11.81	6.19	0.0062	8.25	4.27	0.0028	5.93	3.63	0.0034
Monday	4.71	3.29	0.0020	7.88	5.21	0.0043	5.46	3.59	0.0020
Tuesday	4.82	3.01	0.0015	5.52	3.41	0.0019	5.58	4.07	0.0040
Wednesday	14.48	7.66	0.0087	11.65	6.31	0.0065	10.10	6.14	0.0088
Thursday	8.28	5.14	0.0040	13.72	8.06	0.0077	10.65	6.32	0.0050
Friday	7.19	4.29	0.0038	9.05	4.61	0.0022	13.71	7.26	0.0070

Table 10. e_{day} (%), e_{FMSE}^{day} and $\sigma_{e,day}^2$ of autumn of 2016, 2015 and 2014 for the SP model, average and naïve models.

Day	Autumn (October 2016)								
	SP			AVG			Naïve		
	e_{day} %	e_{FMSE}^{day}	$\sigma_{e,day}^2$	e_{day} %	e_{FMSE}^{day}	$\sigma_{e,day}^2$	e_{day} %	e_{FMSE}^{day}	$\sigma_{e,day}^2$
Saturday	15.16	8.15	0.0089	15.21	7.98	0.0075	8.26	4.77	0.0041
Sunday	9.24	5.59	0.0065	8.72	6.02	0.0098	7.04	4.89	0.0065
Monday	17.35	12.08	0.0114	13.74	9.64	0.0076	10.52	6.69	0.0016
Tuesday	11.64	7.37	0.0028	7.65	4.67	0.0007	5.40	3.55	0.0008
Wednesday	8.71	5.90	0.0023	6.08	3.88	0.0005	3.24	2.11	0.0002
Thursday	2.56	1.65	0.0001	3.80	2.46	0.0002	2.63	2.10	0.0005
Friday	1.96	1.58	0.0003	5.71	3.58	0.0003	2.35	1.80	0.0003

Day	Autumn (October 2015)								
	SP			AVG			Naïve		
	e_{day} %	e_{FMSE}^{day}	$\sigma_{e,day}^2$	e_{day} %	e_{FMSE}^{day}	$\sigma_{e,day}^2$	e_{day} %	e_{FMSE}^{day}	$\sigma_{e,day}^2$
Saturday	18.66	10.40	0.0110	11.77	6.63	0.0048	5.69	3.48	0.0019
Sunday	6.94	3.78	0.0017	8.37	4.70	0.0030	19.95	10.82	0.0132
Monday	6.95	3.79	0.0014	6.15	3.45	0.0014	14.06	7.64	0.0057
Tuesday	9.92	5.43	0.0019	10.47	5.63	0.0017	6.25	3.84	0.0020
Wednesday	2.38	1.34	0.0001	5.30	2.84	0.0004	2.06	1.21	0.0001
Thursday	8.82	5.35	0.0019	10.59	6.55	0.0034	8.35	5.13	0.0020
Friday	8.71	5.62	0.0018	5.03	3.44	0.0009	8.16	6.88	0.0074

Day	Autumn (October 2014)								
	SP			AVG			Naïve		
	e_{day} %	e_{FMSE}^{day}	$\sigma_{e,day}^2$	e_{day} %	e_{FMSE}^{day}	$\sigma_{e,day}^2$	e_{day} %	e_{FMSE}^{day}	$\sigma_{e,day}^2$
Saturday	7.07	4.94	0.0026	8.65	6.07	0.0040	27.77	16.96	0.0126
Sunday	16.01	10.23	0.0179	11.65	7.42	0.0093	12.76	7.72	0.0085
Monday	6.93	5.87	0.0036	18.27	13.71	0.0124	16.94	14.54	0.0227
Tuesday	8.35	5.64	0.0016	12.59	8.18	0.0023	5.75	4.87	0.0031
Wednesday	7.63	6.22	0.0038	14.81	10.63	0.0061	5.20	4.64	0.0026
Thursday	5.81	4.25	0.0012	9.92	6.95	0.0023	5.19	4.66	0.0028
Friday	12.93	7.25	0.0032	10.08	7.04	0.0086	22.69	12.13	0.0042

The naïve errors presented in the previous tables are neither very high, as in winter (January 2014), nor low, as in autumn (October 2016). The SP model displays a more stable error than the AVG and naïve models, i.e., lower volatility across several errors. Furthermore, the error variance shows lower values. For example, in Table 10, for Friday, autumn (October 2014), the SP model produces a higher

e_{day} (%) error than the AVG model; nevertheless, $\sigma^2_{e,day}$ is more than two-times lower than the AVG error variance. This behavior is present for some days of the representative weeks per season. Table 11 shows the average values of $\bar{e}_{day}$ (%), $\bar{e}^{day}_{FMSE}$ and $\bar{\sigma}^2_{e,day}$ of each season for 2016, 2015 and 2014.

Table 11. Average values of $\bar{e}_{day}$ (%), $\bar{e}^{day}_{FMSE}$ and $\bar{\sigma}^2_{e,day}$ of each season for 2016, 2015 and 2014 in order to evaluate the SP model, average and naïve models.

Season & Year	SP			AVG			Naïve		
	$\bar{e}_{day}$ %	$\bar{e}^{day}_{FMSE}$	$\bar{\sigma}^2_{e,day}$	$\bar{e}_{day}$ %	$\bar{e}^{day}_{FMSE}$	$\bar{\sigma}^2_{e,day}$	$\bar{e}_{day}$ %	$\bar{e}^{day}_{FMSE}$	$\bar{\sigma}^2_{e,day}$
Winter (January, 2016)	**15.46**	8.20	0.0093	18.84	10.26	0.0142	24.28	11.26	0.0146
Winter (January, 2015)	**10.26**	6.67	0.0075	13.18	7.98	0.0084	15.82	9.22	0.0086
Winter (January, 2014)	**25.42**	11.86	0.0184	25.58	12.36	0.0275	51.97	23.97	0.0535
Spring (June, 2016)	**7.10**	3.51	0.0022	7.43	3.57	0.0020	22.07	9.61	0.0067
Spring (June, 2015)	7.67	4.94	0.0034	7.94	5.07	0.0032	**5.38**	3.44	0.0014
Spring (June, 2014)	**8.82**	6.38	0.0050	9.55	6.43	0.0033	22.27	14.04	0.0100
Summer (August, 2016)	6.58	3.29	0.0023	**6.52**	3.23	0.0022	8.30	4.03	0.0025
Summer (August, 2015)	8.95	6.01	0.0048	9.40	6.08	0.0037	**6.81**	4.77	0.0031
Summer (August, 2014)	**9.01**	5.17	0.0044	9.53	5.36	0.0040	**8.61**	5.19	0.0049
Autumn (October, 2016)	9.51	6.04	0.0046	8.70	5.46	0.0038	**5.63**	3.70	0.0020
Autumn (October, 2015)	8.91	5.10	0.0028	**8.24**	4.74	0.0022	9.21	5.57	0.0046
Autumn (October, 2014)	**9.24**	6.34	0.0048	12.28	8.57	0.0064	13.75	9.36	0.0081

Note that the numbers in bold indicate that the SP model has the lowest $\bar{e}_{day}$ (%) value compared to the average and naïve models.

The SP model has lower errors than the other models, and the errors are low in general. The naïve method has very low errors in some cases as a consequence of some days being very similar and with low prices [33].

4.2. Discussion

The forecasted prices of the SP model depend on the input data, where the main idea comes from Figure 10, which shows three lines: real price, forecasted price and forecasted trend of the real price. Thus, if we focus on Hour 13, Figure 10 shows the errors that are affecting the input data for the SP model. Hence, two errors are described, $e^{trend}_{p,s}$ and $e^{price}_{p,s}$. Thus, $\hat{e}^{\pm}_{p,s}$ of Equation (14) is defined in (22). The final error could be increased or reduced; $e^{trend}_{p,s}$ is evaluated from the forecasted trend, and $e^{price}_{p,s}$ comes from the forecasted price. In this way, if both errors tend to reduce the final error, the SP model will have a low final error, whilst if the error goes in the opposite way, in order to reduce the error, the SP model could produce a worse forecast with a high final error, as happens on the third day of the week studied, because the error will be the minimum of the sum of both errors, $e^{trend}_{p,s}$ and $e^{price}_{p,s}$.

$$\hat{e}^{\pm}_{p,s} = e^{trend}_{p,s} + e^{price}_{p,s}. \tag{22}$$

A summary of the input data that affect the final error is presented as follows:

- $p_t > \hat{p}_t$: the real price is higher than the forecasted price:
  Final error > 0: where $\hat{\lambda}_p < p_t$ as a result of $\lambda^{forecasts}_{p,s}$, $\lambda_{p,s}$ and $\hat{e}^{\pm}_{p,s}$ influence $\hat{\lambda}_p$ through $\beta_{p,s}$.
- $p_t < \hat{p}_t$, otherwise:
  Final error < 0: where $\hat{\lambda}_p > p_t$ as a result of $\lambda^{forecasts}_{p,s}$, $\lambda_{p,s}$ and $\hat{e}^{\pm}_{p,s}$ influence $\hat{\lambda}_p$ through $\beta_{p,s}$.

In short, $\hat{\lambda}_p$ follows $\lambda^{forecasts}_{p,s}$, whilst $\lambda_{p,s}$ and $\hat{e}^{\pm}_{p,s}$ influence $\beta_{p,s}$, increasing or decreasing $\beta_{p,s}$, the percentage of each forecasted price of the five ARIMA models.

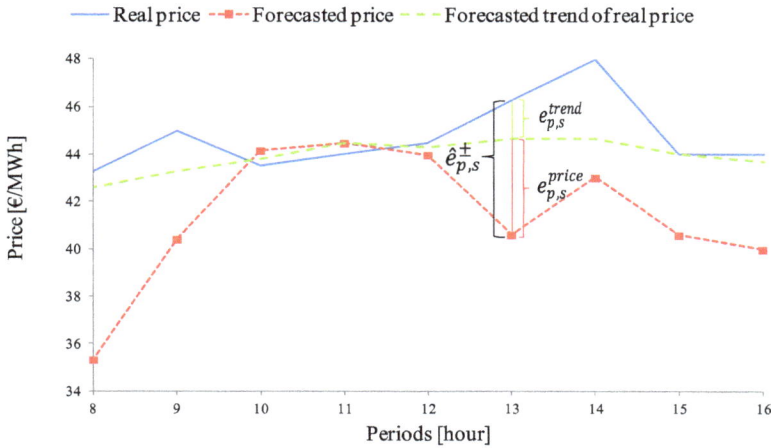

Figure 10. Description of $\hat{e}_{p,s}^{\pm}$.

For this reason, the input data can change the final forecasted price by means of β. Moreover, the $\beta_{p,s}$ value lies in the interval $\beta_{p,s} \in [0,1]$ where each value per period depends on $\lambda_{p,s}^{forecasts}$, $\lambda_{p,s}$ and $\hat{e}_{p,s}^{\pm}$. If $\beta_{p,s}$ is equal to zero for a specific scenario and period, the contribution of that scenario as a forecasted price is null, whilst for $\beta_{p,s} > 0$, it can contribute to the final price, $\hat{\lambda}_p$, and the sum of all $\beta_{p,s}$ terms for all scenarios has to be equal to one, as shown in (13).

5. Conclusions

In this paper, a new approach to improve the day-ahead electricity market price forecasting has been presented, where the error of the portfolio model can considerably reduce the final error, which depends on the quality of input data for the portfolio model. If the input data correctly follow the trend of prices and the error of the models used to create the portfolio, the final error may be very low. In contrast, with the average model, it would not be possible to achieve a perfect forecast. A portfolio approach could attain a zero final error if one input datum or a mix of forecasts closely follow the final error. Due to having more input data than the other models, this approach has higher flexibility. However, a drawback could be the incorrect creation of input data, increasing the final error. Hence, future work could be related to the creation of a better trend, error and forecasting data.

Acknowledgments: This work was supported in part by the Ministry of Economy and Competitiveness of Spain, under Project ENE2015-63879-R (MINECO/FEDER, UE), and the Junta de Comunidades de Castilla-La Mancha, under Project POII-2014-012-P.

Author Contributions: Agustín A. Sánchez de la Nieta and Javier Contreras proposed the methodology. Virginia González performed the simulations. All authors revised the manuscript.

Conflicts of Interest: The authors declare no conflict of interest.

References

1. Larsen, E.R.; Bunn, D.W. Deregulation in electricity: Understanding strategic and regulatory risk. *J. Oper. Res. Soc.* **1999**, *50*, 337–344.
2. Rothwell, G.S.; Gómez, T. *Electricity Economics: Regulation and Deregulation*; Wiley-IEEE Press: New York, NY, USA, 2003; Volume 12.
3. Pinson, P.; Chevallier, C.; Kariniotakis, G.N. Trading wind generation from short-term probabilistic forecasts of wind power. *IEEE Trans. Power Syst.* **2007**, *22*, 1148–1156.

4. Guerrero-Mestre, V.; Sánchez de la Nieta, A.A.; Contreras, J.; Catalão, J.P.S. Optimal bidding of a group of wind farms in day-ahead markets through an external agent. *IEEE Trans. Power Syst.* **2016**, *31*, 2588–2700.
5. Sánchez de la Nieta, A.A.; Contreras, J.; Muñoz, J.I. Optimal coordinated wind-hydro bidding strategies in day-ahead markets. *IEEE Trans. Power Syst.* **2013**, *28*, 798–809.
6. Sánchez de la Nieta, A.A.; Tavares, T.A.M.; Martins, R.F.M.; Matias, J.C.O.; Catalão, J.P.S.; Contreras, J. Optimal generic energy storage system offering in day-ahead electricity markets. In Proceedings of the 2015 IEEE Eindhoven IEEE PowerTech, Eindhoven, The Netherlands, 29 June–2 July 2015; pp. 1–6.
7. Weron, R. Electricity price forecasting: A review of the state-of-the-art with a look into the future. *Int. J. Forecast.* **2014**, *30*, 1030–1081.
8. Haykin, S.; Network, N. *Neural Networks: A Comprehensive Foundation*; Prentice Hall PTR: Upper Saddle River, NJ, USA, 2004; Volume 2.
9. Abedinia, O.; Amjady, N.; Shafie-Khah, M.; Catalão, J.P.S. Electricity price forecast using combinatorial neural network trained by a new stochastic search method. *Energy Convers. Manag.* **2015**, *105*, 642–654.
10. Conejo, A.; Contreras, J.; Espinola, R.; Plazas, M.A. Forecasting electricity prices for a day-ahead pool-based electric energy market. *Int. J. Forecast.* **2005**, *21*, 435–462.
11. Keles, D.; Scelle, J.; Paraschiv, F.; Fichtner, W. Extended forecast methods for day-ahead electricity spot prices applying artificial neural networks. *Appl. Energy* **2016**, *162*, 218–230.
12. Panapakidis, I.P.; Dagoumas, A.S. Day-ahead electricity price forecasting via the application of artificial neural network based models. *Appl. Energy* **2016**, *172*, 132–151.
13. Box, G.E.; Jenkins, G.M.; Reinsel, G.C.; Ljung, G.M. *Time Series Analysis: Forecasting and Control*; John Wiley & Sons: New York, NY, USA, 2015.
14. Pedregal, D.J.; Young, P.C. Statistical approaches to modelling and forecasting time series. In *Companion to Economic Forecasting*; John Wiley & Sons: New York, NY, USA, 2002; pp. 69–104.
15. Wei, W.W.S. *Time Series Analysis*; Addison-Wesley: Boston, MA, USA, 1994.
16. Amjady, N. Day-ahead price forecasting of electricity markets by a new fuzzy neural network. *IEEE Trans. Power Syst.* **2006**, *21*, 887–896.
17. Contreras, J.; Espinola, R.; Nogales, F.J.; Conejo, A.J. ARIMA models to predict next-day electricity prices. *IEEE Trans. Power Syst.* **2003**, *18*, 1014–1020.
18. Garcia, R.C.; Contreras, J.; van Akkeren, M.; Garcia, J.B.C. A GARCH forecasting model to predict day-ahead electricity prices. *IEEE Trans. Power Syst.* **2005**, *20*, 867–874.
19. Gardner, E.S., Jr.; McKenzie, E.D. Forecasting trends in time series. *Manag. Sci.* **1985**, *31*, 1237–1246.
20. Holt, C.C. Forecasting seasonals and trends by exponentially weighted moving averages. *Int. J. Forecast.* **2004**, *20*, 5–10.
21. Harvey, A.C. *Forecasting, Structural Time Series Models and the Kalman Filter*; Cambridge University Press: Cambridge, UK, 1990.
22. Granger, C.W.J.; Ramanathan, R. Improved methods of combining forecasts. *J. Forecast.* **1984**, *3*, 197–204.
23. García-Martos, C.; Rodríguez, J.; Sánchez, M.J. Mixed models for short-run forecasting of electricity prices: Application for the Spanish market. *IEEE Trans. Power Syst.* **2007**, *22*, 544–552.
24. Bessler, D.A.; Brandt, J.A. Forecasting livestock prices with individual and composite methods. *Appl. Econ.* **1981**, *13*, 513–522.
25. Raviv, E.; Bouwman, K.E.; van Dijk, D. Forecasting day-ahead electricity prices: Utilizing hourly prices. *Energy Econ.* **2015**, *50*, 227–239.
26. Osório, G.J.; Gonçalves, J.N.D.L.; Lujano-Rojas, J.M.; Catalão, J.P.S. Enhanced Forecasting Approach for Electricity Market Prices and Wind Power Data Series in the Short-Term. *Energies* **2016**, *9*, 693.
27. Zhao, J.H.; Dong, Z.Y.; Xu, Z.; Wong, K.P. A statistical approach for interval forecasting of the electricity price. *IEEE Trans. Power Syst.* **2008**, *23*, 267–276.
28. Maciejowska, K.; Nowotarski, J.; Weron, R. Probabilistic forecasting of electricity spot prices using Factor Quantile Regression Averaging. *Int. J. Forecast.* **2016**, *32*, 957–965.
29. Sapankevych, N.I.; Sankar, R. Time series prediction using support vector machines: A survey. *IEEE Comput. Intell. Mag.* **2009**, *4*, 24–38.
30. Taylor, J.W.; McSharry, P.E.; Buizza, R. Wind power density forecasting using ensemble predictions and time series models. *IEEE Trans. Energy Convers.* **2009**, *24*, 775–782.
31. Birge, J.R.; Louveaux, F. *Introduction to Stochastic Programming*; Springer: New York, NY, USA, 2011.

32. Information System of the Spanish System Operator (esios). Available online: https://www.esios.ree.es/en (accessed on 13 December 2016).

33. Operator of the Iberian Energy Market, Polo Español S.A. (OMIE). Available online: http://www.omie.es/files/flash/ResultadosMercado.swf (accessed on 13 December 2016).

34. Pedregal, D.J.; Contreras, J.; Sánchez de la Nieta, A.A. ECOTOOL: A general MATLAB forecasting toolbox with applications to electricity markets. In *Handbook of Networks in Power Systems I*; Springer: New York, NY, USA, 2012; pp. 151–171.

35. The Mathworks Inc., Matlab. Available online: http://www.mathworks.com (accessed on 13 December 2016).

energies

Article

Ensemble Prediction Model with Expert Selection for Electricity Price Forecasting

Bijay Neupane [1], Wei Lee Woon [2] and Zeyar Aung [2,*]

[1] Department of Computer Science, Aalborg University, Fredrik Bajers Vej 5, 9100 Aalborg, Denmark; bj21.neupane@gmail.com
[2] Department of Electrical Engineering and Computer Science, Masdar Institute of Science and Technology, Block 1A Masdar City, Abu Dhabi 54224, UAE; wwoon@masdar.ac.ae
* Correspondence: zaung@masdar.ac.ae; Tel.: +971-50-131-0423

Academic Editor: Javier Contreras
Received: 7 September 2016; Accepted: 4 January 2017; Published: 10 January 2017

Abstract: Forecasting of electricity prices is important in deregulated electricity markets for all of the stakeholders: energy wholesalers, traders, retailers and consumers. Electricity price forecasting is an inherently difficult problem due to its special characteristic of dynamicity and non-stationarity. In this paper, we present a robust price forecasting mechanism that shows resilience towards the aggregate demand response effect and provides highly accurate forecasted electricity prices to the stakeholders in a dynamic environment. We employ an ensemble prediction model in which a group of different algorithms participates in forecasting 1-h ahead the price for each hour of a day. We propose two different strategies, namely, the Fixed Weight Method (FWM) and the Varying Weight Method (VWM), for selecting each hour's expert algorithm from the set of participating algorithms. In addition, we utilize a carefully engineered set of features selected from a pool of features extracted from the past electricity price data, weather data and calendar data. The proposed ensemble model offers better results than the Autoregressive Integrated Moving Average (ARIMA) method, the Pattern Sequence-based Forecasting (PSF) method and our previous work using Artificial Neural Networks (ANN) alone on the datasets for New York, Australian and Spanish electricity markets.

Keywords: electricity price forecasting; ensemble model; expert selection

1. Introduction

Deregulated electricity markets are becoming increasingly common, in line with the evolution of a range of smart grid initiatives. In a traditional fixed-priced electricity market, consumption of electricity follows a distinct and more-or-less regular peak demand curve. This peak demand forces the supplier to use resources to meet the peak demand, and those resources are redundant for rest of the time. To overcome this inefficiency, the concept of demand management is put forward as a part of the smart grid initiative [1]. A smart grid utilizes the information about the behaviors of supplier and consumers of electricity and tries to optimize the production and the distribution of electricity.

The smart grid enables two-way peer-to-peer communications between the energy supplier (e.g., a retailer) and the consumers. This distributed information flow, which takes an Internet-like form, will enable the supplier to price the energy based on the consumption feedback from the consumer. On the other hand, the consumers can also schedule their consumption behavior to achieve optimal utilization at the lowest possible cost. In addition, nowadays, a substantial portion of energy is generated from renewable resources, like wind and solar, which are naturally less predictable than the traditional resources, like fossil fuel. All of these factors create a dynamism in the electricity market, under which the main concern for the supplier is to manage a healthy ratio between demand and supply. The general idea of demand management is to design a pricing mechanism that decides the hourly prices that can persuade the consumers to change their usage patterns in order to lower the peak demand, with the expectation that the consumers will respond to it. Another objective of this mechanism is to eliminate fluctuations in the demand beyond a defined threshold.

Under a dynamic pricing scheme, users of electricity will depend on the price per unit at a particular time of the day. A consumer has access to a retail electricity market, and he/she can make a decision on the time to buy the desired amount of electricity from the market. Thus, a cost-conscious consumer will be interested in the possible electricity prices in the coming hours, days or even weeks and will try to optimize his/her utilization and minimize the total bill through smart usage of electricity. With dynamic pricing systems where the consumers would pay based on their time of consumption and the amount of load they consume, it is essential for the consumers to have some price prediction mechanism to assist in scheduling their energy consumption strategy in advance.

In addition to the end consumers, price forecasting is equally important for other stakeholders in the deregulated electricity markets, like the wholesalers, traders and retailers. The ability to accurately forecast the future wholesale prices will allow them to perform effective planning and efficient operations, leading to ultimate financial profits for them.

The problem of electricity price forecasting is related, yet distinct from that of electricity load (demand) forecasting [2–5]. Although the load and the price are correlated, the relationship is non-linear. The load is influenced by various factors, such as the non-storability of electricity, consumers' usage patterns, weather conditions, social factors (like holidays) and general seasonality of demand. On the other hand, the price is affected by all of those aforementioned factors plus additional macro- and micro-economic factors, like the government's regulations, competitors' pricing, market dynamism, etc. As a consequence, the electricity price is much more volatile than the load, thus leading to occasional price spikes. A number of research works has been performed on electricity price forecasting [6–9]. However, to the best of our knowledge, none of them is able to provide adequately accurate results consistently for all of the cases for the respective experimental data of their target market. Thus, a more accurate price forecasting system is necessary to facilitate all of the stakeholders, where the consumers' consumption patterns will depend on the future electricity prices and so are the businesses of the wholesalers, the traders and the retailers.

A good price forecasting system should consider different factors associated with the dynamic pricing scheme in the smart grid and should be able to tackle it in an efficient manner. One of the main challenges in price forecasting under a dynamic pricing scheme is to overcome the aggregate demand response effect from the consumers, which causes sharp rises in peak demand, triggering sharp changes in prices. Different consumers have different priorities regarding the utilization of electricity under a dynamic pricing scheme; thus, their responses to a certain price value might vary substantially. This unpredictable behavior of the consumer might cause high fluctuations in the demand curve, which in turn causes higher fluctuations in electricity prices in a circular manner.

Our Contributions

In order to address the above challenges in electricity price forecasting, we propose an ensemble forecasting solution with the following contributions:

- It carefully engineers a set of features from information, such as past electricity price data (various price data from multiple viewpoints), weather data (temperature) and calendar data (days of the week and holidays). However, all holidays are treated equally, and special days, like Christmas, are not treated differently.
- It presents a wrapper method for feature selection that trains and automatically updates the algorithms to select the set of features best suited for the particular algorithm.
- It offers two different ensemble models, namely the Fixed Weight Method (FWM) and the Varying Weight Method (VWM), that iteratively evaluate the weights of the selected learning algorithms (denoted as an "expert"), and the final predictions are based on the assigned weights.

- It presents the fallback mechanism to tackle the fluctuation and aggregate demand response effect and ensures that the prediction accuracy lies within a desirable range.
- It performs an extensive evaluation of the proposed model with different datasets and experimental configurations.
- The experimental results show that the proposed ensemble model automatically selects the set of features and experts that are tailored to the particular market and best captures the trends, seasonality and patterns in the energy prices.

The performance of the proposed model is evaluated and compared with the results of the standard statistical time-series forecasting method called the Autoregressive Integrated Moving average (ARIMA) [10], the state-of-the-art symbolic forecasting method called Pattern Sequence-based Forecasting (PSF) [11] and our own previous work [12] on the same three datasets—New York (NYISO, New York Independent System Operator) [13], Australian (ANEM, Australian Energy Market Operator) [14] and Spanish (OMEL, Operador del Mercado Ibérico de Energía, Polo Español, S.A.) [15] markets. It is observed that our proposed ensemble learning model uses engineered features and expert selection to provide superior results. Previously, ensemble models for price prediction have been proposed in different fields, e.g., crude oil price [16] and carbon price [17]. However, to our best knowledge, our proposed model is the first to utilize ensemble learning involving different participating algorithms for the purpose of electricity price forecasting.

The remainder of the paper is organized as follows. Section 2 present the related works in energy price and demand forecasting. Section 3 presents our proposed forecast model with ensemble learning. Section 4 discuss the experimental setup, dataset and evaluation metrics. Sections 5 and 6 present the experimental results, discussion and analysis. Finally, Section 7 concludes the paper and provides directions for future research.

2. Related Work

Electricity price forecasting has become one of the most significant aspects in deregulated electricity markets for planning, production and trading. The positive economic consequences have attracted many stakeholders to invest time and money for the development of new methods for precise price prediction. This financial aspect has drawn immense interest to many researchers and has produced many significant research works and contributions in electricity price forecasting. This research thrust gains more momentum with the introduction of smart grid. The papers [6–9] provide good surveys on various methods of electricity price forecasting. We will discuss some of the existing electricity price forecasting methods below.

In [18], the authors proposed an Autoregressive Integrated Moving Average (ARIMA)-based statistical model of electricity price forecasting. The model was based on wavelet transformation where final forecasted results were obtained by applying inverse wavelet transformation. In [19], the proposed method was an augmented ARIMA model, which was an enhancement of the Box and Jenkins [20] model. Tan et al. [21] performed electricity price forecasting using wavelet transform combined with ARIMA and another statistical model, namely Generalized Autoregressive Conditional Heteroskedasticity (GARCH). The work in [22] applied a mixture of wavelet transform, linear ARIMA and nonlinear neural network models to predict normal prices and price spikes separately.

In [23–25], the authors proposed different prediction models using the Artificial Neural Network (ANN). Each proposed model utilized different sets of features created using historical market clearing price, system load and fuel price. The range of model varies from a simple three-layer architecture to combination models, including the Probability Neural Network (PNN) and Orthogonal Experimental Design (OED). In [25], the author implemented PNN as a classifier, which showed the advantage of a fast learning process, as it requires a single-pass network training stage for adjusting weights. OED was used to find the optimal smoothing parameter, which helps to increase prediction accuracy.

In [26,27], the authors proposed price prediction models using Support Vector Regression (SVR). The work in [26] used projected Assessment of System Adequacy (PASA) data as one of the inputs for the model, and that in [27] implemented the Artificial Fish Swarm Algorithm (AFSA) for choosing the parameter of SVM models.

Several ensemble-based forecasting models can be found in the literature [28–33]. For most of the previously proposed models, the final forecast is made as a simple or weighted average of the output of the participating algorithms, i.e., perform a collaborative or voted decision. However, this paper proposes a method where the final forecast is the output from a single expert system, i.e., the one with the highest weight, where the weights are periodically updated based on the models' accuracy. The main idea behind this approach is to find an expert that can reliably capture the current trends in prices rather than making decisions based on the group of amateurs.

Other diverse approaches for electricity price and load forecasting include Self-Organizing Map (SOM) [34], hybrid Principal Component Analysis (PCA) [35], Data Association Mining (DAM) [36], the Bayesian Method [37], Fuzzy Inference [38], Multiple Regression [36], Kernel Machine [39], Neural Networks [32,40], Particle Swarm Optimization (PSO) [41], etc.

Martínez-Álvarez et al. [11] presented the Pattern Sequence-based Forecasting (PSF) algorithm to produce one step-ahead forecasts of the electricity prices based on pattern sequence similarity. K-means clustering was first applied before the sequence similarity research. Experiments were conducted on three different electricity markets, namely New York (NYISO) [13], Australia (ANEM) [14] and Spain (OMEL) [15], for the years 2004–2005, while testing is carried out using data from 2006. Experimental results showed that PSF provided better accuracy than other methods (like ARIMA, naive Bayes, ANN, WNN, etc.) did. As this work is relatively recent and the three datasets used are publicly available, we use the results described in this research as benchmarks in order to evaluate those obtained by our proposed ensemble-based method.

We also compared the proposed model with our previous work [12], where we have implemented the Artificial Neural Network (ANN) model for price forecasting for the same publicly available 2004–2006 datasets from the three electricity markets as in PSF [11], as well as the more recent 2008–2012 datasets from the same markets. The ANN-based model showed promising results and was able to obtain higher forecasting accuracy compared to PSF. The results of our newly proposed ensemble-based model are also compared with those obtained from this ANN-only approach.

3. Proposed Prediction Model

Our objective is to develop a robust model that can sustain its good performance irrespective of various uncertainty factors. For that, we propose an ensemble prediction model that provides flexibility in choosing the type of algorithm for price prediction. This flexibility enables the user to choose the algorithm based on available resources, time constraints and computational complexity.

We believe that incorporating the modified ensemble learning [42] scenario into the well-known prediction methods will help to improve the performance of the prediction model. With the current research on price prediction, machine learning algorithms, like Artificial Neural Network (ANN) [43], Support Vector Regression (SVR) [44] and Random Forest (RF) [45] showed promising results. Hence, this paper proposes an ensemble learning strategy with ANN, SVR and RF as the members of the expert algorithm that learn from the environment and update their parameters based on the information they have collected.

3.1. Model Formulation

Consider a wholesale electricity market where a retailer proposes an hour-ahead bidding price based only on information that is available at the present moment. Once the actual price is known, the retailer is able to evaluate the validity of the predictions and will seek to minimize the difference between the actual and predicted market prices.

Now, let us consider an ensemble forecasting model involving a number of forecasting algorithms. Let $A = \{a_1, a_2, a_3, \ldots, a_n\}$ denote a set of participating forecasting algorithms, where n is the number of participating algorithms and $a_i (1 \leq i \leq n)$ is an individual participating algorithm.

Each hour of the day is treated separately, such that, in total, 24 separate ensemble forecasting models are constructed, one for each hour $h \in \{1, \ldots, 24\}$. For the sake of simplicity, we omit this hour parameter in our description of the proposed model below. Therefore, unless stated otherwise, all variables used below belong to an individual hour h of the day.

The prediction error of an algorithm $x \in A$ is defined as follows:

$$\text{prediction error: } E_x = |P'_x - P| \tag{1}$$

where P is the actual price and P'_x is the price predicted by algorithm x. Let W_i ($1 \leq i \leq n$) be the past performance "weight" of algorithm a_i (the performance weights are calculated using one of two algorithms, the Fixed Weight Method (FWM) or the Varying Weight Method (VWM), which will be explained in detail later).

The algorithm whose past performance weight W_i is the highest on a given day is selected as the "expert" algorithm for that day and is denoted as $\hat{a}$.

$$\text{expert algorithm: } \hat{a} = \underset{a_i \in A}{\arg \max} \, (W_i) \tag{2}$$

In most cases, we will use the forecasts produced by the expert algorithm as our final prediction result. This is due to our expectation that the algorithm with the highest performance weight (i.e., the one that has performed the best recently) will also give us the best prediction result for today. However, this expectation might not always be realistic. For example, if the best performing algorithm is rotating among all of the participating algorithms, the recent best performer may not be today's best performer, and consequently, the expert algorithm we have selected for today may not be actually optimal for today. Thus, in order to alleviate this effect and to make sure that the our prediction result of our ensemble algorithm on average is at least as good as that of the best individual algorithm, we include the following fallback mechanism.

Suppose we make the observations of our forecasting process for m number of days. Then, we have a list $\hat{A}$ of containing m expert algorithms.

$$\text{list of experts: } \hat{A} = [\hat{a}^{(1)}, \hat{a}^{(2)}, \ldots, \hat{a}^{(m)}] \tag{3}$$

Our expectation is that over m days, the overall performance of the list $\hat{A}$'s expert algorithms on their corresponding days should be superior to that of any individual participating algorithm acting alone. In order words, the cumulative prediction error incurred by our selected expert algorithms should be less than that of any individual algorithm. Formally, we should have:

$$\sum_{j=1}^{m} E_{\hat{a}^{(j)}} \leq \sum_{j=1}^{m} E_{x^{(j)}}, \ \forall x \in A \tag{4}$$

Therefore, in the proposed algorithms, the constraint in Equation (4) above is checked at every round. If the past cumulative prediction error of the selected expert algorithms over m days exceeds that of any of the individual algorithms, the forecasts produced by the best individual algorithm are then used as the final prediction result. In addition, all participating algorithms are constantly re-trained using all available data to date to address the problem of concept drift [46], which often occurs in time-series data like electricity prices.

3.2. Model Architecture

The proposed model for electricity price prediction is presented in Figure 1. We have to recruit prediction algorithms that exhibit promising results when utilized separately to participate in our ensemble model. Here, we show three participating algorithms for demonstration purposes (in theory, any number of different algorithms can be used under this model depending on the processing power and time available). The proposed model performs feature engineering on the price data along with the corresponding temperature and calendar data collected from a de-regulated electricity market, which is followed by feature selection, learning, predicting and model updating steps.

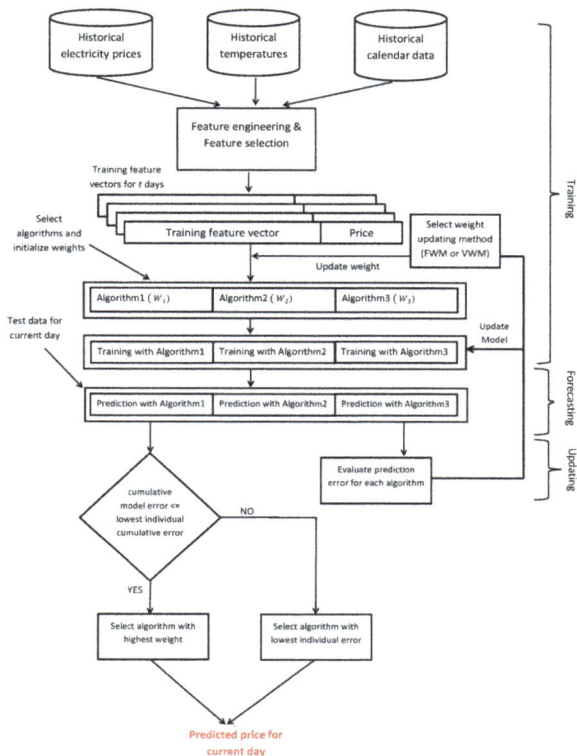

Figure 1. Overview of the proposed ensemble model demonstrated with three participating algorithms.

All of the participating algorithms will generate predictions, but the actual decision will be made by the algorithm whose performance was best recently in the previous days. On the first day of the model deployment, the expert algorithm will be chosen randomly. From the second day onward, for each hour of the day, the performance of each algorithm will be evaluated, and the best algorithm will be chosen based on the past prediction accuracy. The best predictor will be the expert predictor, whose predicted value for the next day will be the decisive value. At the end of each day when the actual price becomes available, every algorithm will analyze its own performance for that day. If the performance is within the range of the threshold, the models are maintained. Otherwise, the models are updated by re-training them with all of the available data up to date. The proposed ensemble models, using fixed weights and varying weights respectively, are described in Algorithms 1 and 2.

It should be noted that both algorithms are designed for each individual hour of the day. Therefore, for each algorithm, we need to run 24 separate instances of it to forecast the electricity prices for 24 h.

3.2.1. Algorithm 1: Fixed Weight Method

The Fixed Weight Method (FWM) is described in Algorithm 1. Briefly, FWM is initialized by assigning a weight of zero to all participating algorithms, except for one randomly-chosen expert algorithm, whose weight is set to one. Models for each participating algorithm are then built using the training dataset and are used to predict the target electricity prices for the unseen test data. Once we obtain the prediction from each model, the constraint in Equation (4) is checked as a fallback procedure, and the final predicted value is decided. Then, the weight of each participating algorithm is updated based on the respective prediction accuracy. The weight for the model with the highest prediction accuracy is set to one, and zero is assigned to the weights of the rest of the models. The model with weight equal to one will be the expert model for the next day. If the performance of the expert model is below that of an individual model acting alone, a re-train signal is sent to the system, which will initiate the retraining of all of the individual models. Newly built models will replace the old models, but will maintain the weights of the previous models.

Algorithm 1 Fixed weight method ($TrainSet[1..t]$, $TestSet[1..m]$)

1: choose the set A of n participating algorithms: $A = \{a_1, a_2, a_3, \ldots, a_n\}$
2: $\hat{A} \leftarrow [\,]$; /* initialize list of experts */
3: $e \leftarrow random(1..n)$; $W_e \leftarrow 1$; /* randomly select an algorithm as expert and assign its weight as 1 */
4: **for** $i \leftarrow 1$ to n **do**
5:   **if** $i \neq e$ **then**
6:     $W_i \leftarrow 0$; /* initialize weights of all algorithms except expert's as 0 */
7:   **end if**
8:   $mo_i \leftarrow Train(a_i, TrainSet[1..t])$; /* build prediction model mo_i by training algorithm a_i using data in the training set for days $1..t$ */
9:   $CumuError[i] \leftarrow 0$; /* initialize cumulative prediction error of all algorithms as 0 */
10: **end for**
11: $ExpertCumuError \leftarrow 0$; /* initialize cumulative prediction error of list of experts as 0 */
12: $retrain \leftarrow$ FALSE;
13:
14: **for** $j \leftarrow 1$ to m **do**
15:   /* for all m number of days in $TestSet$, do the following three steps */
16:
17:   /* Step 1: carrying out prediction */
18:   **for** $i \leftarrow 1$ to n **do**
19:     /* apply prediction model mo_i of algorithm a_i using data in the training set plus those in the testing set up until previous day */
20:     $P'_{a_i} \leftarrow GetPrediction(mo_i, a_i, TrainSet \cup TestSet[1..j-1])$;
21:     **if** $W_i = 1$ **then**
22:       $\hat{a} \leftarrow a_i$; /* if weight is 1, then it is the current expert */
23:       add $\hat{a}$ to list $\hat{A}$; /* insert the current expert into list of experts */
24:       $PredictedPrice \leftarrow P'_{\hat{a}}$; /* tentatively, predicted price will be output of the current expert */
25:     **end if**
26:   **end for**
27:
28:   /* Step 2: fallback procedure and reporting */
29:   $l \leftarrow arg\,min_{i=1..n}\,CumuError[i]$; /* l is index of algorithm with lowest cumulative error */
30:   **if** $CumuError[l] < ExpertCumuError$ **then**
31:     $PredictedPrice \leftarrow P'_{a_l}$; /* take algorithm l's output instead of expert algorithm's */
32:     $retrain \leftarrow$ TRUE; /* all models must be updated later */
33:   **end if**
34:   report $PredictedPrice$; /* report predicted price for j-th day as our output */
35:
36:   /* Step 3: prediction error calculation and updating — after actual price P of day j is known */
37:   $E_{\hat{a}} \leftarrow |P'_{\hat{a}} - P|$; /* calculate expert algorithm's prediction error */
38:   $ExpertCumuError \leftarrow ExpertCumuError + E_{\hat{a}}$;
39:   **for** $i \leftarrow 1$ to n **do**
40:     $E_{a_i} \leftarrow |P'_{a_i} - P|$; /* calculate all algorithms' prediction errors */
41:     $CumuError[i] \leftarrow CumuError[i] + E_{a_i}$;
42:   **end for**
43:   $W_e \leftarrow 0$; /* reset weight of old expert to 0 */
44:   $e \leftarrow arg\,min_{i=1..n}\,P'_{a_i}$; /* select algorithm with lowest error for j-th day as new expert for $(j+1)$-th day */
45:   $W_e \leftarrow 1$; /* set weight of new expert as 1 */
46:   /* re-train models if required */
47:   **if** $retrain =$ TRUE **then**
48:     **for** $i \leftarrow 1$ to n **do**
49:       $mo_i \leftarrow Train(a_i, TrainSet \cup TestSet[1..j])$; /*update all models using the latest available data till now */
50:     **end for**
51:     $retrain =$ FALSE; /* reset $retrain$ flag */
52:   **end if**
53: **end for**

3.2.2. Algorithm 2: Varying Weight Method

Algorithm 2 describes the steps for the Varying Weight Method (VWM). VWM follows a similar approach as proposed in FWM, with few changes in updating the weight of participating algorithm. In this model, the weight of each participating algorithm varies based on the prediction accuracy achieved by the respective algorithm in all of the previous predictions made by it, whereas in FWM, the weight of the algorithm is either zero or one based on its previous day performance. At the beginning, the weight for all participating algorithms is set to one, and randomly, one algorithm is chosen to be the expert algorithm for the first day. These models are used to predict the electricity prices for the unseen test data. Once we obtain the prediction value from each model and come to know the actual price, we evaluate the performance of each model and update the weight based on the function of their prediction accuracy and learning rate λ. The algorithm with the highest accuracy (lowest prediction error) will have its weight increased, and the other algorithms with lower accuracy will have their weight decreased based on the prediction accuracy achieved by them. The main benefit of this model is that it considers the individual prediction error value when updating the weight. Therefore, the weight of any algorithm is dependent on the cumulative error and number of times it has been the best predictor.

3.3. Data Preprocessing

As presented in Figure 1, we need to perform feature engineering and feature selection prior to carrying out model building and prediction themselves.

3.3.1. Feature Engineering

The electricity market data follow a time series pattern and provide the information about the daily electricity prices over a period of time. The information in its raw form does not contain any specific features (attributes) that can be used in electricity price prediction. Therefore, from the time series data, we need to generate relevant features to be used in prediction models as an input. Previous research works have shown that prediction models are often affected by higher variance in time series data. Thus, feature generation, also known as feature engineering, is one of the important aspects in building the prediction model, where the features are carefully created to reduce over-fitting of the model and accurately capture the target value. In our previous work, we have shown that generating relevance features from single or a few sources improves the prediction accuracy of the model by a significant margin [12]. In this research, we have engineered 47 different features to capture various hidden trends in the electricity market.

In order to predict the hour h's electricity price, we extract the hourly price data for the past 24 h $(h - 1$–$h - 24)$ window, yielding 24 different features. The features that can best represent the short-term trend in the electricity market are the previous 24 h data, as observed in [47]. These data provide us a good insight for short-term trends, but fail to capture seasonal and long-term trends. In order to build a robust prediction model, both short and long term and the seasonal effect should be captured efficiently. A sudden high fluctuation in electricity price might occur due to the seasonal behavior and other factors. In order to capture these uncertain behaviors in electricity price, we created putatively relevant features based on a historical time series electricity price dataset. Therefore, 20 additional features, like last year same day same hour price, last year same day same hour price fluctuation, last week same day same hour price, last week same day price fluctuation, etc., were created.

In order to achieve an even better forecasting accuracy, we also introduce various features that are not directly associated with price data. We explore various other factors that can affect the electricity load and the price of the market. We have found that according to [7], the temperature, the day of the week and the occurrence of holidays can all affect the electricity load and price. Therefore, we also incorporate these three non-price features into our generated feature set. For the temperature features, we use historical and forecasted temperature data provided by Weather Underground [48]. For the holiday data, we use predefined holiday information in the geographical area of the target electricity market.

Algorithm 2 Varying weight method ($TrainSet[1..t]$, $TestSet[1..m]$, λ)

1: choose the set A of n participating algorithms: $A = \{a_1, a_2, a_3, \ldots, a_n\}$
2: $\hat{A} \leftarrow [\ ]$; /* initialize list of experts */
3: $e \leftarrow random(1..n)$; /* randomly select an algorithm as expert */
4: **for** $i \leftarrow 1$ to n **do**
5:    $W_i \leftarrow 1$; /* initialize weights of all algorithms as 1 */
6:    $mo_i \leftarrow Train(a_i, TrainSet[1..t])$; /* build prediction model mo_i by training algorithm a_i using data in the training set for days $1..t$ */
7:    $CumuError[i] \leftarrow 0$; /* initialize cumulative prediction error of all algorithms as 0 */
8: **end for**
9: $ExpertCumuError \leftarrow 0$; /* initialize cumulative prediction error of list of experts as 0 */
10: $retrain \leftarrow$ FALSE;
11:
12: **for** $j \leftarrow 1$ to m **do**
13:    /* for all m number of days in $TestSet$, do the following three steps */
14:
15:    /* Step 1: carrying out prediction */
16:    **for** $i \leftarrow 1$ to n **do**
17:       /* apply prediction model mo_i of algorithm a_i using data in the training set plus those in the testing set up until previous day */
18:       $P'_{a_i} \leftarrow GetPrediction(mo_i, a_i, TrainSet \cup TestSet[1..j-1])$;
19:    **end for**
20:    **if** $j \neq 1$ **then**
21:       $e \leftarrow \arg\max_{i=1..n} W_i$ /* if weight is the highest, then it is the current expert */
22:    **end if**
23:    $\hat{a} \leftarrow a_e$; add $\hat{a}$ to list $\hat{A}$; /* insert the current expert into list of experts */
24:
25:    $PredictedPrice \leftarrow P'_{\hat{a}}$; /* tentatively, predicted price will be output of the current expert */
26:
27:    /* Step 2: fallback procedure and reporting */
28:    $l \leftarrow \arg\min_{i=1..n} CumuError[i]$; /* l is index of algorithm with lowest cumulative error */
29:    **if** $CumuError[l] < ExpertCumuError$ **then**
30:       $PredictedPrice \leftarrow P'_{a_l}$; /* take algorithm l's output instead of expert algorithm's */
31:       $retrain \leftarrow$ TRUE; /* all models must be updated later */
32:    **end if**
33:    report $PredictedPrice$; /* report predicted price for j-th day as our output */
34:
35:    /* Step 3: prediction error calculation and updating — after actual price P of day j is known */
36:    $E_{\hat{a}} \leftarrow |P'_{\hat{a}} - P|$; /* calculate expert algorithm's prediction error */
37:    $ExpertCumuError \leftarrow ExpertCumuError + E_{\hat{a}}$;
38:    **for** $i \leftarrow 1$ to n **do**
39:       $E_{a_i} \leftarrow |P'_{a_i} - P|$; /* calculate all algorithms' prediction errors */
40:       $CumuError[i] \leftarrow CumuError[i] + E_{a_i}$;
41:    **end for**
42:    $s \leftarrow \arg\min_{i=1..n} E_{a_i}$ /* s is index of algorithm with smallest prediction error for current day j*/
43:    /* update weights */
44:    **for** $i \leftarrow 1$ to n **do**
45:       **if** $i = s$ **then**
46:          $W_i \leftarrow W_i \times (E_{a_i} \times \lambda)$ /* increase weight of algorithm a_i by factor of λ and E_{a_i} */
47:       **else**
48:          $W_i \leftarrow W_i / (E_{a_i} \times \lambda)$ /* decrease weight of algorithm a_i by factor of λ and E_{a_i} */
49:       **end if**
50:    **end for**
51:    /* re-train models if required */
52:    **if** $retrain =$ TRUE **then**
53:       **for** $i \leftarrow 1$ to n **do**
54:          $mo_i \leftarrow Train(a_i, TrainSet \cup TestSet[1..j])$; /*update all models using the latest available data till now */
55:       **end for**
56:       $retrain =$ FALSE; /* reset $retrain$ flag */
57:    **end if**
58: **end for**

It should be noted that oil and gas prices, and other factors, like load and types of resources used for electricity generation, might also affect the pricing. However, for the three target markets in out studies (namely, New York, Australia and Spain), these data are not easily accessible to us, and we leave them to be considered in our future work.

Normalization is one of the best approaches to deal with the input data where the attributes are of different measurements and scales. In our case, as we use various input data with different scales, we need to normalize all 47 attributes to achieve consistency. We use the mapminmax function in MATLAB [49] to normalize the input attributes into the range $(-1.0, 1.0)$.

3.3.2. Feature Selection

Though 47 features were created using the historical electricity price, calendar and weather data, using all of the created features for building the model poses the threat of over-fitting. All of the generated features are analyzed to remove redundant, irrelevant and loosely-coupled features. Thus, the feature selection process is used to select the most relevant features from the original feature set.

Feature selection is a crucial step in building a robust prediction model. Here, we utilize the wrapper method [50] using WEKA [51] for subset selection from a pool of features. A wrapper involves a search algorithm for finding the optimal subset of features in the feature space and evaluating the subset using the learning algorithm. Using cross-validation, it evaluates the estimated accuracy obtained from the learning algorithm by adding or removing features from the features subset in hand. In our case, we use the Artificial Neural Network (ANN) as the learning algorithm; 10-fold cross-validation is carried out for the training set.

In selecting the optimal feature set, among the 47 features, the wrapper method is applied to the 23 features apart from those for the past 24 h. Technically, the resultant training accuracy may be less than the best possible accuracy since 24 features are omitted. However, due to the verified importance of the past 24 h data [47], we choose to exclude them for the sake of saving expensive computation of the wrapper process, whose running time cost is exponential to the number of input features.

The final feature set obtained for the New York (NYISO) dataset after the feature selection process is shown as an example in Table 1.

Table 1. Selected features for the NYISO (New York Independent System Operator) dataset.

Attribute No.	Description	Notation
1–24	Previous 24 h's prices	p_{h-1} to p_{h-24}
25	Previous year same hour price	p_{y-1}
26	Previous year same day average price	$p_{avg(y-1)}$
27	Previous week same hour price	p_{d-7}
28	Previous hour price increase/decrease	$\|p_{h-1} - p_{h-2}\|$
29	Previous year same day, same hour, price increase/decrease	$\|p_{(y-1)_h} - p_{(y-1)_{h-1}}\|$
30	Temperature of the day	T_d
31	Day of the week	W_d
32	Holiday (Y/N)?	H_d

4. Experimental Setup

4.1. Data

We evaluated our proposed ensemble model by performing experiments with the dataset from three different deregulated electricity markets of New York (NYISO) [13], Australia (ANEM) [14] and Spain (OMEL) [15]. We selected data from these markets to compare the results of our proposed model with those in the previous works. As mentioned in [11], a vast amount of research has been carried out using the data from these markets. The NYISO electricity market contains data from various areas from New York and provides data for hourly electricity price. From NYISO, we selected "Capita" as the reference area to benchmark our results with those of the previous works [11,12]. ANEM represents the market clearing data in the Australian market since its deregulation with half hour resolution. Again, we selected the data from the "Queensland" area to be consistent with the experiments in those two previous works. Likewise, for the Spanish (OMEL) market, we also used the same data as those previous works.

4.2. Evaluation Metrics

Following are the performance measures used to validate our proposed model. These measures are used in order to facilitate direct comparison with the results obtained in the other similar studies.

Mean Error Relative to the mean actual price (MER):

$$MER = 100 \times \frac{1}{N} \sum_{i=1}^{N} \frac{|P_i - P_i'|}{\bar{P}} \qquad (5)$$

where P_i defines the actual price and P_i' defines the predicted price. $\bar{P}$ is the mean price for the period of interest, and N is the number of predicted hours. This indicator is irrespective of the absolute values.

Mean Absolute Error (MAE):

$$MAE = \frac{1}{N} \sum_{i=1}^{N} |P_i - P_i'| \qquad (6)$$

The indicator is dependent on the absolute range of the electricity price.

Mean Absolute Percentage Error (MAPE):

$$MAPE = 100 \times \frac{1}{N} \sum_{i=1}^{N} \frac{|P_i - P_i'|}{P_i} \qquad (7)$$

This indicator is irrespective of the absolute values. If the range of the electricity price is vast, a prediction may give a high MAE value, but a low MAPE.

5. Experimental Results

Two sets of experiments (Experiments I and II) were performed using two different time periods, 2004–2006 (on the NYISO, ANEM and OMEL datasets) and 2008–2012 (on the NYISO and ANEM datasets only), respectively.

5.1. Experiment I: 2004–2006

In Experiment I, the NYISO, ANEM and OMEL datasets for the time period of 2004–2006 are used. For all of the datasets, data from March 2004–March 2006 were used as the training set and April 2006–December 2006 as the testing set. We use the exact same experimental protocol as in Martínez-Álvarez et al. [11]. The following five methods are compared.

1. Ensemble learning using the Fixed Weight Method (FWM) (Algorithm 1) using three participating machine learning algorithms: Artificial Neural Network (ANN), Support Vector Regression (SVR) and Random Forest (RF) [52],
2. Ensemble learning using the Varying Weight Method (VWM) (Algorithm 2) using the same participating algorithms as in FWM,
3. Artificial Neural Network only (ANN-only) (published results as presented in our previous work [12]),
4. Pattern Sequence-based Forecasting (PSF) (published results as presented in Martínez-Álvarez et al. [11]) and
5. Autoregressive Integrated Moving Average (ARIMA) [10] (Note: the auto.arima() function in R's [53] Forecast package is used.)

5.1.1. Experiment I-A: NYISO Dataset

Table 2 shows the MER (Equation (5)), MAE (Equation (5)) and MAPE (Equation (7)) of the four methods for New York (NYISO) data for the testing period of the year 2006. We can see that the results obtained from FWM (VWM in the brackets) have average MER of 3.92% (3.86%) with the SD (Standard Deviation) of 1.03 (1.10), MAE of USD 2.25/MWh (USD 2.20/MWh) with SD of 0.53 (0.52) and MAPE of 3.97% (3.93%) with SD of 0.75 (0.77). The worst month is December 2006, where MER is 5.61% (5.70%), MAE is 3.02 (3.07) and MAPE is 5.46% (5.55%).

For both FWM and VWM, the ANN predictor appears to dominate the other algorithms where more than 60% of its predictions were selected as the final prediction in both cases. Generally, VWM offers slightly better results than FWM with the improvements (decreases in error) of 0.06% of MER, USD 0.05/MWh of MAE and 0.04% of MAPE.

Table 2. MER (Mean Error Relative to the mean actual price), MAE and MAPE evaluation metrics for the NYISO market for the year 2006 (Experiment I-A). VWM, Varying Weight Method; FWM, Fixed Weight Method; PSF, Pattern Sequence-based Forecasting.

Month	MER (%)					MAE (USD/MWh)					MAPE (%)			
	VWM	FWM	ANN-Only	PSF	ARIMA	VWM	FWM	ANN-Only	PSF	ARIMA	VWM	FWM	ANN-Only	ARIMA
January				4.45	6.51				2.25	4.75				6.17
February				5.53	6.28				3.02	4.23				5.97
March				6.30	6.87				3.97	3.80				5.95
April	**3.61**	3.79	3.99	4.94	5.56	2.23	**2.22**	2.34	3.51	3.32	**3.66**	3.82	4.02	5.67
May	**2.67**	2.79	2.94	7.59	4.92	**1.62**	1.64	1.72	4.63	2.73	**3.02**	3.07	3.23	5.09
June	**3.32**	3.47	3.65	3.34	4.70	**1.96**	2.03	2.14	2.31	2.73	**3.64**	**3.64**	3.83	5.02
July	3.95	**3.89**	4.09	3.93	5.31	**2.22**	2.28	2.40	2.28	3.39	3.82	**3.75**	3.95	5.56
August	**4.56**	4.58	4.82	5.37	5.17	**2.59**	2.68	2.82	3.49	3.55	3.80	**3.76**	3.96	5.25
September	**2.64**	2.72	2.86	6.24	4.83	**1.58**	1.59	1.68	4.49	2.24	**3.27**	3.42	3.60	5.15
October	**3.05**	3.19	3.36	7.43	6.18	**1.78**	1.87	1.97	4.23	3.23	**3.85**	3.89	4.09	6.35
November	5.24	5.25	5.53	**5.19**	9.64	2.77	**2.90**	3.05	3.53	6.06	**4.76**	4.91	5.17	9.04
December	5.70	**5.61**	5.90	6.04	12.42	3.07	**3.02**	3.18	3.08	5.79	5.55	**5.46**	5.75	9.43
Mean	3.86	**3.92**	4.13	5.53	6.53	**2.20**	2.25	2.37	3.40	3.82	**3.93**	3.97	4.18	6.22
SD	1.10	**1.03**	1.09	1.29	2.29	**0.52**	0.53	0.55	0.84	1.19	0.77	**0.75**	0.79	1.47

Notes: (1) For the NYISO dataset, The results for the period January–March 2006 cannot be presented for VWM, FWM and ANN-only because the feature vectors of their corresponding training data during the period of January–March 2005 cannot be constructed. Construction of those training feature vectors in turn requires the data before March 2004, which is not readily available to us. For PSF and ARIMA, such a feature vector construction is not required, and neither are the data before March 2004. (2) MAPE results for PSF are not presented because it was not used as an evaluation criterion in [11]. (3) The Standard Deviation (SD) results provided here are computed using the stdev (estimated standard deviation based on a sample) function in Microsoft Excel. They are different from the ones reported in [11] and [12], which are computed by a different standard deviation function.

5.1.2. Experiment I-B: ANEM Dataset

The Australian (ANEM) dataset is particularly challenging. It is highly volatile with a large number of unexpected abnormalities and outliers. There were large fluctuations in the electricity prices with the highest price of AUD 9739/MWh in January 2006 and the lowest value of AUD 7.81/MWh in February 2004. The variance and skewness for each market datum will be discussed in the following Section 6.

Due to these highly fluctuating values and outliers, forecasted price for this market has a high range of error. From Table 3, we can see that the performance FWM (VWM in the brackets) of on ANEM data is 10.06% (9.24%) of average MER with SD of 4.32 (3.83), AUD 3.25/MWh (AUD 3.11/MWh) MAE with SD of 1.40 (1.34) and 8.70% (7.93%) MAPE with SD of 3.23 (2.74).

Bad performances are observed for the months of January, June, July and August 2006. The worst performance of 18.47% (17.14%) MER and 14.53% (13.15%) MAPE are obtained July 2006. The best performance of 4.86% (4.44%) MER and 5.22% (4.75%) MAPE are in the month of March 2006.

Though the results on this ANEM dataset by FWM and VWM have higher error percentages than those on the previous NYISO dataset, the performance of VWM is still better than those of the other methods (ANN-only and PSF) on the same dataset. On the other hand, the accuracy of FWM is found to be slightly lower than that of the ANN-only method, but still higher than that of PSF. The final predictions for ANEM dataset also follow the same trend as in the NYISO dataset where a majority of the predictions were based on ANN as the expert algorithm for both FWM and VWM.

5.1.3. Experiment I-C: OMEL Dataset

The results obtained from the Spanish (OMEL) market are shown in Table 4. We can see that the average MER of FWM (VWM in the brackets) is 5.34% (5.26%) with SD of 0.54 (0.62), which indicates that the monthly errors are not much different from the average error. MAE for the Spanish data is EURc 0.34/kWh (EURc 0.35/kWh) with the SD of 0.03 (0.05) and MAPE of 5.75% (5.62%) with SD of 1.25 (1.07). MAE for the OMEL dataset is very low compared to other markets because the prices are in a different unit of measurement, which is EUR cent per kWh instead of USD/AUD per MWh in the NYISO/ANEM datasets.

It can also be observed that, unlike the previous two cases of NYISO and ANEM, VWM is not always better than FWM for all three evaluation criteria. Whilst VWM is slightly better than FWM in terms of the average MER and MAPE, it is slightly worse than FWM in terms of MAE.

5.2. Experiment II: 2008–2012

To further verify these encouraging results, a second set of experiments were performed using the more recent dataset. Electricity price data from the June 2008–May 2011 period were extracted from the NYISO and ANEM datasets and used as the training set, while data from June 2011–May 2012 were used as the testing set (note: the OMEL dataset is not available for the period 2008–2012, and neither are the experimental results of PSF for that time period). Therefore, only FWM (with ANN, SVR and RF participating algorithms), VWM (with the same participating algorithms), ANN-only and ARIMA are compared. It is observed that the results of our proposed model for this Experiment II are even slightly better than those in Experiment I.

5.2.1. Experiment II-A: NYISO Dataset

From Table 5, we can see that the overall performances of both FWM and VWM for this 2008–2012 NYISO dataset are even slightly better than those for the Experiment I-A (NYISO's 2004–2006 dataset) despite the fact that the 2008–2012 data contain many spikes and outliers. For this dataset, FWM (VWM in the brackets) provides average MER of 3.86% (3.85%) with SD of 0.57 (0.57), MAE of USD 1.48/MWh (USD 1.48/MWh) with SD 0.54 (0.54) and MAPE 3.99% (3.94%) with SD of 1.45 (1.42). Noticeable improvement can be observed in MAE with a decrease of USD 0.77/MWh (USD 0.72/MWh) when compared to Experiment I-A.

Table 3. MER, MAE and MAPE evaluation metrics for the ANEM (Australian Energy Market Operator) market for the year 2006 (Experiment I-B).

Month	ANEM (2006)													
	MER (%)					MAE (AUD/MWh)					MAPE (%)			
	VWM	FWM	ANN-Only	PSF	ARIMA	VWM	FWM	ANN-Only	PSF	ARIMA	VWM	FWM	ANN-Only	ARIMA
January	14.03	16.43	13.96	**5.80**	42.92	4.61	5.32	4.66	**1.51**	11.43	**12.71**	15.20	13.34	49.17
February	**7.04**	7.93	7.26	8.59	55.51	**2.37**	2.57	2.42	5.15	36.85	**6.98**	7.67	7.19	55.21
March	**4.44**	4.86	4.59	7.84	37.79	**1.47**	1.57	1.53	1.73	7.37	**4.75**	5.22	4.79	37.87
April	**5.53**	5.56	5.60	9.92	40.01	**1.78**	1.80	1.87	1.98	8.32	**5.56**	5.96	5.82	42.81
May	10.09	**10.03**	10.07	12.85	49.25	3.22	3.24	3.36	**3.21**	11.72	9.43	**9.27**	9.59	44.43
June	**11.20**	12.01	11.00	22.04	27.23	3.72	3.89	**3.67**	6.81	8.56	**7.54**	9.01	7.85	28.09
July	17.14	18.47	17.97	**17.11**	77.21	6.02	**5.98**	5.99	8.16	37.33	**13.15**	14.53	13.84	41.73
August	**11.12**	12.18	11.28	11.71	25.94	3.81	3.94	**3.76**	3.32	7.78	**7.52**	9.13	7.55	25.56
September	8.36	9.32	8.77	**8.23**	24.20	2.83	3.02	2.93	**2.34**	6.86	7.08	7.03	**7.03**	27.42
October	**5.59**	6.15	5.70	7.66	25.47	**1.89**	1.99	1.90	1.92	6.47	**5.23**	5.77	5.40	29.92
November	10.41	11.61	10.59	**6.76**	28.19	3.54	3.76	3.53	**2.09**	8.88	**9.11**	9.18	9.25	27.00
December	**5.94**	6.15	6.12	6.42	39.70	2.00	1.99	2.04	**1.41**	7.26	**6.06**	6.41	6.35	41.44
Mean	**9.24**	10.06	9.41	10.41	39.45	**3.11**	3.25	3.14	3.30	13.24	**7.93**	8.70	8.17	37.55
SD	**3.83**	4.32	3.92	4.87	15.61	1.34	1.40	**1.31**	2.23	11.26	**2.74**	3.23	2.90	9.82

Table 4. MER, MAE and MAPE evaluation metrics for the Spanish OMEL (Operador del Mercado Ibérico de Energía, Polo Español, S.A.) market for the year 2006 (Experiment 1-C).

	OMEL (2006)													
	MER (%)					MAE (EURc/kWh)					MAPE (%)			
Month	VWM	FWM	ANN-Only	PSF	ARIMA	VWM	FWM	ANN-Only	PSF	ARIMA	VWM	FWM	ANN-Only	ARIMA
January	5.91	**5.65**	6.00	7.26	8.31	0.40	**0.36**	0.40	0.53	0.61	5.00	**4.56**	5.07	8.91
February	5.20	5.34	5.25	**4.93**	8.05	**0.34**	0.35	0.35	0.36	0.58	**3.49**	3.88	3.66	8.02
March	**5.05**	5.48	5.28	5.88	10.80	**0.34**	0.35	0.35	0.43	0.55	6.28	6.33	**6.19**	11.26
April	5.68	6.11	5.90	**3.62**	11.44	0.39	0.39	0.39	**0.28**	0.58	**5.55**	5.88	5.73	10.67
May	**5.88**	5.60	6.00	8.11	10.42	0.40	**0.36**	0.40	0.64	0.51	5.88	**5.69**	6.11	10.38
June	5.12	5.36	5.14	**3.76**	8.92	0.33	0.35	0.34	**0.29**	0.42	**5.05**	5.68	5.27	8.47
July	4.68	4.59	4.59	**4.30**	8.19	**0.29**	0.30	0.30	0.33	0.41	4.59	**4.52**	4.55	8.16
August	**5.10**	5.07	5.17	5.37	9.36	**0.33**	0.33	0.34	0.42	0.43	5.50	**5.39**	5.57	8.78
September	**5.61**	5.44	5.66	6.41	10.64	0.36	**0.35**	0.37	0.50	0.56	5.36	**5.03**	5.42	10.06
October	**6.04**	5.78	6.08	7.89	11.87	0.41	**0.37**	0.40	0.58	0.53	6.31	**6.23**	6.64	10.60
November	**3.83**	4.10	3.94	8.30	7.53	**0.25**	0.27	0.26	0.64	0.28	7.13	7.91	7.25	10.61
December	**5.06**	5.50	5.04	8.02	12.06	**0.33**	0.36	0.33	0.59	0.41	7.33	7.94	7.61	12.32
Mean	**5.26**	5.34	5.34	6.15	9.80	**0.35**	0.34	0.35	0.47	0.49	**5.62**	5.75	5.76	9.85
SD	0.62	**0.54**	0.64	1.76	1.60	0.05	**0.03**	0.04	0.13	0.10	**1.07**	1.25	1.11	1.36

Table 5. MER, MAE and MAPE evaluation metrics for the NYISO market for the years 2011–2012 (Experiment II-A).

Month	NYISO (2011–2012)											
	MER (%)				MAE (USD/MWh)				MAPE (%)			
	VWM	FWM	ANN-Only	ARIMA	VWM	FWM	ANN-Only	ARIMA	VWM	FWM	ANN-Only	ARIMA
June 2011	4.25	**4.24**	4.27	5.42	**1.92**	1.97	2.01	2.36	**5.19**	5.32	4.31	5.19
July 2011	4.40	**4.36**	4.66	6.46	2.71	**2.68**	2.97	3.59	**7.12**	7.25	4.72	5.82
August 2011	3.39	**3.34**	3.54	5.04	1.50	**1.48**	1.56	2.21	**3.87**	3.99	3.51	5.13
September 2011	**3.56**	3.61	3.86	5.23	**1.38**	**1.38**	1.47	2.05	**3.72**	3.73	3.76	5.41
October 2011	**3.49**	3.50	3.55	5.98	1.33	**1.32**	1.33	2.31	3.60	**3.56**	3.48	5.89
November 2011	3.57	**3.56**	3.61	6.73	1.31	**1.28**	1.30	2.38	**3.41**	3.47	3.62	6.36
December 2011	**3.73**	3.80	4.01	7.97	**1.35**	1.36	1.44	2.81	**3.68**	**3.68**	4.16	6.98
January 2012	**4.82**	4.87	4.92	9.72	2.19	**2.15**	2.32	3.90	**5.71**	5.82	5.14	8.26
February 2012	**3.22**	3.25	3.70	6.67	1.06	**1.05**	1.20	2.03	2.84	**2.83**	3.81	6.05
March 2012	**4.74**	4.76	4.83	8.44	**1.27**	1.29	1.31	2.25	**3.48**	3.48	4.86	7.93
April 2012	3.84	**3.80**	4.18	7.21	0.96	**0.95**	1.06	1.88	**2.44**	2.56	4.09	7.20
May 2012	**3.22**	3.23	3.45	5.41	0.84	**0.83**	0.88	1.54	**2.18**	2.25	3.27	5.80
Mean	**3.85**	3.86	4.05	6.69	**1.48**	1.49	1.57	2.44	**3.94**	3.99	4.06	6.34
SD	0.56	0.57	**0.52**	1.44	**0.54**	**0.54**	0.59	0.68	1.42	1.45	**0.60**	1.04

237

The worst forecasting results obtained were 4.87% (4.82%) of MER in January 2012, USD 2.68/MWh (USD 2.71/MWh) of MAE and 7.25% (7.12%) of MAPE both in July 2011. The main reason for the higher forecasting error in July 2011 and January was due to the higher numbers of spikes and outliers in those months in the New York market.

5.2.2. Experiment II-B: ANEM Dataset

The performances of FWM and VWM in this period for the ANEM dataset are noticeably better for those of Experiment I-B (ANEM's 2004–2006 dataset). For this dataset, FWM (VWM in the brackets) provides average MER of 7.16% (6.98%) with SD of 3.83 (3.98), MAE of AUD 2.09/MWh (AUD 2.05/MWh) with SD 1.37 (1.36) and MAPE of 6.22% (6.16%) with SD of 3.01 (2.98). The improvements (deceases in error) over ANEM's 2004–2006 dataset by FWM (VWM in the brackets) are 2.90% (2.26%) in MER, AUD 1.16/MWh (AUD 1.06/MWh) in MAE and 2.48% (1.77%) in MAPE.

There are also decreases in SD for both FWM and VWM, which indicate that the errors in the years 2011–2012 are lessdeviated from the mean error when compared to those for the years 2004–206 in the ANEM dataset.

The worst performance of FWM (VWM in the brackets) was in the months of January–March 2012 with 14.36% (13.72%) MER in February 2012. The best performance of 3.72% (3.66%) MER was in May 2012. The best and the worst MAPEs were 3.83% (3.76%) in May 2012 and 12.72% (12.61%) in February 2012, respectively.

6. Analyses and Discussions

6.1. Variance and Skewness vs. Accuracy

The statistical distributions of the price data can significantly affect the model's prediction accuracy. In this section, we analyze different properties of the electricity price data for all three markets. One key aim is to find correlations between the data distribution and prediction error and to justify the requirement for an independent prediction model for each hour of the day as proposed in our approach. Figures 2 and 3 show overall variance and average price for 2004–2006 and 2008–2012 training and testing data for all three electricity market along with average forecasting accuracy. From both figures, we can see that the NYISO and ANEM data are of high variance. However, when we compare the value with the respective average price, ANEM shows a higher variance in price with a lower average price.

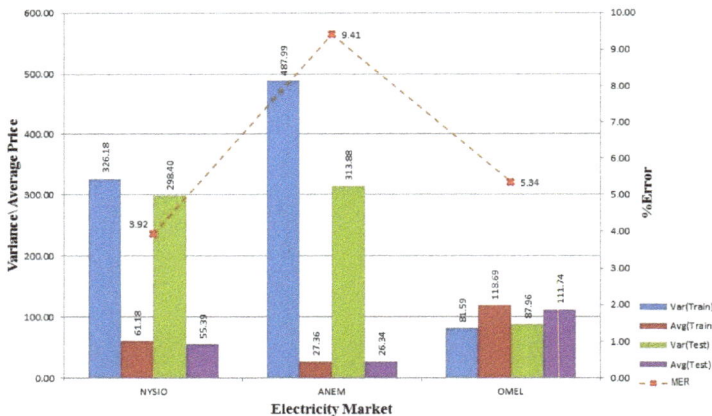

Figure 2. Variance, average and error in training and testing data for NYISO, ANEM and OMEL (2004–2006).

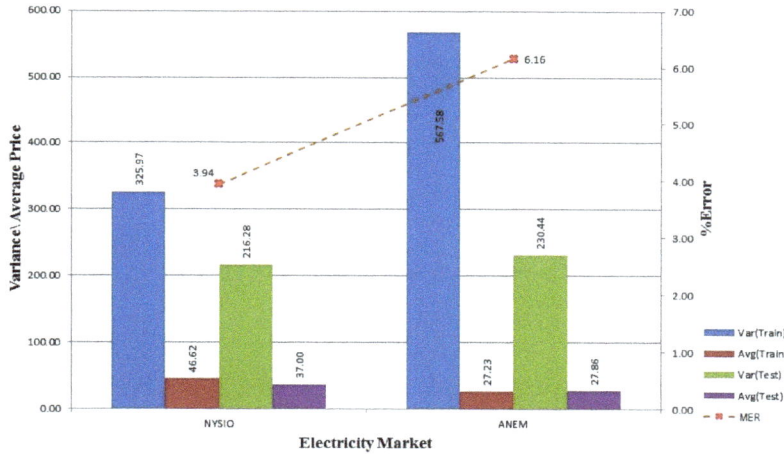

Figure 3. Variance, average and error in training and testing data for NYISO and ANEM (2008–2012).

Higher variance in price and higher deviation in hourly training and testing data might be the reason for higher error for ANEM's 2004–2006 dataset as shown in Figure 2. From Figure 3, we can see for the 2008–2012 dataset, ANEM continues to have higher variance in the training and the testing data, but Figure 8 shows that the hourly variance deviation between the training and the testing data is less, which helps to lower the forecasting error for ANEM. In Figure 6, we can see that the OMEL dataset shows a different response to hourly variance in the dataset. As opposed to NYISO and ANEM, the OMEL 2004–2006 dataset is of a lower variance to average price ratio, but still, its prediction error is higher than that of NYISO. The main cause behind this prediction error is due to the presence of a few outliers, where the ratio between the maximum the the minimum electricity price is more than 1000 fold (the data values for the OMEL dataset in Figure 2 are proportionally adjusted to make them comparable to those of the other markets because, unlike the others, the OMEL market represents electricity price in Euro cent per kWh).

For every electricity market, if we inspect the variances in prices within the same market for different hours of the day, we can observe that there is a great fluctuation in variances. Figures 4–6 show the hourly variance in 2004–2006 electricity price data for NYISO, ANEM and OMEL, respectively. Similarly, Figures 7 and 8 show the hourly variance in 2008–2012 electricity prices for NYISO and ANEM, respectively. From these five figures, we can see that each market exhibits different distributions of data over different hours of the day. For example, if we look at Figure 5, we can see that the ANEM 2004–2006 dataset shows price variances ranging from 30–1400 AUD/MWh for the training dataset and from 32–1950 AUD/MWh for the testing dataset. We can see similar fluctuations in price variances for the NYISO and OMEL datasets, as well. These markets are also of different variances in price over hours of the day, but both the training and the testing data follow a similar trend. On the other hand, the ANEM market shows a large difference in the training vs. testing curves for the 2004–2006 dataset, but somewhat more consistent curves for the 2008–2012 dataset.

From Figure 9, we can see that the ANEM 2004–2006 dataset exhibits high skewness along with higher forecasting error when compared to NYISO and OMEL. This high skewness continues in the ANEM 2008–2012 dataset shown in Figure 10.

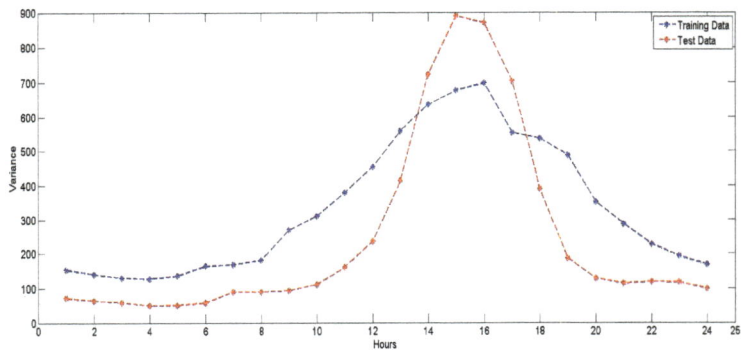

Figure 4. Hourly variance in training and testing data for NYISO (2004–2006).

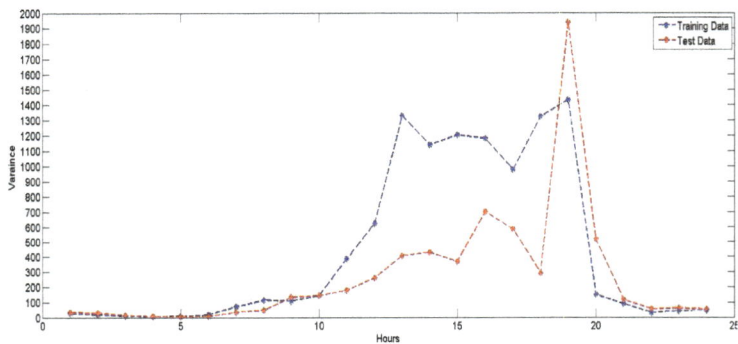

Figure 5. Hourly variance in training and testing data for ANEM (2004–2006).

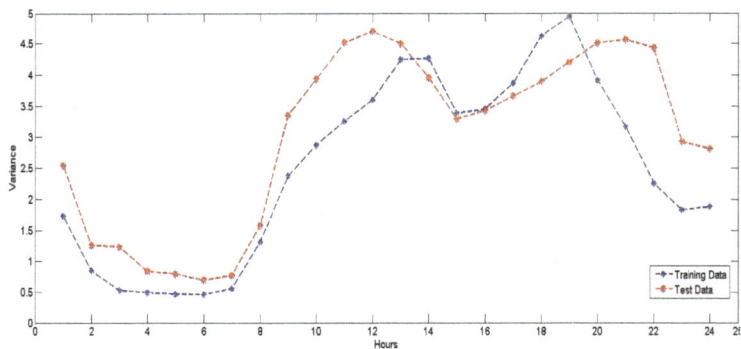

Figure 6. Hourly variance in training and testing data for OMEL (2004–2006).

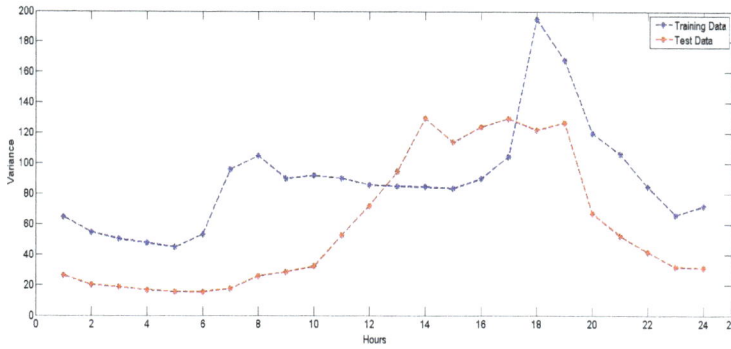

Figure 7. Hourly variance in training and testing data for NYISO (2008–2012).

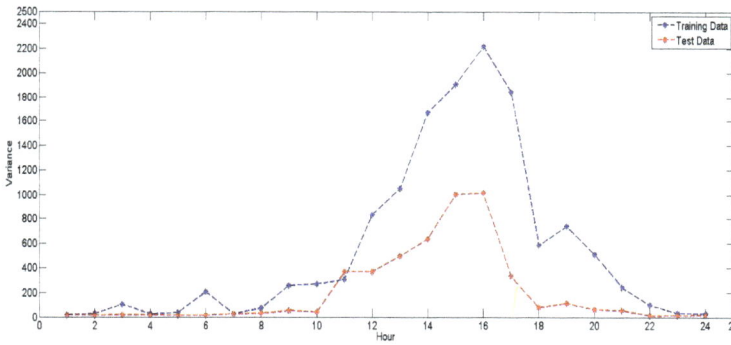

Figure 8. Hourly variance in training and testing data for ANEM (2008–2012).

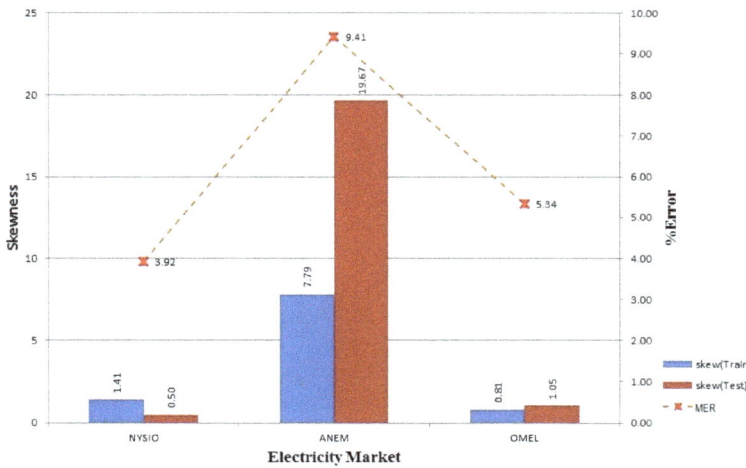

Figure 9. Skewness in training and testing data for NYISO, ANEM and OMEL (2004–2006).

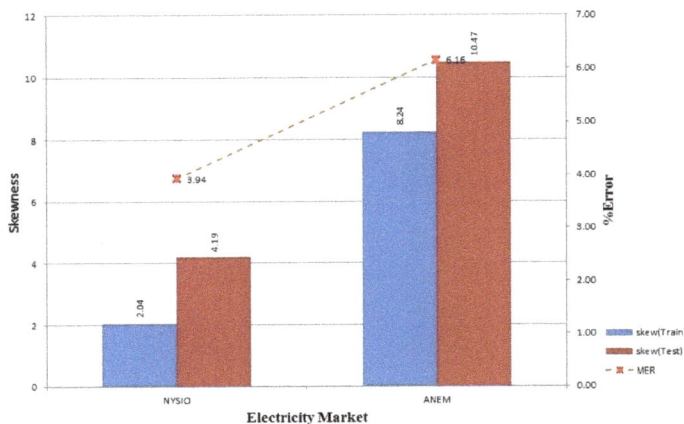

Figure 10. Skewness in training and testing data for NYISO and ANEM (2008–2012).

From all of the above observations, we can infer that electricity markets are of different price distributions, which are highly dependent on the hour of the day, with some hours having higher variance in price and some lower variance. This distribution is further influenced by the deployment of the smart grid, where the electricity price depends on various factors, like the load, user behavior, demand-response, etc. Under this circumstance, it very difficult to find an approach that can offer consistent performance over different electricity markets. This issue is somehow resolved by our proposed model, which provides flexibility on participating algorithms and tries to adapt to the changes in the data distribution. Our proposed model captures the variation in price with carefully-engineered features and builds varying forecasting models separately, one for each hour of the day. The model automatically adjusts itself to certain changes in the environment by evaluating the performance of the model at the end of each day and makes necessary adjustments if required.

6.2. Ensemble Model

Two different experiments were performed with different approaches (FWM and VWM) for updating the weights of the participating algorithms. From the results, we can see that the performance of VWM was slightly better than that of FWM. This was because in VWM, the weights are adjusted based on the prediction error, i.e., the change in weight is higher if the difference between the real and the predicted value is higher. Whereas in FWM, changes in weight do not depend on the level of accuracy of the algorithm, and it just looks at which algorithm performs the best. In VWM, each algorithm is evaluated based on all of its previous errors, which is directly correlated to the final prediction accuracy of the model. VWM is better (incurs less error) than FWM by 0.06%–0.18% of MER, 0.03–0.04 of MAE (USD/MWh, AUD/MWh or EURc/kWh) and 0.05%–0.24% of MAPE. Both approaches show high improvement in accuracy compared to our benchmark method (PSF) and some improvement over our previous work (ANN-only). During the experiments, no single scenario (models for each hour of the day and dataset) had the same expert throughout the test period. These periodic changes of expert show that all of the participating algorithms continually learn from the mistakes and adjust their parameters to capture the current price trends and external impact on prices, eventually showing their presence in the ensemble, i.e., being selected as an expert. Thus, the proposed ensemble approach was able to automatically create a tailored model for each scenario and achieved a higher accuracy.

Analyzing the results, among the three participating algorithms, we found the performance of ANN to be comparatively better than SVR and RF when the price is of higher fluctuation, whereas the performances of SVR and RF were better when the price is of lower fluctuation. ANN shows consistent performance with both VWM and FWM, whereas SVR fails to perform well in VWM, as the prediction error for SVR was very high during the peak price, which decreased the weight of SVR by a large margin. RF also shows consistent performance, but it also degrades in some cases of the peak price. In general, high and sudden variations in the electricity price, which are influenced by many unforeseen factors, cause degradation in performance of our ensemble model (both for VWM and FWM). We can see that our model's performance for a few months is below its average performance due to many sharp price increases in those months.

6.3. Comparisons with ARIMA, PSF and ANN-Only Methods

To validate the performance of FWM and VWM, the results obtained were compared with those obtained using other methods, namely ARIMA, PSF and ANN-only.

6.3.1. Proposed Methods vs. ARIMA

Both FWM and VWM outperformed the standard time series forecasting method, ARIMA [10], by large margins in all five test cases, as shown in Tables 2–6. It can be observed that ARIMA performs quite poorly, particularly on the ANEM dataset (2004–2006), which contained some huge price spikes.

6.3.2. Proposed Methods vs. PSF

In [11], Pattern Sequence-based Forecasting (PSF) was reported to have outperformed other contemporary works, namely ARIMA, naive Bayes, ANN (this ANN implementation is distinct from our previous work (ANN-only) [12] because of different feature engineering approaches), WNN (Weighted Nearest Neighbor) [54], the Structural model (STR) [55] and other mixed models. As testing was performed on the 2004–2006 data from all three markets in the PSF paper, we also perform the testing on same data and achieve the better results shown in Tables 2–4.

It can be observed that FWM (VWM in brackets) forecasts with 1.61% (1.67%) improved accuracy in terms of MER in the NYISO dataset. There are improvements (i.e., decreases in error) of average MER by 0.35% (1.17%) for the ANEM dataset and 0.81% (0.89%) for the OMEL dataset.

Similar accuracy improvements by FWM/VWM over PSF can be seen for the MAE criterion, as well. FWM (VWM in brackets) offers higher accuracy over PSF in average MAE by USD 1.15/MWh (USD 1.20/MWh), AUD 0.19/MWh (AUD 0.05/MWh) and EURc 0.13/kWh (EURc 0.12/kWh) for NYISO, ANEM and OMEL respectively.

Furthermore, we claim that the performances of the two proposed methods are more stable as the standard deviations of both the MER and MAE obtained by FWM/VWM are smaller than those obtained by PSF.

6.3.3. Proposed Methods vs. ANN-Only

Finally, we compare the proposed techniques with an existing ANN-only method, which we had presented in an earlier work [12] and that was shown to produce higher forecasting accuracy than PSF.

For the 2004–2006 datasets, both FWM and VWM provided better results than ANN-only for NYISO. FWM (VWM in the brackets) provides improvements of 0.21% (0.27%) in MER, USD 0.12/MWh (USD 0.17/MWh) in MAE and 0.21% (0.25%) in MAPE.

For ANEM (2004–2006), FWM turns out to be inferior to ANN-only by −0.65% MER, −0.11 AUD/MWh MAE and −0.53% MAPE. However, VWM is still better than ANN-only by 0.17% MER, AUD 0.03/MWh MAE and 24% MAPE. For the OMEL dataset, both FWM and VWM are slightly better than or perform equally as ANN-only.

The improvements of FWM (VWM in the brackets) for OMEL (2004–2006) are 0% (0.08%) MER, EURc 0/kWh (EURc0.01/kWh) MAE and 0.01% (0.14%) MAPE. In terms of SD, FWM provides smaller SD values than ANN-only in five out of nine test cases (i.e., 3 datasets × 3 evaluation criteria), and VWM provides smaller SD than ANN-only in six out of nine test cases.

Table 6. MER, MAE and MAPE evaluation metrics for the ANEM market for the years 2011–2012 (Experiment II-B).

| | ANEM (2011–2012) | | | | | | | | | | | |
| Month | MER (%) | | | | MAE (AUD/MWh) | | | | MAPE (%) | | | |
	VWM	FWM	ANN-Only	ARIMA	VWM	FWM	ANN-Only	ARIMA	VWM	FWM	ANN-Only	ARIMA
June 2011	**4.11**	4.32	4.60	11.08	**1.15**	1.20	1.28	2.81	4.42	**4.34**	4.62	10.99
July 2011	**6.85**	7.01	7.46	15.63	1.98	**1.95**	2.07	4.13	6.59	**6.52**	6.94	15.26
August 2011	**4.73**	4.76	5.06	9.60	**1.30**	1.32	1.41	2.77	**4.07**	4.18	4.44	9.29
September 2011	**4.05**	4.17	4.43	8.56	**1.13**	1.16	1.23	2.35	4.02	**3.99**	4.25	8.61
October 2011	**4.19**	4.36	4.64	7.87	**1.15**	1.21	1.29	2.15	**4.08**	4.29	4.57	7.98
November 2011	6.90	**6.89**	7.33	12.61	**1.88**	1.91	2.03	3.95	**5.52**	5.73	6.10	11.63
December 2011	**4.65**	4.71	5.01	7.18	**1.30**	1.31	1.39	1.84	**4.38**	4.43	4.72	7.88
January 2012	**12.53**	13.05	13.88	24.09	**2.97**	3.10	3.30	6.82	8.58	**8.54**	9.09	19.20
February 2012	**13.72**	14.36	15.28	24.58	**5.29**	5.29	5.63	7.44	**12.61**	12.72	13.53	19.84
March 2012	**13.04**	13.17	14.01	24.47	**4.04**	4.17	4.44	6.75	**10.96**	11.17	11.89	19.98
April 2012	5.40	**5.34**	5.68	9.27	**1.42**	1.48	1.58	2.71	4.93	**4.90**	5.21	9.05
May 2012	**3.66**	3.72	3.95	7.93	**1.00**	1.03	1.10	2.03	**3.76**	3.83	4.07	8.32
Mean	**6.98**	7.16	7.61	13.57	**2.05**	2.09	2.23	3.81	**6.16**	6.22	6.62	12.34
SD	**3.83**	3.98	4.24	6.92	**1.36**	1.37	1.46	2.05	**2.98**	3.01	3.20	4.88

Similar trends are also observed for the NYISO and ANEM (2008–2012) datasets, thus confirming the effectiveness of the proposed VWM and FWM methods.

7. Conclusions

Electricity price forecasting in the deregulated electricity market is essential to facilitate the decision making processes of the stakeholders. Although extensive research has been carried out in this field, the accuracy of existing techniques is not consistently high, especially in volatile and complex market conditions. In this paper, an ensemble-based model that combined three different electricity price forecasting algorithms was proposed. Also presented were two different approaches for updating the weights of the participating algorithms and for selecting the final expert algorithm, whose prediction will be adopted as the final model prediction.

Comparative experimental studies were performed to benchmark the proposed model against a number of existing techniques; these were: ARIMA, which is the standard statistical time series model, PSF, a recent highly-regarded method, which was superior to many other existing methods, as well as a method which we had presented in an earlier work, which used a single ANN regressor. The experiments were conducted on data collected from three different electricity markets and for time periods ranging from 2006–2012, and the results showed that our model outperforms the conventional approaches and produces robust and accurate forecasts, even with a variety of different datasets and over a long period of time.

However, there is still room for improvement, and we plan to carry out the following tasks in the future:

- Further testing on other electricity markets.
- Inclusion of other exogenous features, such as oil/gas prices, electricity generation modalities, etc.
- Incorporation of features to model dynamics associated with the smart grid, like demand response and load balancing.
- Development of better weighting schemes to further improve the accuracy.

Acknowledgments: This research work was funded by Masdar Institute of Science and Technology, Abu Dhabi, United Arab Emirates.

Author Contributions: Bijay Neupane designed and implemented the system, performed the experiments and wrote the initial draft of the paper. Zeyar Aung and Wei Lee Woon supervised the project, provided technical guidelines and insights, performed benchmarking experiments and carried out the final write-up of the paper.

Conflicts of Interest: The authors declare no conflict of interest.

References

1. Simmhan, Y.; Agarwal, V.; Aman, S.; Kumbhare, A.; Natarajan, S.; Rajguru, N.; Robinson, I.; Stevens, S.; Yin, W.; Zhou, Q.; et al. Adaptive energy forecasting and information diffusion for smart power grids. In Proceedings of the 2012 IEEE International Scalable Computing Challenge (SCALE), Ottawa, ON, Canada, 13–16 May 2012; pp. 1–4.
2. Shawkat Ali, A.B.M.; Zaid, S. Demand forecasting in smart grid. In *Smart Grids: Opportunities, Developments, and Trends*; Springer: Berlin, Germany, 2013; pp. 135–150.
3. Shen, W.; Babushkin, V.; Aung, Z.; Woon, W.L. An ensemble model for day-ahead electricity demand time series forecasting. In Proceedings of the 4th ACM Conference on Future Energy Systems (e-Energy), Berkeley, CA, USA, 22–24 May 2013; pp. 51–62.
4. Jurado, S.; Nebot, A.; Mugica, F.; Avellana, N. Hybrid methodologies for electricity load forecasting: Entropy-based feature selection with machine learning and soft computing techniques. *Energy* **2015**, *86*, 276–291.
5. Papaioannou, G.P.; Dikaiakos, C.; Dramountanis, A.; Papaioannou, P.G. Analysis and modeling for short- to medium-term load forecasting using a hybrid manifold learning principal component model and comparison with classical statistical models (SARIMAX, Exponential Smoothing) and artificial intelligence models (ANN, SVM): The case of Greek electricity market. *Energies* **2016**, *9*, 635.

6. Li, G.; Liu, C.C.; Lawarree, J.; Gallanti, M.; Venturini, A. State-of-the-art of electricity price forecasting. In Proceedings of the 2005 CIGRE/IEEE PES International Symposium, New Orleans, LA, USA, 5–7 October 2005; pp. 110–119.

7. Aggarwal, S.K.; Saini, L.M.; Kumar, A. Electricity price forecasting in deregulated markets: A review and evaluation. *Int. J. Electr. Power Energy Syst.* **2009**, *31*, 13–22.

8. Weron, R. Electricity price forecasting: A review of the state-of-the-art with a look into the future. *Int. J. Forecast.* **2014**, *30*, 1030–1081.

9. Martínez-Álvarez, F.; Troncoso, A.; Asencio-Cortés, G.; Riquelme, J.C. A survey on data mining techniques applied to electricity-related time series forecasting. *Energies* **2015**, *8*, 13162–13193.

10. Shumway, R.H.; Stoffer, D.S. *Time Series Analysis and Its Applications: With R Examples (Springer Texts in Statistics)*, 3rd ed.; Springer: Berlin, Germany, 2011.

11. Martínez-Álvarez, F.; Troncoso, A.; Riquelme, J.C.; Aguilar-Ruiz, J.S. Energy time series forecasting based on pattern sequence similarity. *IEEE Trans. Knowl. Data Eng.* **2011**, *23*, 1230–1243.

12. Neupane, B.; Perera, K.S.; Aung, Z.; Woon, W.L. Artificial neural network-based electricity price forecasting for smart grid deployment. In Proceedings of the 2012 IEEE International Conference on Computer Systems and Industrial Informatics (ICCSII), Dubai, UAE, 18–20 December 2012; pp. 1–6.

13. NYISO: New York Independent System Operator. Available online: http://www.nyiso.com/public/markets_operations/market_data/pricing_data/index.jsp (accessed on 1 November 2012).

14. ANEM: Australian National Electricity Market. Available online: http://www.aemo.com.au/en/Electricity/NEM-Data/Price-and-Demand-Data-Sets (accessed on 1 November 2012).

15. OMEL: Operador del Mercado Ibérico de Energía, Polo Español, S.A. (Operator of the Iberian Energy Market, Spanish Public Limited Company). Available online: http://www.omel.es (accessed on 1 November 2012).

16. Yu, L.; Wang, S.; Lai, K.K. An EMD-based neural network ensemble learning model for world crude oil spot price forecasting. In *Soft Computing Applications in Business*; Studies in Fuzziness and Soft Computing; Springer: Berlin/Heidelberg, Germany, 2008; Volume 230, pp. 261–271.

17. Zhu, B. A novel multiscale ensemble carbon price prediction model integrating empirical mode decomposition, genetic algorithm and artificial neural network. *Energies* **2012**, *5*, 355–370.

18. Contreras, J.; Espinola, R.; Nogales, F.J.; Conejo, A.J. ARIMA models to predict next-day electricity prices. *IEEE Trans. Power Syst.* **2003**, *18*, 1014–1020.

19. Jakasa, T.; Androcec, I.; Sprcic, P. Electricity price forecasting—ARIMA model approach. In Proceedings of the 8th International Conference on the European Energy Market (EEM), Zagreb, Croatia, 25–27 May 2011; pp. 222–225.

20. Box, G.E.P.; Jenkins, G.M.; Reinsel, G.C. *Time Series Analysis: Forecasting and Control*, 3rd ed.; Prentice Hall: Upper Saddle River, NJ, USA, 1994.

21. Tan, Z.; Zhang, J.; Wang, J.; Xu, J. Day-ahead electricity price forecasting using wavelet transform combined with ARIMA and GARCH models. *Appl. Energy* **2010**, *87*, 3606–3610.

22. Voronin, S.; Partanen, J. Price forecasting in the day-ahead energy market by an iterative method with separate normal price and price spike frameworks. *Energies* **2013**, *6*, 5897–5920.

23. Singh, N.K.; Tripathy, M.; Singh, A.K. A radial basis function neural network approach for multi-hour short term load-price forecasting with type of day parameter. In Proceedings of the 6th IEEE International Conference on Industrial and Information Systems (ICIIS), Kandy, Sri Lanka, 16–19 August 2011; pp. 316–321.

24. Singhal, D.; Swarup, K.S. Electricity price forecasting using artificial neural networks. *Int. J. Electr. Power Energy Syst.* **2011**, *33*, 550–555.

25. Lin, W.M.; Gow, H.J.; Tsai, M.T. Electricity price forecasting using enhanced probability neural network. *Energy Convers. Manag.* **2010**, *51*, 2707–2714.

26. Gong, D.S.; Che, J.X.; Wang, J.Z.; Liang, J.Z. Short-term electricity price forecasting based on novel SVM using artificial fish swarm algorithm under deregulated power. In Proceedings of the 2nd International Symposium on Intelligent Information Technology Application (IITA), Shanghai, China, 21–22 December 2008; Volume 1, pp. 85–89.

27. Sansom, D.C.; Downs, T.; Saha, T.K. Support vector machine based electricity price forecasting for electricity markets utilising projected assessment of system adequacy data. In Proceedings of the 6th International Power Engineering Conference (IPEC), Singapore, 27–29 November 2003; pp. 783–788.

28. Shrivastava, N.A.; Panigrahi, B.K. A hybrid wavelet-ELM based short term price forecasting for electricity markets. *Int. J. Electr. Power Energy Syst.* **2014**, *55*, 41–50.
29. Li, S.; Goel, L.; Wang, P. An ensemble approach for short-term load forecasting by extreme learning machine. *Appl. Energy* **2016**, *170*, 22–29.
30. He, K.; Yu, L.; Lai, K.K. Crude oil price analysis and forecasting using wavelet decomposed ensemble model. *Energy* **2012**, *46*, 564–574.
31. Alamaniotis, M.; Bargiotas, D.; Bourbakis, N.G.; Tsoukalas, L.H. Genetic optimal regression of relevance vector machines for electricity pricing signal forecasting in smart grids. *IEEE Trans. Smart Grid* **2015**, *6*, 2997–3005.
32. Lahmiri, S. Comparing variational and empirical mode decomposition in forecasting day-ahead energy prices. *IEEE Syst. J.* **2015**, in press.
33. Ferruzzi, G.; Cervone, G.; Monache, L.D.; Graditi, G.; Jacobone, F. Optimal bidding in a day-ahead energy market for micro grid under uncertainty in renewable energy production. *Energy* **2016**, *106*, 194–202.
34. He, D.; Chen, W.P. A real-time electricity price forecasting based on the spike clustering analysis. In Proceedings of the 2016 IEEE/PES Transmission and Distribution Conference and Exposition (T&D), Dallas, TX, USA, 3–5 May 2016; pp. 1–5.
35. Hong, Y.Y.; Wu, C.P. Day-ahead electricity price forecasting using a hybrid principal component analysis network. *Energies* **2012**, *5*, 4711–4725.
36. Motamedi, A.; Zareipour, H.; Rosehart, W.D. Electricity price and demand forecasting in smart grids. *IEEE Trans. Smart Grid* **2012**, *3*, 664–674.
37. Bracale, A.; De Falco, P. An advanced Bayesian method for short-term probabilistic forecasting of the generation of wind power. *Energies* **2015**, *8*, 10293–10314.
38. Mori, H.; Itagaki, T. A fuzzy Inference net approach to electricity price forecasting. In Proceedings of the 54th IEEE International Midwest Symposium on Circuits and Systems (MWSCAS), Seoul, Korea, 7–10 August 2011; pp. 1–4.
39. Alamaniotis, M.; Bourbakis, N.; Tsoukalas, L.H. Very-short term forecasting of electricity price signals using a Pareto composition of kernel machines in smart power systems. In Proceedings of the 2015 IEEE Global Conference on Signal and Information Processing (GlobalSIP), Orlando, FL, USA, 14–16 December 2015; pp. 780–784.
40. Longo, M.; Zaninelli, D.; Siano, P.; Piccolo, A. Evaluating innovative FCN Networks for energy prices' forecasting. In Proceedings of the 2016 International Symposium on Power Electronics, Electrical Drives, Automation and Motion (SPEEDAM), Druskininkai, Lithuania, 13–15 October 2016; pp. 315–320.
41. Kintsakis, A.M.; Chrysopoulos, A.; Mitkas, P.A. Agent-based short-term load and price forecasting using a parallel implementation of an adaptive PSO trained local linear wavelet neural network. In Proceedings of the 12th International Conference on the European Energy Market (EEM), Lisbon, Portugal, 19–22 May 2015; pp. 1–5.
42. Zhou, Z.H. *Ensemble Methods: Foundations and Algorithms*; Chapman and Hall/CRC: Boca Raton, FL, USA, 2012.
43. Heaton, J. *Introduction to the Math of Neural Networks*; Heaton Research, Inc.: St. Louis, MO, USA, 2012.
44. Smola, A.J.; Schölkopf, B. A tutorial on support vector regression. *Stat. Comput.* **2004**, *14*, 199–222.
45. Breiman, L. Random forests. *Mach. Learn.* **2001**, *45*, 5–32.
46. Ditzler, G.; Roveri, M.; Alippi, C.; Polikar, R. Learning in nonstationary environments: A survey. *IEEE Comput. Intell. Mag.* **2015**, *10*, 12–25.
47. Azadeh, A.; Ghadrei, S.F.; Nokhandan, B. One day-ahead price forecasting for electricity market of Iran using combined time series and neural network model. In Proceedings of the 2009 IEEE Workshop on Hybrid Intelligent Models and Applications (HIMA), Nashville, TN, USA, 30 March–2 April 2009; pp. 44–47.
48. Weather Underground. Weather Forecasts and Reports. Available online: https://www.wunderground.com/ (accessed on 1 December 2016).
49. Matlab-Matworks. Available online: http://www.mathworks.com/products/matlab/ (accessed on 1 December 2016).
50. Kohavi, R.; John, G.H. Wrappers for feature subset selection. *Artif. Intell.* **1997**, *97*, 273–324.
51. Hall, M.; Frank, E.; Holmes, G.; Pfahringer, B.; Reutemann, P.; Witten, I.H. The WEKA data mining software: An update. *SIGKDD Explor. Newsl.* **2009**, *11*, 10–18.

52. Neupane, B. Ensemble Learning-based Electricity Price Forecasting for Smart Grid Deployment. Master's Thesis, Masdar Institute of Science and Technology, Abu Dhabi, UAE, 2013. Available online: http://www.aungz.com/PDF/BijayNeupane_Master_Thesis.pdf (accessed on 1 December 2016).

53. R: A Language and Environment for Statistical Computing. Available online: http://www.R-project.org/ (accessed on 1 December 2016).

54. Lora, A.T.; Santos, J.M.R.; Exposito, A.G.; Ramos, J.L.M.; Santos, J.C.R. Electricity market price forecasting based on weighted nearest neighbors techniques. *IEEE Trans. Power Syst.* **2007**, *22*, 1294–1301.

55. Chen, J.; Deng, S.J.; Huo, X. Electricity price curve modeling and forecasting by manifold learning. *IEEE Trans. Power Syst.* **2008**, *23*, 877–888.

MDPI AG

St. Alban-Anlage 66

4052 Basel, Switzerland

Tel. +41 61 683 77 34

Fax +41 61 302 89 18

http://www.mdpi.com

Energies Editorial Office

E-mail: energies@mdpi.com

http://www.mdpi.com/journal/energies

www.ingramcontent.com/pod-product-compliance
Lightning Source LLC
Chambersburg PA
CBHW051726210326
41597CB00032B/5620